THE ALGAE

THE ALGAE

V. J. CHAPMAN, M.A., Ph.D. (Camb.)

Professor of Botany in the University of Auckland
New Zealand

and

D. J. CHAPMAN, B.Sc. (Auck.), Ph.D. (Cal.)

Ass. Professor of Botany, University of Chicago

MACMILLAN

(In BCL)

First published January 1962
Reprinted June 1962, 1964, 1968, 1969

Second edition 1973

Published by
MACMILLAN AND CO LTD
London and Basingstoke
Associated companies in New York Toronto
Melbourne Dublin Johannesburg and Madras

Printed in Great Britain by
ROBERT MACLEHOSE AND CO LTD
The University Press, Glasgow

PREFACE TO SECOND EDITION

In preparing this new edition substantial changes have been made in both the arrangement and the material. The underlying system in the taxonomic portion is the same as in the first edition as we are convinced that this is the best way to introduce students to different plants. A great deal of work has been carried out on the flagellates and allied organisms in recent years and this has thrown new light on algal phylogeny. The original chapters on these groups have therefore been greatly expanded and we have also included the Euglenophyta in this new edition. In view of the importance of phytoplankton in the sea and in fresh waters an entirely new chapter on this aspect of the algae has been added.

The book is intended as a brief introductory text only. Because of this the choice of material and references for inclusion must be highly selective. This is especially true of certain chapters, for example those on physiology and symbiosis, for which very extensive works are now available. The references are intended as a guide to further detailed study.

In preparing this new edition we want to express our thanks to the following experts who have very kindly read certain chapters and made very useful and valuable comments: Professor G. F. Papenfuss (Chlorophyta and Phaeophyta); Professor R. E. Norris (Rhodophyta); Professor I. Manton (Prasinophyceae and Haptophyceae); Professor Harlan Johnson (Fossil Algae); Professor J. E. Morton (Rocky coast Ecology); Professor H. Clemençon (Xanthophyceae); Dr. J. R. Taylor (Phytoplankton); Dr. U. V. Cassie (Bacillariophyta); Dr. A. R. Loeblich III (Pyrrhophyta). Whilst we are greatly indebted to the above specialists we accept full responsibility for the contents of this book and the way in which the material is set out.

We would like to thank the authorities at the Cranbrook Institute of Science who made Professor Chapman's visit to the United States possible and consequent personal collaboration on this book.

V. J. CHAPMAN   *Cranbrook Institute of Science and Auckland University*

D. J. CHAPMAN   *University of Chicago*

PREFACE TO FIRST EDITION

This book has been prepared in response to many requests that I have received from colleagues. Although the present volume follows the general lay-out of my first book *An Introduction to the Study of Algae*, there has been so much rearrangement, the complete rewriting of many chapters and the addition of new ones, that it cannot be considered a second edition of the first book. I am particularly grateful to the Syndics of the Cambridge University Press for releasing the original copyright so that it has been possible to use some portions of the first book. In recent years considerable advances have been made in the study of the algae and this has affected, not only our knowledge of their life histories, but also our views on phylogeny. These new results have necessitated much rearrangement of material. There has also been extensive new work on the ecology of the algae, so that the chapters on marine algal ecology and ecological factors are more or less completely new. The same is true also of algal physiology. In order that the survey should be more complete there is a chapter briefly describing the historical aspect of the subject, and also a chapter on the economic uses of algae.

The purpose of the book remains the same as my first volume, namely to provide a general survey of all aspects of the algae such as may be required by a University undergraduate, with selected portions (marked by an asterisk) that can be used by first year students or which are suitable for use in schools at the sixth form level. The type method of instruction has again been employed because continued experience has not changed my belief in it. I still feel, however, that no study of the algae is complete without reference to the other aspects which have been included. I am aware that not everyone will agree with what has been included and material that has been excluded, nor with all the views expressed, but where there are divergencies of opinion, I have tried to see that the other view is also presented or made known. This is perhaps particularly true in the chapter on Evolution.

There are a number of excellent works available to advanced students and research workers, and the present volume is not intended to compete with them in any way. Apart from those works to which I expressed my indebtedness in the preface to *An Introduction to the Study of Algae*, there are some new books that have appeared since, and which have provided valuable reference material. I include here the second volume of Fritsch's *Structure and Reproduction of the Algae*, the *Manual of Phycology*, edited by the late Gilbert Smith, and the recent extensive work on the Rhodophyceae by the late Prof. Kylin. The sources of the new illustrations are acknowledged in the legends.

At the end of each chapter certain references are provided which can be consulted for further specific information. No attempt has been made to provide anything approaching a complete reference list and the choice of the various references is entirely my own. I am aware that this choice will not necessarily please everyone, but I hope it will prove useful.

I am grateful to my colleagues, Dr. J. A. Rattenbury for critical reading of the manuscript and to Dr. J. M. A. Brown for his reading and criticism of the two chapters on ecological factors and algal physiology. Dr. A. B. Cribb of the University of Brisbane has also read the proofs and made valuable suggestions for which I am most appreciative. My thanks go to Mrs. J. Rutherford for assistance in proof-reading.

V. J. CHAPMAN

London 1960

CONTENTS

GLOSSARY

Acronematic: Smooth flagellum with terminal hair.

Akinete: Thick walled resting spore, generally derived from vegetative cell.

Aplanospore: Non-motile spore.

Autospore: Aplanospore resembling a mature cell.

Auxospore: Cell resulting from syngamy (Bacillariophyceae).

Carpogonium: Female gametangium (Rhodophyceae).

Carposporangium: Sporangium resulting from division of zygote nucleus (Rhodophyceae).

Carpospore: Spore produced by carposporangium (Rhodophyceae).

Carposporophyte: Product of fertilized carpogonium (Rhodophyceae).

Coenocyte: Multinucleate cell lacking cross walls.

Coenobium: A group of independent cells enclosed in a common envelope or united together so that the colony functions as a single individual.

Consocies: An ecological term used to denote a community in a succession in which only a single species is dominant.

Cryptostomata: Thallus cavities from which hairs arise (Phaeophyceae).

Ecad: A term used to designate a morphological form of a species that is related to a specific ecological niche.

Endospore: Thin walled spore within the cell (Cyanophyceae).

-etum: Termination used in ecology to distinguish an Association.

Exospore: Spore formed, one at a time, from apex downwards (Cyanophyceae).

Flimmergeissel: Flagellum with lateral flagellar hairs ('tinsel flagellum').

Frustule: One half of a cell (Bacillariophyceae).

Girdle: Transverse groove (Dinophyceae). Region where frustules overlap (Bacillariophyceae).

Gonimoblast: Gonimoblast filaments are the cells bearing the carposporangia (Rhodophyceae).

Haptonema: Flagella-like organ used for attachment to substrate (Haptophyceae).

Heterocyst: Spore-like structure (Cyanophyceae).

Heterokont: Condition in which two flagella are of dissimilar morphology.

Heteromorphic: Morphologically dissimilar.

Heterotrichous: Combination of erect filament and prostrate thallus portion.

Hormogone: Multicellular segment capable of gliding (Cyanophyceae).

Hypnospore: A resistant spore of the same size and dimensions as the cell
from which it is formed.

Isokont: Condition in which two flagella are of similar morphology.

Isomorphic: Morphologically similar.

Intercalary: Applied to describe a meristem in the middle of a filament or
thallus.

Mastigoneme: Hair-like thread occurring on flagella.

Meiospore: Spore formed at meiosis.

Monosporangium: Vegetative cell that changes to produce a single spore
called monospore (Rhodophyceae and Phaeophyceae).

Oospore: Zygospore resulting from oogamous fusion.

Pantonematic: Arrangement describing flagellar hairs on both sides of
flagellum.

Paraphyses: Sterile hairs or threads within sporangium or gametangium
(Phaeophyceae).

Peitschengeissel: Flagellum lacking lateral hairs: 'whiplash' or 'smooth'
flagellum.

Pellicle: Thin membrane around the protoplast.

Pericarp: Gametophyte tissue surrounding the carposporophyte (Rhodophyceae).

Pleuronematic: Flagellum with hairs = Flimmergeissel.

Plurilocular: Multichambered gametangia or sporangia (Phaeophyceae).

Polyspore: Spore of a group in which more than four spores are formed from
mother cell.

Pneumatocyst: Gas bladders or enlarged hollow area of stipe (Phaeophyceae).

Socies: A term used to designate a small community in a consocies dominated
by a subordinate species.

Sorus: Cluster of sporangia (Phaeophyceae).

Statospore: Resting cell or spore formed within vegetative cell, usually with
ornamented cell wall.

Stichonematic: Flagellum with lateral hairs on one side only.

Synzoospore: Compound zoospore.

Tetraspore: Spore from a group of four formed at meiosis.

Tetrasporophyte: Plant producing tetraspores (Rhodophyceae).

Trichoblast: Colorless hair-like branch (Rhodophyceae).

Trichocyst: Cytoplasmic organelle, often dart or thread-like and often
ejected by cell (Dinophyceae, Cryptophyceae and Chloromonadophyceae).

Trichogyne: Hair-like extremity of female gametangium (Rhodophyceae).

Trichome: Filament (Cyanophyceae).

Trichothallic: Describes growth at the base of uniseriate filament (Phaeophyceae).

Unilocular: Describing a single-chambered sporangium (Phaeophyceae).

Zoochlorellae: Chlorophyceae endosymbiotic with invertebrates.

Zooxanthellae: Dinophyceae endosymbiotic with invertebrates.

Zoospore: Motile flagellated spore.

Zygospore: Thick walled resting stage resulting when zygote from isogamy ceases motility.

Zygote: Diploid cell resulting from syngamy.

I

CLASSIFICATION

The student is probably familiar with the seaweeds of the seashore and may even have observed green skeins in stagnant freshwater ponds and pools. These plants, which are among the simplest in the plant kingdom, belong to the group known as Algae.

In recent years, mainly as a result of biochemical and ultrastructural studies, there has been considerable debate over the major divisions of the plant kingdom. It is commonly agreed that the group of organisms, characterized by a lack of a distinct nuclear membrane and hence of a nucleus, should be separated from all other plants. One school of thought suggests that these organisms, which comprise the blue-green algae (Cyanophyta) and the bacteria, should be termed the Monera (Dougherty and Allen, 1960, Whittaker, 1969), while another school terms them the Procaryota (Stanier and van Niel, 1962). This latter school includes all other plants in one major subdivision, the Eucaryota, and within this subdivision the simplest photosynthetic members are those plants known as the Algae. Those who recognise the Monera sometimes group the Algae, Protozoa and Fungi as the Protista; sometimes the fungi are segregated into a separate sub-division, the Mycota. The terrestrial plants and aquatic higher plants with more complex organization are classed as the Metaphyta.

Whittaker's more recent proposal involves five kingdoms, three of which are the Fungi, Plantae and Protista and the eucaryote algae are therefore distributed between the two latter kingdoms. This seems an artificial arrangement and will not be used here.

It is important that any classification should be as broadly based as possible and with our present knowledge it can be a matter of opinion which of the two major subdivisional classifications one employs. We have employed the Procaryota-Eucaryota concept, assigning the algae (except the blue-greens) to 10 divisions within the Eucaryota.

The principal criteria for the primary classification of Algae are differences in pigmentation, other biochemical characteristics and the structure of organelles such as flagella (Fig. 1.1). While there are one or two exceptions to the general basis of differentiation, these are so few (and even then some-

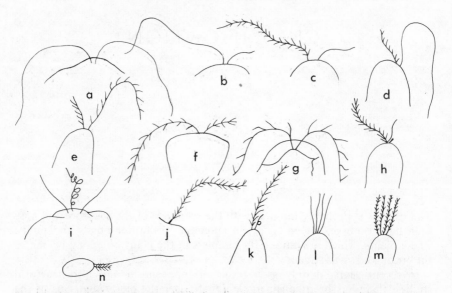

Fig. 1.1 Flagellar types. a, *Chlamydomonas*; b, *Anisomonas*; c, *Thallasomonas*; d, *Chilomonas*; e, *Cryptomonas*; f, *Bipedinomonas*; g, *Haematococcus*; h, *Synura*; i, *Chrysochromulina*; j, *Micromonas*; k, *Mallomonas*; l, *Draparnaldia* zoospore; m, *Platymonas*; n, male gamete of *Vaucheria*. (after Prescott)

what uncertain) that they can be neglected. A study of Table 1.1, which summarizes the pigmentation differences between the respective algal classes, indicates that there are also resemblances between some groups, e.g. Chrysophyceae, Bacillariophyceae and resemblances between these as a whole and the Phaeophyceae; and between the Rhodophyceae and Cyanophyceae. These relationships are, however, discussed more fully later (see pp. 328, 339).

Chlorophyll *a*, one of the green pigments, is the most abundant chlorophyll in all types of algae, and β-carotene is widely distributed. There are a large number of carotenoids (yellow pigments) and many algal classes possess their

TABLE 1.1

GENERALIZED DISTRIBUTION OF CHLOROPLAST PIGMENTS OF SYSTEMATIC SIGNIFICANCE WITHIN THE ALGAE.

	Cyanophyceae	Rhodophyceae	Cryptophyceae	Prasinophyceae	Chlorophyceae	Charophyceae	Euglenophyceae	Xanthophyceae	Chloromonadophyceae	Phaeophyceae	Bacillariophyceae	Chrysophyceae	Haptophyceae	Desmophyceae	Dinophyceae
CHLOROPHYLLS															
Chlorophyll a	+	+	+	+	+	+	+	+	+	+	+	+	+	+	+
Chlorophyll b				+	+	+	+								
Chlorophyll c			+						+	+	+	+	+	+	+
CAROTENOIDS															
α-carotene		±	+	±	±										
β-carotene	+	+	Tr	+	+	+	+	+	+	+	+	+	+	+	+
Flavacene	+														
Zeaxanthin	±	+		+	+	+									
Lutein		+		+	+	+									
Violaxanthin				+	+	+				+					
Echinenone	+				b										
Canthaxanthin					b										
Astaxanthin					b										
Isozeaxanthin	+														
Myxoxanthophyll	+														
Oscillaxanthin	±														
Siphonaxanthin					a										
Micronone				+											
Loroxanthin					+										
Heteroxanthin								+							
Neoxanthin				+	+	+									
Fucoxanthin										+	+	+	+		
Diatoxanthin							c	c	c		±	±	±		
Diadinoxanthin							+	+	+		±	±	±	+	+
Alloxanthin			+												
Crocoxanthin			+												
Monadoxanthin			±												
Dinoxanthin														+	+
Peridinin														+	+
BILIPROTEINS															
Allophycocyanin	+	+													
c-Phycocyanin	+	+													
r-Phycocyanin		±													
c-Phycoerythrin	±														
b-Phycoerythrin		±													
r-Phycoerythrin		±													
Cryptomonad biliproteins			+												
Phytochrome				+		+									

Very minor pigments with very limited species distribution and isolated exceptions to this general distribution have been omitted.

  + = Present in all (or the great majority of) species.
  ± = Present in some, but not all species.
  Tʀ = Present in trace amounts.
  a = Siphonaceous green algae only.
  b = Secondary carotenoids.
  c = In the Euglenophyceae, Xanthophyceae, Chloromonadophyceae, the identification of the xanthophylls other than diadinoxanthin and heteroxanthin is in doubt.

modified from D. Chapman,
1965 (Thesis)

own characteristic types. The Siphonales, an order of the Chlorophyceae, is treated separately because of the carotenoid differences.

The Rhodophyta and Cyanophyta possess biliproteins that are similar to each other, but which are different, with regard to their protein constituents, from those of the Cryptophyta. The phycocyanins (blue pigments) usually predominate in the Cyanophyta and the phycoerythrins (red pigments) are usually dominant in the Rhodophyta. Either may be dominant in the Cryptophyta. There are other biochemical features that appear to be specific to certain algal classes. These include the storage polysaccharides, the chemical constituents of the cell walls, the characteristic sterols and some other miscellaneous compounds that will be mentioned later. Colourless forms are known in all classes except the Rhodophyceae, Phaeophyceae, Charophyceae and Prasinophyceae, but their morphology is such that they are classified in the same class as their photosynthetic counterparts.

Using a combination of biochemical characters together with morphological differences, the algae can be classified as follows:

	Division	*Class*
PROCARYOTA	CYANOPHYTA (blue-green 'algae')	CYANOPHYCEAE
EUCARYOTA	1. RHODOPHYTA (red algae)	RHODOPHYCEAE
	2. CHLOROPHYTA (green algae)	CHLOROPHYCEAE PRASINOPHYCEAE (temporary name) CHAROPHYCEAE
	3. EUGLENOPHYTA (Euglenoids)	EUGLENOPHYCEAE
	4. CHLOROMONADOPHYTA* (Chloromonads)	CHLOROMONADO- PHYCEAE
	5. XANTHOPHYTA* (yellow-green algae)	XANTHOPHYCEAE
	6. BACILLARIOPHYTA (diatoms)	BACILLARIOPHYCEAE
	7. CHRYSOPHYTA (golden-brown algae)	CHRYSOPHYCEAE HAPTOPHYCEAE (temporary name)
	8. PHAEOPHYTA (brown algae)	PHAEOPHYCEAE
	9. PYRRHOPHYTA (dinoflagellates)	DINOPHYCEAE DESMOPHYCEAE
	10. CRYPTOPHYTA (Cryptomonads)	CRYPTOPHYCEAE

* Chapman, D. J. and F. T. Haxo (1966) have established that the Chloromonads possess the same pigment composition as the Xanthophyceae. This might indicate a relationship between these two classes. Scagel *et al.* (1965) have in fact included these two classes in a division, Xanthophyta.

Although the Cyanophyta should not now be included as algae in the strict sense they will be considered here because they have certain similarities (e.g. pigments, gross morphology) to other algae. While there is substantial agreement among taxonomists as to the number and composition of the classes, there is much divergence of opinion as to the groupings of these classes into divisions. One school of thought (e.g. Papenfuss, 1955; Fott, 1959; Smith, 1955; Prescott, 1968; Round, 1965; Dawson, 1967) favours classifications similar to that above, while others (Chadefaud, 1960; Christensen, 1962; Feldmann, 1963; Bourelly, 1968) favour a classification involving only three divisions: Aconta or Rhodophyta (Rhodophyceae) Chlorophyta (Chlorophyceae, Charophyceae, Prasinophyceae and including in some cases the Euglenophyceae) and the Chromophyta (all other classes).

These latter suggestions merit serious consideration because they emphasise the major evolutionary groups within the algae. Our divisions 1, 2 and 3–10 could certainly be grouped together into three taxa corresponding to the Aconta (Rhodophyta), Chlorophyta and Chromophyta of Christensen. Grouping the 14 classes into 3 or 10 divisions is thus a matter of opinion.

Ultimately the classification adopted must depend upon the view taken of the inter-relationships between these groups and other plants and animals. Dodge (1969) has recently surveyed the eyespot structure in flagellated cells. He has divided the algae (with eyespots) into three main 'eyespot groups' that coincide well with other characteristics:

(a) Eyespot within the chloroplast and unrelated to the flagella (Chlorophyceae, Prasinophyceae, Cryptophyceae);

(b) Eyespot within the chloroplast and in close relation to the flagella (Chrysophyceae, Phaeophyceae, most Xanthophyceae);

(c) Eyespot independent of the chloroplast (Euglenophyceae, Dinophyceae).

The above schemes can be expected to undergo modifications as more information becomes available, especially from certain flagellate groups. There are a number of fossil algae not represented among present day forms, and some, the Nematophyceae, that are generally regarded as algae. These are considered in Chapter 13 and are not included in the above classification. A brief summary of the principal class characteristics is as follows.

PROCARYOTA

CYANOPHYTA

CYANOPHYCEAE

The plants in this class show less internal differentiation. Like the bacteria, and unlike the algae, they lack the characteristic nuclear and mitochondrial membranes and hence there are no distinct nuclei and mitochondria, although the nuclear material does appear to be localized.

A distinct membrane-bounded chloroplast is also lacking, and the photosynthetic lamellae are distributed usually in the peripheral region of the cytoplasm.

The product of photosynthesis is Cyanophycean starch, present in granular form. The colour of the cells is commonly blue-green, sometimes olive-green or red, the colour being due to the varying proportions of chlorophyll a, carotenoids and biliproteins. The cell walls lack cellulose and instead comprise mainly amino acids and amino sugars. Contrary to accepted fact, blue-green algae have recently been shown to possess the sterols sitosterol and cholesterol. There is no known sexual reproduction, propagation taking place by simple division, by non-motile endospores, by spores or else by vegetative fragmentation.

EUCARYOTA

CHLOROPHYTA

CHLOROPHYCEAE, PRASINOPHYCEAE, CHAROPHYCEAE

This group used to comprise four great subclasses, the Isokontae (equal cilia), Stephanokontae (ringed cilia), Akontae (no cilia) and Heterokontae (unlike cilia). These subdivisions, as their names imply, were based upon the type of flagella of the motile reproductive bodies. The last has since been renamed the Xanthophyceae and removed to a separate division. The Chlorophyta encompasses a morphological range from simple unicells to quite complex multicellular plants. The pigments are characteristically those of the higher plants: chlorophyll a, b and β-carotene, pigments lutein, violaxanthin and neoxanthin as the usually predominant xanthophylls, but certain exceptions are known: Siphonaxanthin in Siphonales, Micronone in Prasinophyceae. The cell wall is composed commonly of cellulose, except in the Siphonales (see p. 103), plus mannans and xylans with sitosterol as the predominant sterol. In certain members (Dasycladales and Siphonales) calcification is often present. Although the chloroplasts vary considerably in shape and size, the final product of photosynthesis is mostly starch. The motile cells of most members of the Chlorophyceae commonly possess two or four simple and equal flagella that are smooth and lacking lateral appendages. In the Oedogoniales (the former Stephanokontae) there is a ring of flagella, while in the Conjugales (the former Akontae) flagellated reproductive cells are absent. Recent work has shown that certain unicellular algae, formerly in the Chlorophyceae, possess a rather different flagellar structure. These flagella are distinguished by such features as minute scales, lateral hairs, terminal tufts, or bipartite arrangement with a narrower terminal extremity (Fig. 1.1). Genera with these features are removed to a new class, Prasinophyceae.

Sexual reproduction is common and ranges from isogamy to anisogamy and oogamy. Asexual reproduction normally takes place by means of motile zoospores, but a variety of nommotile spores (p. 34) may also be produced.

The Charophyceae comprise one predominantly freshwater order, the Charales, whose members differ from other Chlorophyta in their remarkable morphological features and elaborate oogamy.

EUGLENOPHYTA

EUGLENOPHYCEAE

This class consists of unicellular flagellates of varying shapes and a few palmelloid colonial forms. The organisms lack a cell wall, but are surrounded by a pliable periplast or pellicle. A lateral row of hairs is present on one of the two flagella that are commonly present. The other flagellum is greatly reduced. The cells characteristically possess a gullet with a reservoir, contractile vacuoles and eyespot. While many members of the class are colourless, those that are not possess discoid chloroplasts with chlorophylls *a* and *b*, β-carotene, diadinoxanthin and other xanthophylls. The storage reserve is paramylum, a polysaccharide very similar to laminarin (see pp. 9, 184) and the principal sterol is ergosterol. Reproduction is commonly by longitudinal division, although in some forms cyst formation has been recorded.

CHLOROMONADOPHYTA

CHLOROMONADOPHYCEAE

This is probably the smallest and least understood class of algae. Non-flagellated forms are unknown and about half the described genera lack chloroplasts. When chloroplasts are present they are discoid, small and very numerous, possessing chlorophyll *a* as the only chlorophyll and carotenoids similar to those found in the Xanthophyceae. The cells are biflagellated with heterokont flagella one of which usually trails. Other characteristic cytological features include a gullet and numerous trichocysts (see p. 140). Storage reserves appear to be oil. Under certain conditions the cells may often become encysted or 'amoeboid'.

XANTHOPHYTA

XANTHOPHYCEAE

The plants are usually simple but their lines of morphological development frequently show an interesting parallel with those observed in the Chlorophyceae and Chrysophyceae. In the non-flagellated forms the cell wall, which contains cellulose, is frequently of two unequal or equal halves which overlap each other. The motile cells commonly possess two unequal heterokont flagella, the longer of which is a flimmergeisel (pleuronematic) type, the shorter a peitschengeissel (smooth) type. Reproduction is by zoospores, aplanospores or occasionally statospores. Sexual reproduction is rare and when present is isogamous, though in one genus (*Vaucheria*) there is a well-developed oogamy. This group characteristically possesses chlorophyll *a* as the sole chlorophyll, and carotenoids, identical to some in the Chrysophyta, Euglenophyta and Bacillariophyta, while the normal storage reserve is oil or leucosin, and ergosterol has been reported as the principal sterol.

EUSTIGMATOPHYCEAE

Certain Xanthophyceae differ from the majority in the organization of their motile cells. Hibberd and Leedale (1970) have proposed that such genera should be assigned to a new class, Eustigmatophyceae. This class includes those genera in which the eyespot is independent of the chloroplast; the zoospores have one emergent flagellum; the chloroplast lacks a girdle lamella; golgi bodies and pyrenoids are present in the vegetative cell, absent in zoospores.

Biochemical distinctions have yet to be determined, although the pigments of the Eustigmatophyceae and Xanthophyceae appear to be similar.

BACILLARIOPHYTA

BACILLARIOPHYCEAE (Diatoms)

One of the characteristic features of these plants is their cell wall which is composed of silica and partly of pectin. The wall is always in two halves and frequently ornamented with delicate maskings. Each cell contains one or more variously shaped, golden-brown chloroplasts, containing the same pigments as those of the Chrysophyta. Like the Chrysophyceae, the members of this class produce chrysolaminarin as the photosynthetic food reserve. They are divided into two major groups, the representatives of one group being radially symmetrical, those of the other being bilaterally symmetrical. Propagation is by cell division, giving rise to two cells of slightly differing size. In the former group, the centric diatoms, there is a sexual reproduction by oogamy, in which the male gamete possesses a single pleuronematic (flimmergeissel) flagellum. The other group, the pennate diatoms, reproduce sexually by a form of amoeboid isogamy, the gametes fusing in pairs. Members of this class are unicellular, but often occur as colonial forms.

CHRYSOPHYTA

CHRYSOPHYCEAE, HAPTOPHYCEAE

These form a group, mostly of unicellular organisms, many of which are naked and lack a cellulose wall, although others show high degrees of silicification or calcification. The principal pigments are chlorophyll a and c and fucoxanthin. The principal photosynthetic product is chrysolaminarin, a polysaccharide very similar to laminarin of the Phaeophyceae. The main sterols so far identified are fucosterol and porifasterol.

CHRYSOPHYCEAE: The members of this class are genera whose flagellate stages are characteristically heterokontic. One flagellum is of the flimmergeisel type (pantonematic), while the other is smooth (peitschengeissel $\varDelta$). Flagella scales if present are mineralized and certain genera also possess silicified cysts.

HAPTOPHYCEAE: The flagellate stages of this class are isokontic and characterized by flagella that are always smooth and scales, if present, that are not mineralized. Certain genera of this class also possess a third filiform appendage, or anchorage organ, called a haptonema.

Reproduction is commonly by cell division, by zoospores or by statospores, which are silicified cysts that generally have a small aperture closed by a small plug.

PHAEOPHYTA

PHAEOPHYCEAE

This group comprises the common brown algae of the seashore, and it is worth noting that the majority are wholly marine. There are a few, rare freshwater species. The brown colour is due to the fucoxanthin protein which masks chlorophylls *a* and *c* (see Table 1). The principal product of photosynthesis is the sugar alcohol, mannitol, together with the polysaccharide laminarin, while the cell walls not only contain cellulose but also specific polysaccharides such as alginic acid and fucoidin. In some of the larger and more advanced brown algae these compounds are present in sufficient quantity to be of commercial importance. The glucans or glucose polymers (e.g. laminarin), like paramylum and chrysolaminarin are characterized by predominantly β 1:3 linkages instead of the α 1:4 linkages found in the starches of the Chlorophyta and Rhodophyta.

The simplest forms are filamentous, but there are examples of all stages of development and differentiation particularly in the large seaweeds of the Pacific and Arctic shores with their great size and complex internal and external differentiation. There is also a number of greatly reduced parasitic and epiphytic forms. The pyriform motile reproductive cells are heterokont with two laterally inserted flagella. The longer anteriorly directed flagellum is a flimmergeissel (except in *Fucus* where it is the shorter), while the posteriorly directed flagellum is a peitschengeissel. The motile reproductive cells are commonly produced in special organs that are either unilocular or plurilocular. Some genera also reproduce by non-motile monospores, tetraspores and aplanospores. Sexual reproduction ranges from isogamy to oogamy.

PYRRHOPHYTA

DESMOPHYCEAE, DINOPHYCEAE

Most of the members of these two classes are motile unicells, but there has been an evolutionary tendency towards an 'algal' form and the development of short filaments, e.g. *Dinothrix, Dinoclonium*.

In the Dinophyceae, the motile cells are laterally biflagellate. One flagellum lies in the girdle or transverse groove and encircles the cell, while the other trails behind from the sulcus or ventral groove. Within this class are the 'armoured' and 'unarmoured' dinoflagellates. In the former, the motile cell possesses a cellulose wall with elaborately sculptured plates, while, as the name implies, the unarmoured forms lack the cellulose wall and sculptured plates.

The Desmophyceae are characterized by motile cells that lack a cellulose wall, and are divided vertically into two halves. There is no girdle or sulcus and the flagella are apically inserted.

Both classes are noted for their large nucleus, containing distinct moniliform chromatin threads, even in the resting stage. Those genera possessing chloroplasts contain chlorophylls *a* and *c* and a number of xanthophylls, of which the predominant one, peridinin, is found only in these two classes. The so-called colourless forms that is, these which lack chloroplasts, are often characterized by variously coloured cytoplasmic pigments. The products of photosynthesis are starch or fat. Some of the autotrophic forms can feed heterotrophically, and saprophytic genera are also known. Cell division is longitudinal (Desmophyceae) or obliquely longitudinal (Dinophyceae) and asexual reproduction is by zoospores. Sexual reproduction in those genera in which it has been recorded is anisogamous or isogamous. Characteristic resting cysts are also produced by many of the forms.

CRYPTOPHYTA

CRYPTOPHYCEAE

With the two exceptions of coccoid *Bjornbergiella* and *Tetragonidium* the members of this class are flagellates. The motile cells possess two slightly unequal flagella, which appear to be flimmergeissel. The cells are asymmetrical and have a very distinctive morphology with a gullet, a trichocyst lined groove and a complex vacuolar system. One small group is noted for its reniform cells. Biochemically this class is very interesting in that it is the only group that possesses chlorophylls *a* and *c* and biliproteins. Additionally it is one of the few algal groups that synthesises α-carotene instead of β-carotene. The xanthophylls, although similar to others in chlorophyll *c*-containing algae, are restricted to this class. The storage product appears to be a starch-like compound. Cell division is longitudinal, and asexual reproduction can occur by zoospores. Sexual reproduction is unknown.

RHODOPHYTA

RHODOPHYCEAE

This group is often referred to as the red algae and the majority of its members are the conspicuous red seaweeds of the seashore. The colour is produced by the phycoerythrins and phycocyanins, which mask the carotenoids and chlorophyll *a*. Chlorophyll *d*, often considered characteristic of this group is probably an artefact of extraction. The products of photosynthesis are a starch-like compound (floridean starch), floridoside and mannoglycerate. There is also a tendency to form polysaccharide sulphate esters and certain groups produce large amounts of polysaccharide mucilages. Motile reproductive stages are unknown. While primitive members are coccoid unicells or simple filaments, all stages up to a complex thallus can be found, although there is not quite the same degree of complexity as in the Phaeophyceae. Despite variations in form there are only two basic types of thallus construction. Protoplasmic connections exist between the cells of all forms except most of those that comprise a small group, the Bangiophycidae (see p. 262). Sexual reproduction is by a complex oogamy, the ovum being retained within the gametangium, and although the subsequent development of the zygote, which is retained on the parent plant, varies to a certain extent, it usually gives rise to filaments that bear special reproductive bodies known as carpospores; these latter are normally responsible for the production of a tetraspore-bearing diploid individual.

The extent to which the various algal classes occur in the marine and fresh-water environments, the latter including aerial and soil dwelling algae is of considerable interest and a general summary of present information is provided in Table 1.2.

TABLE 1.2 (Modified from Yale-Dawson, 1966)

OCCURRENCE OF ALGAL CLASSES IN MARINE AND FRESHWATER HABITATS

Division	Approx. no. of living species	Pre-dominantly	Main Occurrence
Cyanophyta	7,500 but probably much less	M & FW	Benthic, terrestrial Epipelic, planktonic
Chlorophyta			
Chlorophyceae ⎫	7,000	FW 13% M	Benthic, epiphytic planktonic
Prasinophyceae ⎬	250	FW	Epiphytic
Charophyceae ⎭			Benthic
Euglenophyta	400	FW	Epipelic
Chloromonadophyta	7	1 sp M	Plankton
Baccillariophyta	6,000	30–50% M	Plankton, epiphytes Epipelic
Xanthophyta	450	15% M	Benthic, epipelic Plankton
Chrysophyta	650	± 20% M	Plankton
Phaeophyta	1,500	99.7% M	Benthic
Pyrrophyta	1,100	93% M	Plankton
Rhodophyta	4,000	98% M	Benthic
Cryptophyta	100	M FW	Plankton

REFERENCES

Aaronson, S, and Hutner, S. H. (1968). *Q. Rev. Biol.*, **41**, 13.
Bourelly, P. (1968). *Les algues d'eau douce*, Tomes **1**, **2** and **3**, (Boubée et Cie, Paris).
Chadefaud, M. (1960). *Traité de Botanique*, Tome **1**, (Masson et Cie, Paris).
Chapman, D. J. and Haxo, F. T. (1966). *J. Phycol.*, **2**, 98.
Christensen, T. (1962). In *Botanik*, Edit. by T. W. Bocher, M. Lange and T. Sorenson, (Copenhagen).
Christensen, T. (1964). In *Algae and Man*, Edit. by D. F. Jackson, pp. 1–434, (Plenum Press, New York).
Dawson, E. Y. (1966). *Marine Botany, An Introduction*, (Holt, Rinehart and Winston, New York).
Dodge, J. D. (1969). *Br. Phycol. J.*, **4**, 199.
Dougherty, E. G. and Allen, M. B. (1960). In *Comparative Biochemistry of Photoreactive Systems*, Edit. by M. B. Allen, pp. 129–143, (Academic Press, New York and London).
Feldmann, G. (1963). In *Précis de Botanique*, Edit. by M. Chadefaud and M. Emberger, pp. 83–249, (Masson et Cie, Paris).
Fott, B. (1959). *Algenkunde*, (Jena, Stuttgart).
Fott, B. (1965). *Preslia*, **37**, 117.

Fritsch, F. E. (1935). *Structure and Reproduction of the Algae*, Vol. I, pp. 1–12, (Cambridge University Press).

Hibberd, D. J. and Leedale, G. F. (1970). *Nature, Lond.*, **225**, 758.

Joyon, L. and Mignot, J-P. (1969). *Ann. Biol.*, **8**, 1.

Klein, R. M. and Cronquist, A. (1967). *Q. Rev. Biol.*, **42**, 108.

Papenfuss, G. F. (1955). In *A century of progress in the Natural Sciences*, pp. 115–224, (Californian Academy of Sciences).

Prescott, G. F. (1968). *The Algae — A Review*, (Houghton Mifflin Co., Boston).

Pringsheim, E. G. (1965). *Farblose Algen*, (Gustav Fischer Verlag, Stuttgart).

Pringsheim, E. G. (1966). *Z. PflPhys.*, **54**, 99.

Round, F. E. (1965). *Biology of Algae*, (Arnold, London).

Scagel, R. F., Bandoni, R. J., Rouse, G. E., Schofield, W. B., Stein, J. R., Taylor, T. M. C. (1965). *An Evolutionary Survey of the Plant Kingdom*, (Wadsworth, Belmont).

Smith, G. M. (1951). *Manual of Phycology*, pp. 13–19, (Chronica Botanica, Waltham).

Smith, G. M. (1955). *Cryptogamic Botany*, 2nd Ed., Vol. 1, (McGraw Hill, New York).

Stanier, R. Y. and Van Niel, C. B. (1962). *Bact. Rev.*, **42**, 17.

Strain, H. H. (1951). In *Manual of Phycology*, Edit. by G. M. Smith, pp. 243–262, (Chronica Botanica, Waltham).

Whittaker, R. H. (1969). *Science, N.Y.*, **163**, 150.

2

CYANOPHYTA

CYANOPHYCEAE

The name Cyanophyceae has recently been generally adopted for this class previously known as the Myxophyceae. Apart from the main feature that the cells lack true nuclei (see p. 1), they show other features which, with the bacteria, distinguish them from the Eucaryota (Echlin and Morris, 1965). The cell envelope in the Cyanophyceae consists of an outer sheath (capsule in the bacteria), an outer membrane and an inner one (present in gram + ve bacteria, but lacking in gram – ve) and a plasma membrane. The cell envelope is composed largely of mucopeptides built up from amino sugars (e.g. glucosamine) and amino acids (e.g. muramic, diaminopimelic), whereas the cell wall of other algae lack such amino constituents.

The bacteria and Cyanophyceae lack mitochondria, true vacuoles and endoplasmic reticula. Sterols have not been detected in bacteria, and the absence of sterols in the Cyanophyceae was also an accepted fact. Recent work has shown the presence of cholesterol and β-sitosterol in *Anacystis* and *Fremyella*. There is no membrane bounded chloroplast and in the blue-green algae the photosynthetic lamellae are usually distributed in the peripheral cytoplasm.

The separate microscopic plants are characterized by a bluish-green colour, occasionally red, which varies greatly in shade, depending upon the relative

13

proportions of chlorophyll *a*, carotenoids, phycocyanin and phycoerythrin.

As well as the photosynthetic lamellae the Cyanophyceae possess small granules resembling ribosomes. The photosynthetic lamellae can be arranged irregularly or in parallel stacks, but they are always separated by a layer of cytoplasm. Fine fibrils of DNA (deoxyribose nucleic acid) are a feature of the nucleoplasm and these are either distributed throughout the cell or concentrated in the central portion. The use of the electron microscope has shown that the former idea of an outer chromatic zone and an inner colourless zone is not valid and that cells of different species have the structural units just described. which can be arranged in a number of different ways (Fig. 2.1).

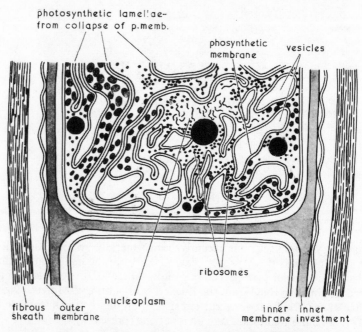

Fig. 2.1   Diagram of cell of *Calothrix*. (after Echlin)

The protoplast is normally devoid of vacuoles, except in old cells, and this is a major factor contributing to the great resistance of the plants to desiccation and of the cells to plasmolysis. In some forms (e.g. *Anabaena*, *Oscillatoria*), and principally in species which are planktonic, pseudovacuoles (gas vacuoles) may be found and it is supposed that these contribute towards their buoyancy by virtue of the gas that they contain. In *Microcystis* these consist of a single layer of particles, a few nanometers in dimension, linearly arranged around the axis of the vacuole which has approximate dimensions of 70×400 nm. The mucilaginous cell sheath, which may surround the whole cell, e.g. *Anacystis*, or form a cylindrical sheath, e.g. *Oscillatoria*, or an interrupted sheath, e.g. *Anabaena*, appears to be composed of cellulose fibrils.

There is considerable variation in the composition of the different cell sheaths, and the amount of material laid down frequently depends upon the external environment. Colour of the sheaths is modified by acidity of the medium and also by the action of fungal parasites. Protoplasmic connections between mature cells are known in one order, the Stigonematales (see p. 22).

The class is characterized by a general absence of well-marked reproductive organs, sexual reproduction or motile reproductive bodies. Fusion of filaments has been reported for a strain of *Nostoc muscorum* (Lazaroff and Vishniac, 1962).

The coccoid forms (spherical cells) multiply by cell division which takes place by means of a progressive constriction, whilst in some other types the cell contents give rise to a number of non-motile bodies that are termed endospores (formerly called gonidia). All stages from simple binary fission to endospores can be found:

(*a*) binary fission, e.g. *Chroococcus turgidus* (= *Anacystis dimidiata*)

(*b*) quadrants and octants formed, e.g. *Anacystis montana*

(*c*) numerous small daughter cells are produced in which there is retention of individual sheaths, e.g. *Anacystis* spp. (These are sometimes called nannocytes or gonidia)

(*d*) the same without individual sheaths

(*e*) abstricted endospores, e.g. *Entophysalis.*

During cell division and at binary fission the photosynthetic lamellae become invaginated at the point of division and a new membrane is formed here by invagination of the inner membrane of the long wall.

Many of the filamentous forms produce specialized cells known as *heterocysts*. These are enlarged cells which possess thickened walls, particularly at their poles, and they usually occur singly, though occasionally they may be formed in rows. They develop from an ordinary vegetative cell, but during development they remain in protoplasmic communication with neighbouring cells by a small cytoplasmic connection. During development the heterocyst increases in size, and there is an apparent increase in the cytoplasmic matrix, disorganization of the photosynthetic lamellae system and disappearance of some of the cytoplasmic inclusions (Wildon and Mercer, 1963). As the heterocyst enlarges a thick outer envelope is laid down outside the cell. There is usually a pore at one end of the terminal heterocyst (at both ends if intercalary) at which the disorganized chloroplast lamellae often reaggregate. There are connections to plasma membranes of adjacent cells, but there does not appear to be any direct protoplasmic connection. It has been shown that heterocyst formation occurs only when the nitrogen level falls below a certain level and that heterocysts are incapable of photosynthesis but may fix nitrogen.

The widespread occurrence of heterocysts, especially terminal ones, and their specialized structure indicates that they must fulfil a significant role. Various suggestions have been made as to their function, and in some cases they do seem to determine the breaking up of the trichomes* or threads into hormogones. These hormogones are short lengths of thread which are cut off,

* The trichome plus the sheath is termed a filament.

thus forming a means of vegetative reproduction among the filamentous types. In the past it has been suggested that the heterocysts represent archaic reproductive organs that are now functionless. Fritsch suggested that the heterocysts, during the vegetative period, secrete substances that stimulate growth and cell division. He also suggested that at the time of reproduction the nature of the secretion changes and as a result akinetes are formed.

Recently Wolk has demonstrated again that under certain conditions (e.g. transfer to ammonium chloride and glucose) heterocysts will 'germinate' to give rise to multicellular germlings. He also found that sporulation (see below) was stimulated by phosphate starvation and that spores could germinate on phosphate agar. Wolk then suggested that these two adaptations are survival mechanisms for deficiencies of nitrogen and phosphate. In other experiments he has shown that sporulation of akinetes adjacent to heterocysts is dependent upon the heterocysts. Further work is, however, needed before the problem of the heterocyst is solved. Recently Stewart (1969) has shown they will fix nitrogen.

Hormogones, besides being cut off by heterocysts, may also be produced by the development of biconcave 'separation discs' at intervals along the filament, or they may break off from the ends of the trichomes. The so-called separation discs are formed by the solidifying of a cell and its contents in the filament. The hormogones, together with certain of the filamentous types, e.g. *Oscillatoria*, and a few Chroococcales, exhibit a slow forward motion and associated with this motion is an active and continual secretion of mucilage along the sides of the filaments. The actual mechanism said to be responsible for locomotion is rhythmic longitudinal waves passing from end to end of the trichomes and caused by changes in the volumes of the protoplasts (see p. 26).

This is probably an over simplification and until further experimental work has been carried out interpretation should be reserved. In four species of *Phormidium* that have been fairly thoroughly studied, it has been found that movement is a response to light, and that it occurs between light intensities of 0·02 lx and 30,000 lx with optimum response at 2,000 lx. It is suggested that there are three different light reaction systems that are independent of each other, so that there is some degree of complexity. The principal system is related to chlorophyll alone but there is another based upon both chlorophyll and the biliproteins. A third system is related to a photoreceptor of unknown character (Nultsch, 1964).

Yet another type of structure, the thick-walled resting spore or akinete, occurs in many of the filamentous forms belonging to the Nostocaceae and Rivulariaceae, and these akinetes normally develop next to a heterocyst, either singly or in a series. On germination they give rise to a new filament.

Many of the forms aggregate into colonies, but in some of the Chroococcaceae the macroscopic plant mass represents an association of such colonies rather than a single colony. The form which any colony may take depends on first, the planes of cell division, and second, the effect of environment, which may determine the consistency of the mucilage. Uneven temperatures, for example, sometimes produce irregular growth. It has been shown experimentally that the environment may affect the shape of colonies of *Microcystis*

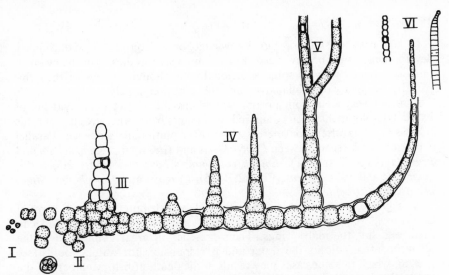

Fig. 2.2 Hapalosiphon types found in all *Mastigocladus* plants. I, *Chroococcus* type; II, *Pleurocapsa* type; III, *Stigonema* type; IV, *Calothrix* type; V, *Scytonema* type; VI, Hormogones, *Anabaena — Oscillatoria* types. (after Schwabe)

and *Chroococcus turgidus* (= *Anacystis dimidiata*) and determines the size of *Rivularia haematites*.

The single floating unicell may not necessarily represent the most primitive condition. The class is exceptionally ancient and must have evolved at a very early stage (see p. 328). It is known that the early waters of the world were not saline as at present and they were probably very much warmer and the fact that many Cyanophyceae are found associated with hot springs may be of significance. Schwabe (1960) has suggested that it is in such places the most primitive forms may be found, and he proposed a radiating system based upon the behaviour of *Mastigocladus laminosus*, a wide-spread thermal alga. In this alga, under different conditions, a wide range of morphological form can develop (Fig. 2.2). This plasticity is regarded as characteristic of a primitive thermal alga and various lines of morphological specialization are proposed as follows:

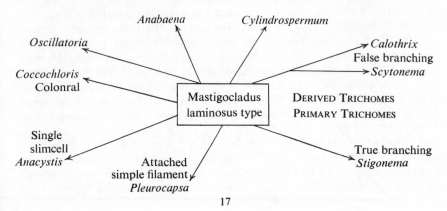

In the present state of our knowledge other courses of evolutionary development could be envisaged. Further intensive studies with the electron microscope may provide some additional data, though in view of the comparative simplicity of cellular structure, this is not very likely.

As may be expected from a primitive and ancient group there is evidence of parallel development when compared with plants from other groups, and this follows what is often considered a typical evolutionary sequence. Parallel evolution can be seen between *Coccochloris* and *Gloeocystis*, *Agmenellum* and *Prasiola* (Figs. 2.4, 3.31), *Entophysalis* and *Characium* (Figs. 2.5, 3.17), *Anacystis* and *Pleurococcus* (Figs. 2.3, 4.5), *Lyngbya* and *Hormidium*, *Stigonema* and *Wittrockiella*.

As a group, the plants are extremely widely distributed under all sorts of conditions, frequently occurring in places where no other vegetation can exist, e.g. hot thermal springs. Their presence in great abundance in the plankton often colours the water and is responsible for water blooms (see p. 396), which in some cases may result in the death of fish. Members of the group also form a large constituent of the soil algae.

One odd site that has been recorded for these algae is in the centre of lake balls composed of the remains of leaves of *Potamogeton pectinatus* found in Lake Vrana, Yugoslavia.

Symbiosis (see p. 385) involving Cyanophyceae is of widespread occurrence. Apart from the very well known example of Lichens (an association of a fungus and either a Chlorophycean or Cyanophycean alga, e.g. *Anacystis*, *Scytonema*), blue-green algae have been recorded in a variety of symbiotic associations, examples of which are given below: *Anabaena azollae* within the leaves of the water fern *Azolla filiculoides*; *Nostoc sphaericum* within cells of the fungus *Geosiphon pyriforme*; *Calothrix chapmanii* within the cells of the green alga *Enteromorpha*. One of the most interesting examples of symbiosis involving blue-green algae is the case of the 'cyanellae' or 'syncyanomes'. These are blue-green algae living within colourless algae or rhizopods. The symbiosis appears to be geologically ancient and unlike other symbiotic Cyanophyceae. The algae have undergone such modification and reduction that some workers have suggested that in certain cases (e.g. *Glaucocystis*) the association is so inseparable that it may be considered as a single entity, and consequently they have suggested the creation of a phylum Glaucophyta to accommodate them. Another possibility is to regard them as only peculiar plastids. Two of the best known examples are *Glaucocystis nostochinearum* (the host is a colourless Chlorophyceae-like organism with vestigial flagella) and *Cyanophora paradoxa*. In this latter case there are two blue-green inclusions, which have been named *Cyanocyta korschikoffiana* (Fam. Cyanocytaceae, order Chroococcales), inside a colourless Cryptomonad host, *Cyanophora*. In the case of *Glaucocystis*, the name *Skujapelta nuda* (Fam. Skujapeltaceae, order: Chroococcales) has been applied to the symbiont, and *Glaucocystis* retained for the host. The symbionts differ from free-living forms in lacking the typical, double-layered cell wall and they also possess a large centrally-located body which may represent a primitive sort of nuclear material (Hall and Claus, 1963). Another of these peculiar forms is the colourless parasite, *Sarcinastrum urosporae* found in *Urospora* and *Ulothrix* (see p. 61).

The Cyanophyceae, besides being primitive, are very ancient and some of the earliest plant fossils known belong to the class (cf. p. 328). It would seem that very little evolution has taken place within the group since they first appeared. Their age and constancy of form, together with their modes of reproduction, are also responsible for the wide distribution of different species, so that there is little or no endemism in the Cyanophycean flora of any region.

Taxonomically the class has been variously treated. Originally it was divided into two orders and in more recent years it has been customary to recognize five orders. Much of the problem in this respect arises from environmentally induced morphological variation. Recent works on *Nostoc* and *Lyngbya* have shown that gross morphology is especially dependent on environment/culture condition. Drouet has shown that some 54 different species (8 genera) are in fact all variations (ecophenes) of *Schizothrix calcicola*. Some authorities, however, have suggested that there should be only one order, the Chroococcales, and eight families. Because of this flux of opinion only the families will be treated in the following pages.

The five orders recognized in many taxonomic works are first, the Chroococcales, which comprises unicellular, free-living forms, living singly or in colonies; second, the Chamaesiphonales, which is a small order consisting of attached unicells occurring singly or in groups and reproducing by means of endospores, which do not occur in the first order. The Pleurocapsales is another very small order with few species that consist of cells that divide in one or two planes to give a crust or cushion. The great majority of the Cyanophyceae are placed in the Nostocales and these are the simple or branched filamentous forms that occur free or aggregated in a common sheath. In some genera the base and apex of the filaments differ, e.g. *Rivularia*, *Calothrix*, and the possession of heterocysts is a feature of many. Reproduction is commonly by means of hormogones but some genera also produce spores. The fifth order is the Stigonematales in which the species form branched threads and there are examples of heterotrichy. Branching is true branching and the threads may be multiseriate. Both hormogones and heterocysts are known.

CHROCCOCCACEAE: *Anacystis* (*ana*, single; *cystis*, bladder) Fig. 2.3

The representatives of this family consist of uni- or multicellular microscopic plants or macroscopic plant colonies. The individual cells are spherical, ovoid, discoid, cylindrical or pyriform in shape, and if colonial can be arranged regularly or irregularly within a gelatinous envelope. In some species of *Anacystis* the colonies are free-floating with a mass of spherical cells: in other, mostly terrestrial, species the cells are single or else united into spherical or flattened colonies each with a small number of cells. The free-floating colonies of *Agmenellum* (Fig. 2.4) form regular plates one cell in thickness which, with increasing age and size, can become curved or twisted. There appears to be considerable variation in the degree of compactness with which the cells are arranged.

B

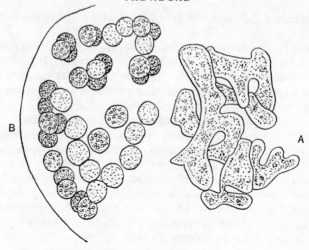

Fig. 2.3 *Anacystis cyanea.* A, colony; B, portion of a colony. (B × 750) (A, after Geitler; B, after Tilden)

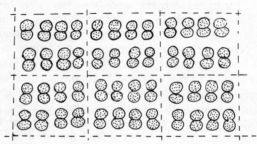

Fig. 2.4 *Agmenellum,* portion of a colony in surface view (× 175) (after Scagel)

Reproduction of colonies is by fragmentation, but multiplication of the cells is by binary fission in two (*Agmenellum*) or three planes (*Anacystis*). In the genus *Johannesbaptista*, which has a linear series of lenticular cells in an elongate gelatinous matrix, division is in one plane only. *Anacystis cyanea* (= *Microcystis aeruginosa*) is a very common water bloom alga and certain strains produce a highly toxic substance (see p. 396). The taxonomy of the family is extremely difficult and one can either recognize a large number of genera and species or only a few. The latter course, based on Drouet and Daily (1956), has been adopted here. The common *Anacystis nidulans*, of physiological use has recently been transferred to a new genus *Lauterbornia nidulans* (Pringsheim, 1968).

CHAMAESIPHONACEAE : *Entophysalis* (*ento*, within; *physalis*, little bladder). Fig. 2.5.

The cells of the sole genus in the family are attached or epiphytic, solitary or arranged in dense clusters on wood, rocks or fresh water or marine plants. The cells are more or less rigid, vary much in shape and are attached at the

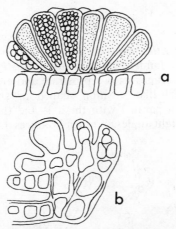

Fig. 2.5 *Entophysalis.* a, colony on host plant; b, filamentous phase. (b, after Tilden)

base by down-growing filaments. In a cushion cell, division is usually horizontal and gives rise to a pseudo-parenchymatous prostrate system, but the upper daughter cells can divide by perpendicular walls, the resulting arrangement of cells being radial. The sheath is thin and hyaline and ultimately opens at the apex. Reproduction is by means of endospores which are abstricted successively from the apex of the surface cells by transverse division. There has been much confusion about this order, but Drouet and Daily consider that species previously placed in different genera, e.g. *Chamaesiphon, Dermocarpa, Xenococcus, Pleurocapsa*, simply represent growth stages of one and the same species. The few species recognized are now placed in the single genus *Entophysalis*.

CLASTIDIACEAE

This family is not well known, the plants consisting of solitary, elongate, epiphytic unicells within a sheath which is thickened at the base to form an attachment pad. There are only two genera, and three species (*Stichosiphon sansibaricus* and *Clastidium setigerun, C. rivulare*) (Fig. 2.6).

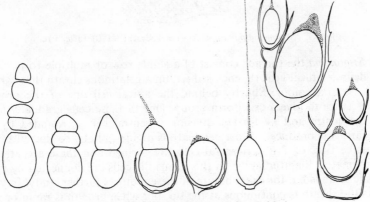

Fig. 2.6 *Clastidium setigerum.* (after Geitler)

STIGONEMATACEAE: *Stigonema* (*stigon*, dotted; *nema*, thread). Fig. 2.7

The members of this family, which includes *Mastigocladus* (p. 391), form branched threads, sometimes with a prostrate and erect portion. The entire plants are floccose, felt-like, cushion shaped or spherical. Branches are formed by cell division parallel with the axis and the daughter cell pushes through the sheath at right angles and then produces its own sheath.

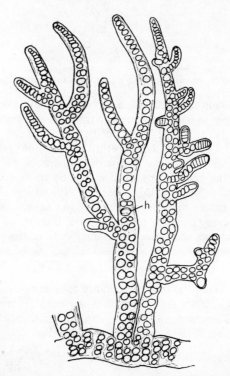

Fig. 2.7   *Stigonema minutum*. (h = heterocyst) (× 170). (after Fremy)

In *Stigonema* the threads consist of a single row or multiple rows of cells with definite apical growth, enclosed within a gelatinous sheath that varies in thickness and colour. Shortly behind the apical cell one of the products divides twice to give a cruciform group. One of these cells gradually moves into the centre and by further division a central cell surrounded by four pericentrals is produced. Most pericentrals divide radially so that ultimately successive tiers of 7–8 pericentral cells are formed and these are arranged approximately longitudinally. In this genus the cells are connected by protoplasmic strands but the exact nature of these requires further study. In young plants the sheath is continuous at the tip, but when growth is rapid or when hormogones are produced and extruded the end bursts. The heterocysts in this genus are commonly lateral, being formed by tangential division of a cell. The filaments are often aggregated to form a macroscopic yellow-green gelatinous mass. The species are most common where there is continual dripping water.

NOSTOCACEAE: *Nostoc* (used by Paracelsus). Fig. 2.8

The gelatinous thallus is solid or hollow, floating, attached or terrestrial and varies much in size and shape. The spheres usually break open when mature and give rise to a flat expanse. There is a dense limiting layer containing numerous intertwined and contorted moniliform filaments with individual hyaline or coloured sheaths, though the sheath may sometimes be

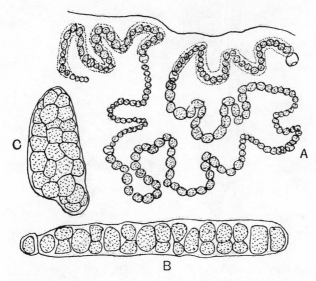

Fig. 2.8 *Nostoc.* A, portion of colony of *N. linckia*; B, C, germinating hormogones of *N. punctiforme.* (A × 400, B × 900) (after Geitler)

absent or very indistinct. Heterocysts are terminal or intercalary and occur singly or in series. Reproduction is by means of hormogones or spores, the latter arising midway between the heterocysts and developing centrifugally.

The closely related genus *Anabaena* differs from *Nostoc* in that no firm colony is formed and it lacks a punctiform stage, whilst in *Wollea* the saccate plants contain trichomes arranged in a parallel manner. A feature of the genus *Cylindrospermum* is the large spore that develops next to the heterocyst. Some species are symbiotic (cf. p. 385), whilst species of both *Nostoc* and *Anabaena* are capable of fixing atmospheric nitrogen (cf. p. 470).

RIVULARIACEAE: *Rivularia* (*rivulus,* a small brook). Fig. 2.9

The colonies form spherical, hemispherical or irregular gelatinous masses that grow on plants, stones or soil, those of *R. atra* being especially frequent on salt marshes. Each colony contains numerous radiating filaments with repeated false branching, each branch terminating in a colourless hair. With the production of mucilage the 'branches' become displaced and their origin is not easy to see. The individual sheaths can be seen near the base of the

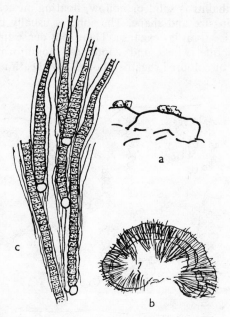

Fig. 2.9 *Rivularia atra*. a, plants on stones; b, transverse section of thallus; c, single trichome in sheath. (a × ⅔, b × 9, c × 300) (a, b, after Newton)

trichomes, but they are diffluent farther up. The heterocysts are basal and in the closely allied genus *Gloeotrichia* spores are produced next to them. In other genera the heterocysts may also be intercalary, e.g. *Calothrix*. In *Calothrix* there is no development of a gelatinous colony and the plant mass is often felt-like, as the filaments are attached at the base only.

SCYTONEMATACEAE: *Scytonema* (*scyto*, leather; *nema*, thread). Fig. 2.10
    The trichomes differ from those of *Oscillatoria* (see below) in the presence of heterocysts. The filaments (trichome and sheath) have distinct basal and apical regions forming little erect tufts. Branching is of the false type, the branches arising either between two heterocysts or else adjoining one as a result of the degeneration of an intercalary cell. The intercalary growth results in strong pressure being applied to the sheath, which finally ruptures so that the trichome forms a loop outside (Fig. 2.10, A–C). Further growth causes this loop to break, thus producing twin branches, one or both of which may subsequently proceed to additional growth, the branch sheaths extending back into the parent sheath (Fig. 2.10, E). False branching may also be

initiated by degeneration of a vegetative cell or heterocyst and subsequent growth of the two filaments on either side.

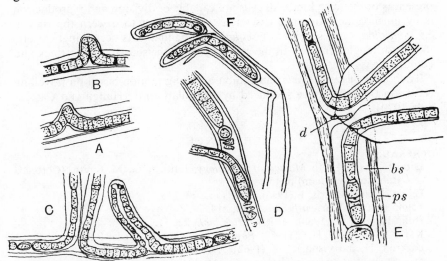

Fig. 2.10 *Scytonema.* A–C, geminate branching in *Scytonema pseudoguyanense*; D, false branching in *Calothrix ramosa*; E, false branching in *Scytonema pseudoguyanense* showing branch sheath (*bs*) terminating at heterocyst. (*ps*=parent sheath, *d*=dead cell); F, hormogones emerging from parent sheath in *S. guyanense*. (A × 470, B, C × 340, D × 570, E × 590, F × 750) (after Bharadwaja)

OSCILLATORIACEAE: *Oscillatoria* (*Oscillare*, to swing). Fig. 2.11
This is the largest family of the class. It differs from the other families in

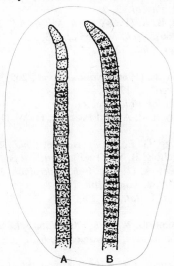

Fig. 2.11 *Oscillatoria.* A, *O. formosa*; B, *O. corallinae.* (A, B × 613) (after Carter)

that the members do not form spores or heterocysts. The trichomes are uniseriate and occur singly or many in the enveloping sheath. Branching when present is of the false kind. All cells are capable of division and reproduction is by hormogones. In some cases it is very difficult to discern the sheath, e.g. *Spirulina*, some species of *Oscillatoria*. The filaments occur singly or else they form tangled masses. In the genus *Lyngbya* the sheath is often thick and coloured.

Trichomes of *Oscillatoria* can exhibit a gliding motion based on travelling waves on the cell surface produced by a fibrillan array (Halfen and Castenholz, 1970).

REFERENCES

GENERAL

Drouet, F. (1951). In *Manual of Phycology*, Edit. by G. M. Smith, (Chronica Botanica, Waltham) p. 159.
Echlin, P. and Morris, I. (1965). *Biol. Rev.*, **40**, 143.
Echlin, P. (1966). *Scientific American*, **214**, 74.
Holm-Hansen, O. (1968). *A. Rev. Microbiol.*, **22**, 57.
Kantz, T. and Bold, H. C. (1969). *Phycol. Stud. IX*, Univ. Texas Publ.
Lazaroff, N. and Vishniac, W. (1962). *J. Gen. Microbiol.*, **28**, 203.
Pringsheim, E. G. (1967). *Öst. bot. Z.*, **114**, 324.
Pringsheim, E. G. (1968). *Planta*, **79**, 1.

SYSTEMATIC

Bourelly, P. (1969). *Öst. bot. Z.*, **116**, 273.
Desikachary, T. V. (1959). *Cyanophyta* (I.C.A.R., New Delhi).
Drouet, F. and Daily, W. A. (1956). *Bot. Stud. But. Univ.*, **12**, 1.
Schwabe, G. H. (1960). *Schweiz. Z. Hydrol.*, **22**, 759.

Development

Jensen, T. E. and Clark, R. L. (1969). *J. Bacteriol.*, **97**, 1494.
Lazaroff, N. (1966). *J. Phycol.*, **2**, 7.
Peat, A. and Whitton, B. A. (1968). *Arch. Mikrobiol.*, **57**, 155.
Pringsheim, E. G. (1966). *Arch. Mikrobiol.*, **55**, 266.
Wolk, C. P. (1965). *Dev. Biol.*, **12**, 15.

Heterocyst

Fay, P. and Walsby, A. E. (1966). *Nature, Lond.*, **209**, 94.
Lang, N. (1965). *J. Phycol.*, **3**, 127.
De Puymaly, A. (1957). *Le Botaniste*, **41**, 209.
Stewart, W. D. P., Haystead, A. and Pearson, H. W. (1969). *Nature, Lond.*, **224**, 226.
Walsby, A. E. and Fogg, G. E. (1968). *Nature, Lond.*, **220**, 810.
Walsby, A. E. and Nichols, B. W. (1968). *Nature, Lond.*, **221**, 673.
Whitton, B. A. and Peat, A. (1967). *Arch. Mikrobiol.*, **58**, 324.
Wolk, C. P. (1965). *Nature, Lond.*, **205**, 201.
Wolk, C. P. (1966). *Am. J. Bot.*, **53**, 260.

Movements

Halfen, L. N. and Castenholz, R. W. (1970). *Nature, Lond.*, **225**, 1163.
Nultsch, W. (1965). *Photochem, Photobiol.*, **4**, 613.

Sterols

Rietz, R. C. and Hamilton, J. G. (1968). *Comp. Biochem. Physiol.*, **25**, 401.
De Souza, N. J. and Nes, W. R. (1968). *Science, N.Y.*, **162**, 363.

Ultrastructure
    Fuhs, G. (1965). *Arch. Mikrobiol.*, **54**, 293.
    Gantt, E. and Conti, S. F. (1969). *J. Bact.*, **97**, 1468.
    Lang, N. J. (1968). *A. Rev. Microbiol.*, **22**, 17.
    Leak, L. V. (1967). *J. Ultrastruct. Res.*, **21**, 61. and *ibid.* (1965), **12**, 135.
    Ris, H. and Singh, R. N. (1961). *J. biophys. biochem. Cytol.*, **9**, 63.
    Wildon, D. C. and Mercer, F. V. (1963). *Aust. J. Biol. Sci.*, **16**, 585.
Gas vacuoles
    Jost, M. and Jones, D. D. (1970). *Canad. J. Microbiol.*, **16**, 159.
    Jost, M. and Matile, P. (1966). *Arch. Mikrobiol.*, **53**, 50.
    Smith, R. V. and Peat, A. (1967). *Arch. Mikrobiol.*, **57**, 111.
    Smith, R. V. and Peat, A. (1968). *Arch. Mikrobiol.*, **58**, 117.
Cyanellae Symbionts
    Chapman, D. J. (1966). *Arch. Mikrobiol.*, **55**, 17.
    Echlin, P. (1967). *Br. phycol. Bull.*, **3**, 225.
    Geitler, L. (1959). *Handb. PflPhys.*, **X1**, 530.
    Mignot, J-P., Joyon, L. and Pringsheim, E. G. (1969). *J. Protozool.*, **16**, 138.
Cyanophora
    Hall, W. T. and Claus, G. (1965). *J. Cell. Biol.*, **19**, 551.
Glaucocystis
    Hall, W. T. and Claus, G. (1967). *J. Phycol.*, **3**, 37.
    Lefort, M. and Pouphile, M. (1967). *C.r.* hebd. Séanc. *Soc. Biol.*, **161**, 992.
    Schnepf, E. and Koch, W. (1967). *Z. PflPhys.*, **55**, 97.
    Schnepf, E. and Koch, W. (1967). *Arch. Mikrobiol.*, **55**, 149.
Rivulariaceae
    Darley, J. (1968). *Le Botaniste*, **51**, 141.
Anacystis
    Echlin, P. (1963). *Protoplasma*, **58**, 439.
    Pringsheim, E. G. (1968). *Arch. Mikrobiol.*, **63**, 1.
Entophysalis
    Ginsburg-Ardre (1966). *Öst. bot. Z.*, **113**, 362.
Oscillatoria
    Drouet, F. (1968). Revision of the Oscillatoriaceae. *Philadelphia Acad. Nat. Sci. Monogr.* 15.
    Hall, W. T. and Claus, G. (1962). *Protoplasma*, **54**, 23.
    Jost, M. (1965). *Arch. Mikrobiol.*, **50**, 211.
Pleurocapsa
    Beck, S. (1963). *Flora*, **153**, 194.

3

CHLOROPHYTA

PRASINOPHYCEAE, CHLOROPHYCEAE

PRASINOPHYCEAE

This class was proposed only recently, mainly as a result of discoveries using electron-microscopy. Much of the work in this field has been carried out by Manton and Parke and their co-workers. The organisms ´are flagellates containing chlorophylls *a* and *b* and green algal xanthophylls, and are now known to be characterized by one or two layers of plate-like scales on the flagella and in some cases (e.g. *Monomastix*, *Micromonas*) on the cells as well. There may be a single flagellum (*Micromonas squamata*), two equal flagella (*Nephroselmis gilva*) or four flagella (*Prasinocladus*) (Figs. 3.1, 3.2). Hairs are also present on the flagella; these are thick and tapering, often curved at the tips, and usually with internal striations. In the genus *Heteromastix* the origin of the hairs has been clearly traced to vesicles derived from golgi bodies. In *Heteromastix*, *Halosphaera* and *Micromonas squamata* the scales also originate from the vesicles and are subsequently deposited on the outside. Rounded plate scales form the inner of two scale layers in *Prasinocladus*, *Pyramimonas*, *Halosphaera* and *Heteromastix*. The outer scales may be small rods (*Prasinocladus*), stellate (*Heteromastix*) or hexagonal (*Halosphaera*). In *Nephroselmis* there is only one layer of scales (Fig. 3.1). In *Tetraselmis* a product of photosynthesis is mannitol, normally produced in the

Phaeophyta, and if this is confirmed for other members, then their segregation will be further justified on physiological grounds.

Asexual reproduction is by fission, usually into two daughter cells. In *Micromonas squamata* a second flagellum develops before fission, whereas in *Nephroselmis* and *Heteromastix* each daughter cell receives one parent flagellum and one newly grown one.

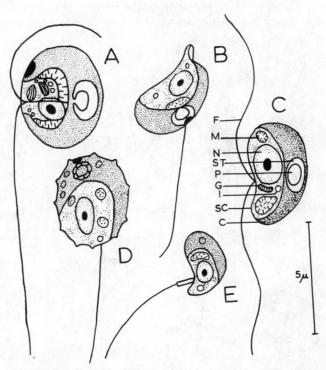

Fig. 3.1 A, *Heteromastic rotunda*; B, *Micromonas squamata*; C, *Nephroselmis gilva*; D, *Pedinomonas tuberculata*; E, *Micromonas pusilla* (after Manton *et al.*) (C = chloroplast; SC = scales; I = lipid body; G = golgi body; P = pyrenoid; ST = starch sheath; N = nucleus; M = mitochondrion; F = flagellum.)

The exact confines of this class have not yet been finally established. Some authorities retain members of the Prasinophyceae (*sensu lato*) as three sub-orders of the Volvocales within the Chlorophyceae. Some would include *Pedinomonas* even though there are no scales on the flagella. Other workers place this genus and *Micromonas pusilla*, where there is a long hair point to the flagella and no scales, in a separate family, the Loxophyceae. The class has been tentatively divided into two orders, the *Pyramimonadales* and the *Halosphaerales*. Two genera that have been studied in some detail are Prasinocladus and Halosphaera.

PYRAMIMONADALES

CHLORODENDRACEAE: *Prasinocladus* (*Prasini*, green coated charioteer; *cladus*, stalked). Fig. 3.2

Species of this genus are found in marine aquaria and on rock surfaces near high-water mark. The motile phase is a quadriflagellate swarmer with a laciniate parietal plastid, whose finger-like projections point anteriorly and posteriorly. The four flagella arise from the bottom of a depression at the anterior end. There is a distinct starch sheath surrounding the pyrenoid and

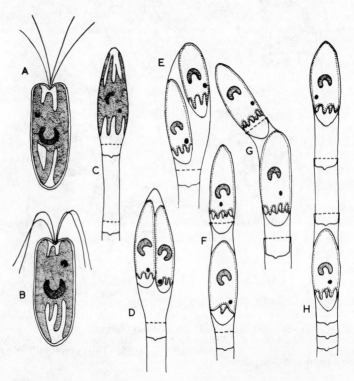

Fig. 3.2 *Prasinocladus lubricus.* A, B, Motile swarmers; C, Cell in theca apex; D–G, Cell division; H, Filament with daughter cell left behind. (after Kylin)

also a 2-layered red eye spot which is more pronounced in summer swarmers than in winter ones. The cells are covered by a periplast and when the organism becomes non-motile this undergoes plastic deformation to form the stalk of the sedentary individual. Because it contains no cellulose, this type of 'wall' may be termed a theca. The theca is composed of three layers, a central homogeneous one and an outer and inner fibrillar layer. Motile bodies are not common. When a cell becomes motile it starts to move out of the parent theca but becomes wedged in the neck, and then a new theca is formed. In this manner a series of empty cups are built up, and the plastic deformation of these empty thecae gives the jointed appearance of a hollow stalk. Division of the non-motile cells takes place by longitudinal fission and this can result in branch formation; sometimes one daughter cell does not emerge but is left behind in the middle of a series of empty thecae. On the few occasions when a swarmer is liberated the species is perpetuated, but at present the conditions under which liberation occurs are not known. Cyst formation may also occur.

The well known symbiont of the platyhelminth *Convoluta roscoffensis* has recently been reassigned to this order as *Platymonas convolutae*.

HALOSPHAERALES

HALOSPHAERACEAE: *Halosphaera* (*halo*, salt; *sphaera*, sphere). Fig. 3.3
The large, free-floating spherical cells possess one nucleus which is suspended either in the central vacuole or in the parietal cytoplasm where it is associated with numerous discoid chloroplasts. A new membrane can be formed internally and then the old one ruptures, but if the outer membrane persists the cells appear to have a multi-layered sheath. Aplanospores and

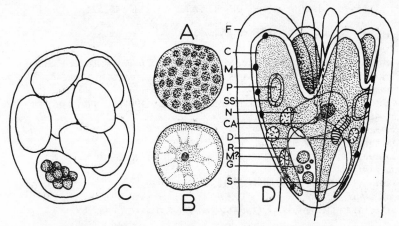

Fig. 3.3 *Halosphaera viridis*. A, mature cell; B, young cell in optical section; C, mature cell with aplanospores; D, '*Pyramimonas*' stage (A–C, after Fritsch; D, after Manton, Oates & Parke)
(F = flagellum; C = chloroplast; P = pyrenoid; SS = Starch sheath; N = nucleus; CA = canal opening into reservoir inside body; D = dictyosome; M = trichocyst or muciferous organelle; R = reservoir connected to exterior by canal; G = carotinoid? granules in reservoir; S = stigma.)

resting cysts have been recorded, but reproduction is primarily by means of quadriflagellate swarmers (Manton, Oates and Parke, 1963). These pear-shaped swarmers very closely resemble known members of the genus *Pyramimonas* both in respect of shape and morphology. There are anterior body lobes which can be outwardly or inwardly curved or which remain vertical. Around the body there are muciferous organelles or trichocysts which can eject their contents as rods, threads or balloons when treated with vital stains. Two or four pyrenoids with starch sheaths are also present. The most distinct structure, however, is a large spheroidal basal reservoir with a canal to the exterior through the frontal lobes. There are two layers of scales but the scales of flagella and body differ in pattern. The outer flagella scales are round or hexagonal plates strictly aligned in nine rows. The second layer consists of much smaller rounded scales, each with a circular groove and a central hub, arranged in a spiral fashion. The cell body scales are more varied but there is a principal diversity between flat lace-like plates and rectangular or circular scales with deeply depressed centres.

REFERENCES

GENERAL
Bourelly, P. (1966). *Les Algues d'eau douce, Tome* 1 (Boubée et Cie, Paris).
Craigie, *et al* 1966.
Manton, I. (1965). *Adv. Bot. Res.*, **2**, 1.
Prasinocladus
Chihara, M. (1963). *Phycologia*, **3**, 19.
Parke, M. and Manton, I. (1965). *J. mar. biol. Ass. U.K.*, **45**, 525.
Platymonas
Parke, M. and Manton, I. (1967). *J. mar. biol. Ass. U.K.*, **47**, 445.
Halosphaera
Manton, I., Oates, K. and Parke, M. (1963). *J. mar. biol. Ass. U.K.*, **43**, 225.
Parke, M. and Den Hartog-Adams, I. (1955). *J. mar. biol. Ass. U.K.*, **45**, 537.
Heteromastix
Manton, I., Rayns, R. G., Ettl, H. and Parke, M. (1965). *J. mar. biol. Ass. U.K.*, **45**, 241.
Micromonas
Manton, I. (1958). *J. mar. biol. Ass. U.K.*, **38**, 319.
Manton, I. and Parke, M. (1959). *J. mar. biol. Ass. U.K.*, **39**, 265.
Monomastix
Manton, I. (1967). *Nova Hedwigia*, **11**, 1.
Nephroselmis
Parke, M. and Rayns, D. G. (1964). *J. mar. biol. Ass. U.K.*, **44**, 209.

CHLOROPHYCEAE

The characteristic cell structure of the Chlorophyceae consists of a large central vacuole surrounded by cytoplasm. In some of the simpler unicellular forms there may also be two or more small, contractile vacuoles whose probable function is excretion. The pigments, which are essentially the same as those in higher plants, are contained in chloroplasts; these vary from one to many per cell and in outline may be discoid, star-shaped, spiral, plate-like or reticulate. There is evidence to show that in certain forms the chloroplasts are capable of movement in response to light stimuli. The proportions of the pigments may change at the time of gamete formation, i.e. in *Ulva lobata* fertile female thalli are olive-green and fertile male thalli are brownish. Cells of fertile thalli contain much more carotene than those of vegetative thalli, and a large proportion is γ-carotene of which only a trace occurs normally. Other colouring matter may also be present, e.g. tannins in some of the Conjugales. Formation of secondary carotenoids is very common in the Chroococcales, and other isolated genera (e.g. *Acetabularia*).

The cells are commonly surrounded by a two-layered wall, the inner, which is often lamellate, being of cellulose (xylan in certain Siphonales), and the outer of pectin. In some forms the outer surface of this pectin sheath is dissolved as fast as it is formed on the inner side. Some of the very primitive forms lack a rigid cell wall. In a few species the outer layer of pectose becomes impregnated with an insoluble substance while in others the pectin layer gradually increases in thickness, and in at least three orders (Siphonales, Dasycladales, Siphonocladales) lime may be deposited on the walls. In *Oedogonium* and *Cladophora* there is said to be a layer of chitinous material on the very outside and a study of the sub-microscopic structure of filamentous species has shown that there may be more than one kind of cellulose in the walls.

The chloroplasts normally contain pyrenoids surrounded by a sheath of starch. The pyrenoids contain fibrillar particles, but the existence of proteins is not substantiated by recent work though RNA may be present (Brown, R. M., *et al.*, 1967). The number of pyrenoids in each cell is frequently a constant specific character and can be used for taxonomic purposes. Very few families of Chlorophyceae lack pyrenoids and those that do are the more highly evolved. In *Derbesia* the pyrenoids disappear when the alga is grown in diffuse light, and in several genera they may disappear during zoospore formation. In such cases they must arise again *de novo*, but more commonly they are perpetuated by simple division at the time of chloroplast division. In those algae with pyrenoids the production of starch granules is associated with their presence.

Each cell is usually uninucleate, but in certain orders a multinucleate condition is found, and in the Siphonales cross walls are only laid down at reproduction so that the vegetative plant is non-septate and multinucleate. The chromosomes are usually small, short and few in number. Where there is only one chloroplast it divides at cell division into two daughter chloroplasts. In many species there is a distinct diurnal periodicity in nuclear and cell division.

The flagella of the motile cells are composed of an axial, cytoplasmic filament or axoneme surrounded, except at the very apex, by a sheath which is probably able to contract. In *Chlamydomonas* and the Volvocaceae the electron microscope has revealed that the axoneme consists of 9 separate fibrils arranged in a ring around 2 central fibrils. This 9 + 2 arrangement is general throughout the algae, except in the haptonema of the Haptophyceae (cf. p. 170). In the motile forms the flagella are associated with other structures which are collectively known as the neuromotor apparatus. At the base of each flagellum is a granular blepharoplast. These are connected by a transverse fibre, or paradesmose, and this, or one of the blepharoplasts, is also connected to the intranuclear centrosome by a thin strand called the rhizo-plast. The motile cells often contain a red-eyespot. In *Platydorina* the eye-spot comprises an assemblage of carotenoid-containing granules situated between the lamellae of the chloroplast (Fig. 3.4).

In other algae the carotenoid granules may have an association with the flagella with (Chrysophyceae) or without plastid association (Euglenophyta), or with flagella and a lamellar body (*Glenodinium foliaceum*), or associated with a retinoid-like structure (Warnowiaceae in Dinoflagellates) (Dodge, 1969).

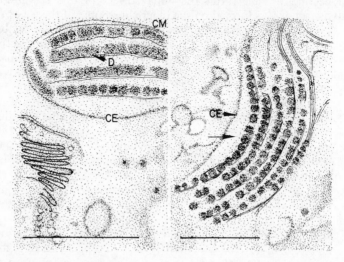

Fig. 3.4 *Platydorina* Eyespot. Stigma of *Platydorina caudata* within chloroplast envelope (CE) and inside cytoplasmic membrane (CM). See *Amer. J. Bot.*, **50** p. 297 for fuller ex-planation of this. (after Lang 1963)

The exact function of the eyespot has not been determined, although it is possible that it acts as a shading device for the parabasal granule.

Vegetative reproduction takes place through fragmentation and ordinary cell division, though in the Chlorococcales and Siphonales vegetative cell division is unknown. Asexual reproduction is by means of biflagellate or quadriflagellate zoospores, except in the Oedogoniales where there is a ring of cilia at the anterior end. The zoospores are often formed during the night and are then liberated in the morning when they may remain motile for as much as three days or for as short a time as three minutes. Their production can sometimes be induced artificially by altering the environmental conditions, e.g. by removing the plant from flowing to still water (*Ulothrix, Oedogonium*), by changing the illumination, or by removing the plant from water for twenty-four hours (*Ulva, Enteromorpha*). Each individual cell may produce one or more zoospores, the number varying with the different species. Liberation depends on: (*a*) lateral or terminal pores, or (*b*) gelatinization of the entire wall, or (*c*) the wall dividing into two equal or unequal portions. In some species non-motile zoospores called aplanospores are formed, but if these then secrete a thick wall they become known as hypnospores. Aplanospores which have the same shape as the parent cell are termed autospores. All these spores develop a new periplast or membrane and therefore differ from a purely resting vegetative cell or akinete.

Sexual reproduction is represented in all the orders and often there is a complete range from isogamy to oogamy, the egg sometimes being retained and fertilized on the parent thallus in the oogamous forms (e.g. *Volvox*, *Oedogonium*, *Coleochaete*, Charales). The isogamous forms are normally dioecious, the two sexes being termed + and –, because as plants and gametes are alike morphologically they are only distinguished by the fact that the gametes can normally only fuse with gametes derived from another plant. In some cases (*Ulva*, *Enteromorpha*, *Chlamydomonas*) relative sexuality has been reported, weak + or – strains fusing with strong + or – strains respectively. Segregation into + and – strains occurs during meiosis, a phenomenon which in many species takes place at the first division of the fertilized egg or zygote. In the Dasycladales (see p. 96) the gametes are produced in special cysts that are liberated from the parent thallus. In the case of oogomous algae, the gametangia (antheridia, the male, and oogonia, the female) are normally of distinctive shape. In some genera gametes are capable of developing into new plants by parthenogenesis.

The occurrence of sexual reproduction can be induced in culture by an abundance or deficiency of food material or by intense insolation. Interspecific hybrids have been recorded in *Spirogyra*, *Ulothrix*, *Stigeoclonium*, *Draparnaldia*, and *Chlamydomonas*. Another striking fact is that characters which may develop in some species under the influence of the external environment are normally found 'fixed' in others. This indicates the plasticity of many members in the class and the phenomenon may also be of importance phylogenetically.

As a group the Chlorophyceae are widespread, occurring in all types of habitat: marine, freshwater, soil and subaerial. Only about 13 per cent are marine and are mainly represented by the Ulvaceae, Siphonocladales, Dasycladales, Derbesiales and Siphonales. Many of the marine species have a definite geographical distribution whereas most of the freshwater and soil algae are cosmopolitan. A few genera, e.g. *Entocladia* (*Endoderma*), *Chlorochytrium*, *Cephaleuros*, *Phyllosiphon*, are parasitic species, whilst species of other genera participate in symbiotic associations, e.g. *Zoochlorella*, *Trebouxia* (cf. p. 385).

VOLVOCALES

This order comprises the simplest members of the Chlorophyceae and is the only order with motile vegetative cells. It ranges from simple unicellular forms of a basic type, e.g. *Chlamydomonas*, to regular colonial aggregates of similar cells, e.g. *Volvox*; in other cases the cells are aggregated in an irregular manner, e.g. *Tetraspora* (p. 47). Throughout the order the motile reproductive swarmer is uniform in character and even when aggregated the individual cells usually retain their power of locomotion. Electron-microscope studies show that the microcellular structure is uniform throughout. Ettl

(1965) has recently proposed subdividing the order into three (Chlamy-domonadales, Volvocales and Tetrasporales) but further detailed studies by the electron-microscope are probably necessary before accepting this change.

CHLAMYDOMONADACEAE: *Chlamydomonas* (*chlamydo*, cloak; *monas*, single). Figs. 3.5, 3.7

The chlamydomonad type of cell characteristically possesses a single chloroplast, a red eye-spot, one pyrenoid, two contractile vacuoles and two flagella, and is often strongly attracted to light (positively phototactic). Variations in the structure of the cell occur throughout the genus which contains about 325 species. Some species lack pyrenoids (*C. reticulata*); some have one (*C. stellata*), or two (*C. de baryana*), or several (*C. gigantea*). The chloroplast is usually basin-shaped but may be reticulate (*C. reticulata*), or axile and stellate (*C. eradians*), or it may be situated laterally (*C. parietaria*). It has been said that under cultural conditions many of the characteristic features can be modified, and that therefore some of the species are not strictly species but are simply phases in the life cycles of other species, or ecophenes of the same species.

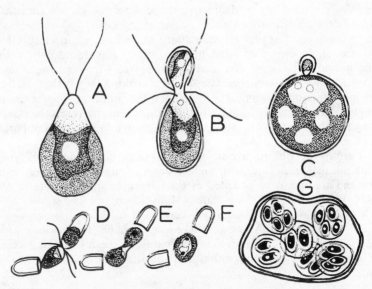

Fig. 3.5 *Chlamydomonas.* A, *C. fonticola*, vegetative cell; B, *C. fonticola*, fusion of naked gametes; C, *C. coccifera*, fusion of gametes; D–F, *C. media*, stages in fusion; G. *C. braunii*, palmelloid stage. (after G. Huber-Pestalozzi)

The motile cells are spherical, ellipsoid, ovoid or pyriform in shape with a thin membrane which sometimes has an outer layer of mucilage. The term naked is often used to describe cells (usually flagellates) which lack a cell wall (e.g. cellulose, pectin). The term is relative as so-called naked cells (e.g. *Chlamydomonas*) often possess an outer membrane, which surrounds the whole cell, and also possess a thin protoplasmic membrane or periplast. In this text the term naked describes cells that lack the cellulose-type wall character-istic of non-flagellates without denoting any particular type of periplastic mem-brane. The two flagella are situated anteriorly and either project through one aperture in the membrane or else through two separate canals. In either case there are two basal granules or blepharoplasts at the point of origin of the flagella (see p. 33). Electron microscope studies have shown that every cell contains mitochondria, dictyosomes, a double cell membrane, endoplasmic reticulum and fenestrated nuclear membrane. Near the dictyosomes there are also large irregular vesicles whose function is unknown.

During asexual reproduction the vegetative cell comes to rest and divides into two or four, more rarely eight or sixteen, zoospores. All divisions at zoospore formation are normally longitudinal. The biflagellate zoospores escape through gelatinization or rupture of the cell membrane, but if this does not occur the colony then passes into a palmelloid state, which is usually of brief duration; in *C. kleinii* however, it is the dominant phase in the life of the organism. *C. kleinii* may thus be regarded as forming a transition to the condition found in *Tetraspora* (cf. p. 47).

In sexual reproduction 8, 16 or 32 biflagellate gametes are formed in each cell. These are normally either + or − in character and fusion takes place between the strains. In *C. agametos* an anticopulatory hormone is produced and so there is no fusion; in *C. monoica* there is physiological anisogamy as the naked contents of one gamete (male) pass into the envelope of the other; in *C. braunii* the anisogamy is more distinct, the female cell producing two or four macrogametes and the male cell 8 or 16 microgametes. In *C. suboogama* each cell gives rise to three macrogametes and an antheridial cell that pro-duces four antherozoids; this is a most interesting condition, because sex segregation must take place at the first division of the mother cell nucleus. In *C. coccifera* there is oogamy, with the female cell producing one macro-gamete enclosed in a wall, while the male cell produces 16 or 32 spherical microgametes. In a related genus, *Chlorogonium oogamum*, one naked ovum is produced and numerous elongate antherozoids. The genus *Chlamydomonas* appears to be polyploid, with eight as the basic haploid chromosome number.

The zygote is at first quadriflagellate but it soon loses the flagella and forms a smooth or spiny wall. It subsequently enlarges in most species and on germination generally gives rise to four swarmers. Meiosis occurs during this segmentation and the normal vegetative cells are therefore haploid. In *C. pertusa* and *C. botryoides*, however, the zygote may remain motile for as long as ten days. In *C. variabilis* the persistent 4-ciliate zygote has for long been known as *Carteria ovata* but it has now been demonstrated that the two organisms represent the haploid and diploid phases respectively of one species. Whether these examples represent a true alternation of generations or whether they must be regarded as possessing a special type of zygote will be con-sidered later (p. 344). In most species the zygote requires a maturation period, which in *C. reinhardii* is 6 days at 25°C. The life history, which probably holds for the majority of the other species, is shown in Fig. 3.6. The species is very easy to grow in culture and the four products of zygote germination readily separate on an agar medium.

Many mutants of *C. reinhardii* have appeared during investigations and it has been possible to work out a number of linkage groups of different genes. Many of the mutant strains are of such a nature that it will be possible to attack fundamental problems in the biochemistry as well as the genetics of these organisms (see p. 472), (Hastings, 1965).

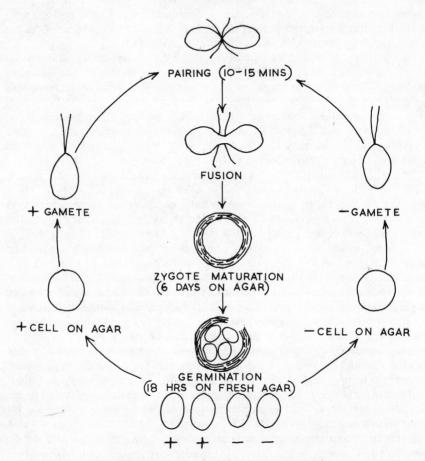

Fig. 3.6   Life history of *Chlamydomonas reinhardii*.

The genus is widespread, the various species occurring principally in small bodies of water, in sewage oxidation lakes (see p. 418) and in the soil.

Within the family Chlamydomonadaceae there is a considerable variety of forms. Among those which lack a true cell wall is the biflagellate *Dunaliella*, one species of which, *D. salina*, is very common in salt ponds. In *Phacotus* the chlamydomonad cell is surrounded by a special outsize hardened envelope which is flattened in side view. *Polytoma* is a colourless member of the family but nevertheless produces starch as its food reserve.

Possible lines of development within the family have recently been discussed by Ettl (1965) and are illustrated in Fig. 3.7.

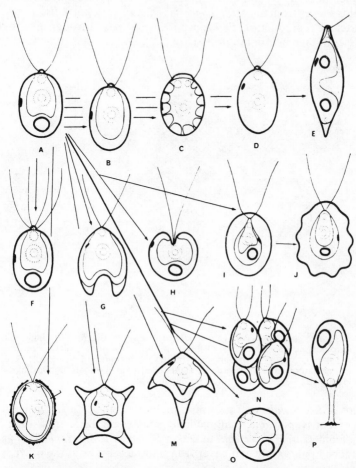

Fig. 3.7 The relationship of *Chlamydomonas* to some closely related genera. The figure only represents morphological relationships; it does not refer to phylogenetic ones. A, *Chlamydomonas*; B, *Chloromonas*; C, *Gloeomonas*; D, *Polytoma*; E, *Chlorogonium*; E, *Carteria*; G, *Selenochloris*; H, *Chlamydonephris*, I, *Sphaerellopsis*; J, *Lobomonas*; K, *Thorakomonas*; L, *Diplostauron*; M, *Brachiomonas*; N, *Pascherina*; O, *Hypnomonas*; P, *Chlorangium*. (after Ettl, 1965)

HAEMATOCOCCACEAE: *Haematococcus* (*haemate*, orange; *coccus*, berry). Fig. 3.8

A characteristic of this genus is the area forming the inner wall between the protoplast, which is very similar to that of *Chlamydomonas*, and the firm, outer cell wall; this region is filled by a watery jelly and crossing it are cyto-plasmic threads which pass from the central protoplast to the cell wall. These are numerous and very fine in *H. pluvialis*, but are fewer and thicker in the other species. The flagella are unusual in possessing a thin tomentum of fine hairs. Contractile vacuoles are not always present and if present they are inconspicuous. The plastid varies in the different species and the number of pyrenoids ranges from one in *H. capensis* to many in *H. pluvialis*.

In asexual reproduction the zoospores may pass through a non-motile aplanospore stage before maturing (Fig. 3.8), and this has developed to such

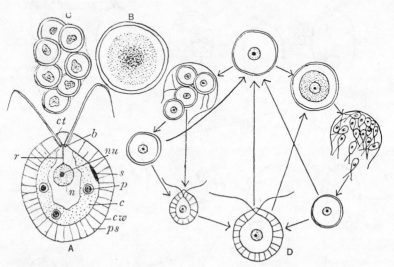

Fig. 3.8 *Haematococcus pluvialis* (*Sphaerella lacustris*). A, diagram of single macrozoid (*b* =blepharoplast, *c* =chloroplast, *ct* =flagellum tube, *cw* =cell wall, *n* =nucleus, *nu* = nucleolus, *p* =pyrenoid, *ps* =protoplasmic strand, *r* =rhizoplast, *s* =stigma.); B, encysted plant with haematochrome in centre; C, eight-celled palmelloid stage; D, diagram illustrating life cycle in bacteria-free cultures. (after Elliott)

an extent in *H. pluvialis* that it forms a major feature of the life history. In all species except *H. pluvialis* small biciliate gametes can be formed, the number ranging from 32 in *H. buetschlii* to 256 in *H. capensis*. In *H. pluvialis* gametes are produced only rarely and never from motile cells, as in the other species, but only from resting cells (Fig. 3.8).

VOLVOCACEAE:

So far the genera considered have comprised only unicells, but a whole series is known in which an increasing number of chlamydomonad cells are combined into colonial forms. The simplest type is probably represented by *Pascheriella* where three or four chlamydomonad cells are attached to each other laterally. The colonies are usually spherical in shape, but in the genus *Gonium* four, eight, or sixteen cells all lying in one plane form a flat quadrangular plate.

In the genus *Pandorina* 4, 8, 16 or 32 cells are closely aggregated in a common spherical envelope, whilst in *Volvulina* the 16 cells are arranged in four

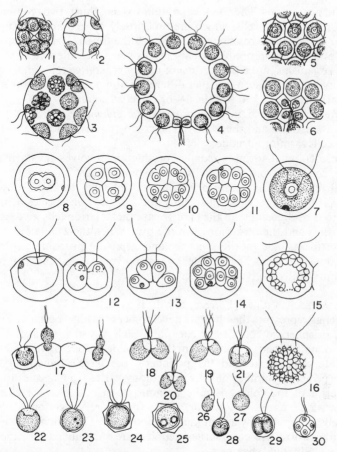

Fig. 3.9 1–3: *Volvulina Steinii*. 1, polar view, anterior pole; 2, optical section, third tier of cells; 3, colony in side view, some cells forming daughter colonies.

4–30: *Astrephomene gubernaculifera*. Structure, figs. 4–7: 4, nearly mature colony, side view, optical section; 5, anterior pole, surface view; 6, posterior pole, surface view, showing rudder; 7, single cell from anterior pole, surface view.

Asexual reproduction, figs. 8–16: (a) Surface view: 8, first nuclear division complete, cleavage beginning; 9, 4-celled stage; 10, third nuclear division complete, cleavage beginning; 11, 8-celled stage; (b) Side view: 12, flattening of outer wall preparatory to division and 2-celled stage; 13, 8-celled stage; 14, 16-celled stage; 15, cell division nearly complete, optical section; 16, daughter colony fully formed, flagella developing; parental eyespot and flagella still present.

Sexual reproduction, figs. 17–25: 17, escape of gametes; 18–20, gametes of various sizes beginning to conjugate; 21, conjugating gametes enclosed in hyaline membrane preparatory to final fusion; 22, planozygote recently formed; 23, planozygote with developing wall, and 24, with wall fully differentiated; 25, zygospore soon after coming to rest, with two pyrenoids.

Germination of zygospore, figs. 26–30: 26, small asymmetric zoospore; 27, larger spherical zoospore; 28, first division in formation of germ-colony beginning; 29, two-celled stage and 30, four-celled stage, flagella of zoospore still present and wall expanded to form vesicle. 1–6 × 260; 7–30 × 500. (after Pocock)

41

tiers (Fig. 3.9). This process has gone further in the genus *Eudorina* where the colonies are spherical or ellipsoidal with the posterior end often marked by mamillate projections. The colonies contain sixteen, thirty-two (commonly) or sixty-four biflagellate cells, which are not closely packed and are sometimes arranged in transverse rows. In *E. illinoiensis* the four anterior cells are much smaller and cannot reproduce. This marks the first differentiation of this series of organisms into a plant soma of purely vegetative cells which die once the colony has reproduced. Further differentiation occurs in *Pleodorina* where the somatic cells form $\frac{1}{3}$ to $\frac{1}{2}$ of the 64- or 28-celled colonies.

A genus comparable to *Pleodorina* is *Astrephomene* which is found in S. Africa, Australia and the U.S.A. The colony consists of 64 cells, the four anterior cells forming a 'rudder'.

In these colonial forms asexual reproduction involves the formation of daughter coenobia from all the cells except somatic cells.

Sexual reproduction in *Pandorina* shows a tendency to anisogamy, but in *Eudorina* it borders between anisogamy and oogamy, the species being either monoecious or dioecious: in the former case the anterior cells give rise to the antherozoids and the posterior form the ova. When the zygote germinates it gives rise to a single motile swarmer and what are probably two or three degenerate swarmers. The motile swarmer eventually comes to rest and divides mitotically to give a new coenobium.

VOLVOCACEAE: *Volvox* (*Volvere*: to roll) Figs. 3.10–3.13

The genus represents the ultimate development that has been reached along this line of spherical colonies or coenobia, each of which behaves as a well-organized unit. Every colony is in the form of a sphere with 500–60,000 biflagellate cells set around the periphery, the flagella emerging through canals. The actual number of cells in a colony varies from species to species. The interior of the colony is mucilaginous, as in *V. aureus*, or else merely contains water, as in *V. globator*, while the entire collection of generally ovoid cells is bounded by a firm mucilaginous wall. The individual cells, each containing two to five contractile vacuoles, are surrounded in some species by distinct individual sheaths.

In some species, too, the cells are united by delicate cytoplasmic threads, or plasmodesmae, and the base of each thread appears to be associated with a mitochondrion. In *V. globator* and *V. rousseletii* the cells are 'sphaerelloid' in nature, each individual cell being enclosed in a separate envelope, while in *V. aureus* a number of individual chlamydomonad cells are enclosed in wedge-shaped prisms (Fig. 3.10). It has therefore been suggested that the volvocine colony has arisen at least twice in the course of evolution, once from a *Haematococcus* and once from a *Chlamydomonas* ancestry.

The majority of the cells, including all those in the anterior quarter, are wholly somatic, and only a few are able to give rise to daughter colonies or gametes. As the plant matures some of the posterior cells enlarge and lose their flagella. The enlarged gonidium then divides longitudinally a number of times and a small hollow sphere is produced with a pore (phialopore) towards the outer edge. These coenobia, which hang down into the parent cavity, then

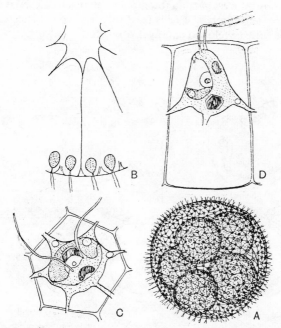

Fig. 3.10 *Volvox*. A, *V. aureus* with daughter colonies; B, structure of *V. aureus* as seen in section; C, surface view of single cell of *V. rousseletii* (× 2000); D, the same in side view (× 2000). (A, B, after Fritsch; C, D, after Pocock)

invert, the process commencing opposite the phialopore. After inversion flagella are formed and the colonies may be liberated into the parental cavity or they may escape to the exterior (Fig. 3.11). If retained they remain until the parent tears open, in some species (*V. aureus*) at the adult phialopore, in other species (*V. globator*) at any place. In *V. africana* it is possible to see as many as four generations in the one original parent colony, because the first parent takes a long time to break down.

In sexual reproduction the plants are either monoecious (*V. globator*) or dioecious (*V. aureus*). Cells which give rise to antherozoids divide to give bowl-shaped or globose colonies containing 16, 32, 64, 128, 256 or 512 spindle-shaped, biflagellate antherozoids. Starr (1968) reports that in eight strains of *Volvox* the liberation of a male proteinaceous hormone induces egg formation. Cells developing into ova enlarge considerably and loose their flagella but do not undergo division. Both bowls and spheres undergo inversion before maturation (cf. Fig. 3.12). The fertilization mechanism is not known for certain, but in *V. aureus* the antherozoids penetrate the female colony and then enter the ovum from the inner side (Holt, 1969). Zygotes do not germinate for a considerable time after liberation. Meiosis occurs at germination but the oospore only produces a single swarmer that is liberated into a vesicle formed by extrusion of the inner oospore wall (endospore). This grows into a 'juvenile' plant of about 128 to 500 cells, which finally inverts. It subsequently reproduces by daughter colonies, each successive generation having an increased number of cells (see p. 55 for comparison with *Hydrodictyon*) until a fully developed colony is produced.

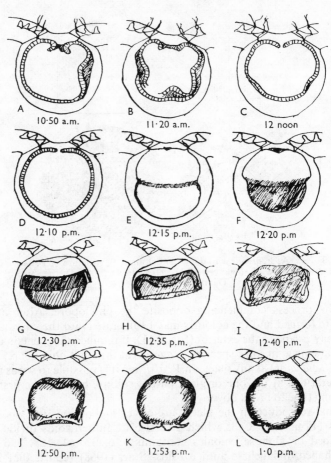

Fig. 3.11 *Volvox capensis* and *V. rousseletii*. A–J, stages in the inversion of a daughter colony: A, denting begins; C, dents smooth out; D, colony round again; E, 'hour-glass' stage; F, posterior half contracts; G, infolding begins; H, infolding complete; I, posterior half emerges through phialopore; J, flask stage begins; K, flask stage ends; L, inversion complete. (All × 150 approx.) (after Pocock)

One of the features of the larger saccate Volvocales (excepting *Astrephomene*) is the inversions that occur at different stages of the life-cycle, and it is difficult to see why they occur or what the conditions were under which they first developed. It may have something to do with the fact that the new cells are formed with the flagella and eye-spot facing the interior, but then the problem arises as to how the individual cells came to be arranged thus.

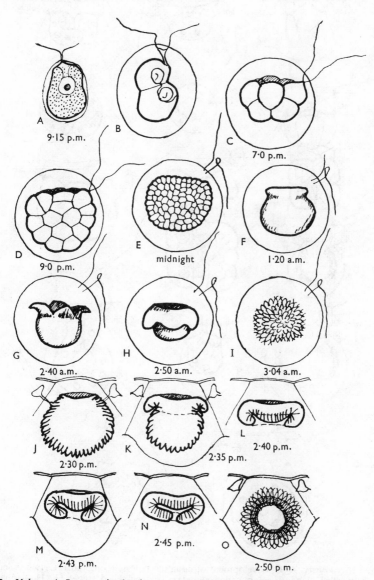

Fig. 3.12 *Volvox*. A–I, stages in the development of the oospore of *V. rousseletii*: A, zoo-spore just after escape; B, first division; F, preparation for inversion; G–I, inversion. (A–I × 375). J–O, stages in the inversion of a sperm bundle of *V. capensis*. (J–O × 750) (after Pocock)

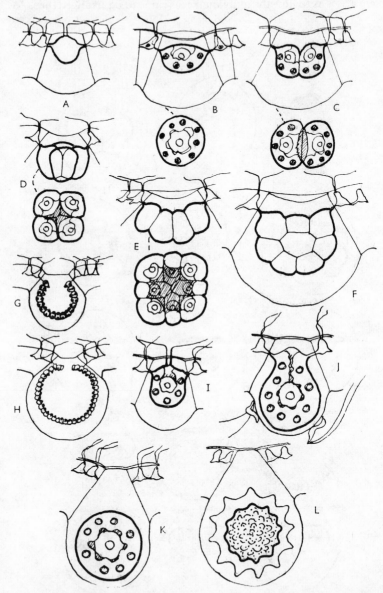

Fig. 3.13 *Volvox*. A–H, development of daughter colony (gonidium): C, two-celled stage; D, four-celled stage; E, eight-celled stage; F, sixteen-celled stage; G, H, formation of phialopore. I–L, development of oospore of *V. rousseletii*: I, flagellar stage; J, mature; K, fertilized; L, exospore formation. (A–H × 750; G, H × 225; I–L × 750) (after Pocock)

The nuclear condition of the Volvocaceae presents some interesting features (Cave and Pocock, 1951). *Gonium*, which would seem generally to be a primitive member of the family, possesses the largest number of chromosomes ($n = 17$). In *Volvox* the species regarded as primitive have more chromosomes ($n = 14$ or 15) than species regarded as advanced ($n = 5$). The facts may necessitate a re-interpretation of development within the family, *Gonium* being regarded as reduced rather than primitive, and similarly with the species of *Volvox*. Colonies with 4, 6, 7, or 8 chromosomes occur in the genus *Astrephomene*. Polypoidy is suggested as clones with 4 chromosomes remain distinct from those with 6, 7, or 8.

TETRASPORACEAE: *Tetraspora* (*tetra*, four; *spora*, spores). Fig. 3.14

Members of this and allied families are placed by some workers in a separate order, the Tetrasporales. They are characterized by non-motile vegetative cells that may temporarily become motile. In so far as the cells are chlamydomonad-like, there seems no justification for the segregation. The species of *Tetraspora* form expanded or tubular, convoluted, light green macroscopic colonies. The colony in these palmelloid forms differs from the organized coenobium of the Volvocaceae because it is merely an unorganized loose assembly of cells. The palmelloid condition (see p. 331) has in this and

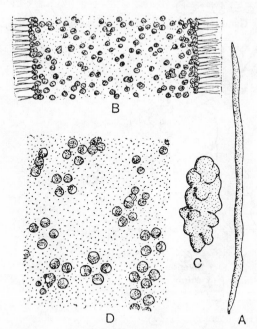

Fig. 3.14 *Tetraspora*. A, *T. cylindrica*; B, portion of colony of *T. cylindrica* showing outer envelope; C, *T. lubrica*; D, portion of colony of *T. lubrica*. (A × ½; B × 155; C × ½; D × 500) (after Smith)

allied genera become the dominant phase, while the motile condition only occupies a brief period in the life cycle.

The colonies are most abundant in the spring when they are attached at first, although later they become free-floating. The spherical to ellipsoidal cells are embedded in the mucilage and are frequently arranged in groups of two or four, each group being enclosed in a separate envelope. Two pseudocilia proceed from each cell to the surface of the main colonial envelope, each thread being surrounded by a sheath of denser mucilage.

Vegetative reproduction takes place by fragmentation of the parent colony,

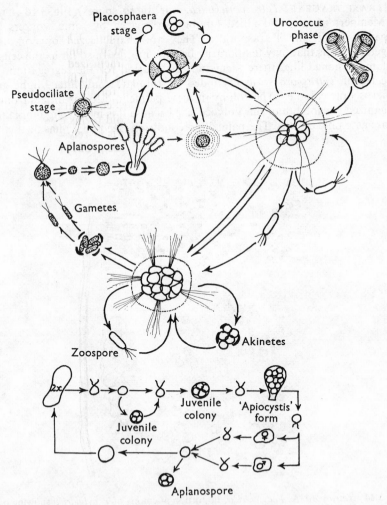

Fig. 3.15   Top: Life history of *Schizochlamys*. (after Thompson)
Bottom: Life history of *Tetraspora gelatinosa*. (after Hirose)

and asexual reproduction is secured by means of biflagellate swarmers that may develop into a new colony or into a thick-walled resting spore that later gives rise to a new colony. The gametophytic colonies are monoecious or dioecious, depending on the species, and give rise to biflagellate isogametes. The zygote gives rise to four to eight aplanospores that are said to grow into new colonies. Work on *T. gelatinosa* suggests that the aplanospores are not always produced and that the zygote, after a resting period, can give rise directly to a new diploid plant which gives rise to asexual swarmers. These develop into juvenile colonies that can perpetuate themselves by biflagellate swarmers. Eventually swarmers are produced that give rise to an attached, pear-shaped colony. The swarmers from this colony give rise to the sexual plant so that in this species there is a definite alternation of morphological phases (Fig. 3.15). It seems probable, though it has not been yet established, that meiosis may well take place at swarmer formation in the attached colonies.

A life cycle has been fully worked out for *Schizochlamys* (Thompson, 1956). Here the adult can reproduce itself by akinetes, zoospores or gametes (Fig. 3.15).

REFERENCES

GENERAL

Bourelly, P. (1966). *Les Algues d'Eau Douce*, Tome **1**, (Boubée et Cie, Paris).
Dodge, J. D. (1969). *Br. Phycol. J.*, **4** (2), 199.
Van Den Hoek, C. (1965). *Nova Hedwigia*, **10**, 367.
Eyespot
Walne, P. L. and Arnott, H. J. (1967). *Planta*, **77**, 325.
Flagella
Jacobs, M., Hopkins, M. and Randall, J. (1969). *Proc. R. Soc.*, **173B**, 61.
Ringo, D. L. (1967). *J. Cell. Biol.*, **33**, 543.
Rosenbaum, J. L., Moulder, J. E. and Ringo, D. L. (1969). *J. Cell. Biol.*, **41**, 600.
Pyrenoid
Brown, Jr., R. M. and Arnott, H. J. (1970). *J. Phycol.*, **6**, 14.
Brown, Jr., R. M. (1967). *J. Phycol.*, **3**, Suppl. 5.
Volvocaceae
Cave, M. S. and Pocock, M. A. (1951). *Am. J. Bot.*, **38**, 800.
Stein, J. R. (1958). *Am. J. Bot.*, **45**, 388.
Volvocales
Ettl, H. (1965). *Nova Hedwigia*, **10**, 515.
Lang, N. J. (1963). *Am. J. Bot.*, **50**, 280.
Astrephomene
Stein, J. R. (1958). *Am. J. Bot.*, **45**, 388.
Chlamydomonas
Brown, Jr., R. M., Johnson, G. and Bold, H. C. (1968). *J. Phycol.*, **4**, 100.
Ettl, H. (1965). *Phycologia*, **5**, 61.
Hastings, P. J. (1965). *Microb. Genet. Bull.*, **23**, 17.
Levine, R. P. and Ebersold, W. T. (1960). *Ann. Rev. Microbiol.*, **14**, 197.
Starr, R. C. (1968). *Proc. Natn. Acad. Sci. U.S.A.*, **59**, 1082.
Eudorina
Doraiswami, S. (1940). *J. Indian bot. Soc.*, **19**, 113.
Goldstein, M. (1964). *J. Protozool.*, **11**, 317.

Gonium
    Stein, J. R. (1958). *Am. J. Bot.*, **45**, 664.
    Stein, J. R. (1966). *J. Phycologia*, **2**, 23.
    Stein, J. R. (1965). *Am. J. Bot.*, **52**, 379.
Haematococcus
    Pocock, M. A. (1960). *Trans. R. Soc. S. Afr.*, **36**, 6.
Pandorina
    Coleman, A. W. (1959). *J. Protozool.*, **6**, 249.
    Friedman, I., Colwin, A. L. and Colwin L. H. (1968). *J. Cell. Sci.*, **3**, 115.
Schizochlamys
    Thompson, R. H. (1956). *Am. J. Bot.*, **43**, 665.
Tetraspora
    Herndon, W. (1958). *Am. J. Bot.*, **45**, 298.
Volvox
    Darden, W. (1966). *J. Protozool.*, **13**, 239.
    Deason, T. R., Darden, W. H. and Ely, S. (1969). *J. Ultrastruct. Res.*, **26**, 85.
    Holt, B. R. (1969). *Ann. Meet. Mich. Acad. Sci. Summ.*
    Pocock, M. A. (1933). *Ann. S. Afr. Mus.*, **16**, 523.
    Pocock, M. A. (1938). *J. Queckett. Micr. Club, Ser.* 4, **1**, 1.

CHLOROCOCCALES

This order contains a number of forms, diverse both in respect of morphology and reproduction. Nearly all are freshwater or terrestrial. Some, such as *Chlorochytrium*, *Chlorococcum*, because of their habit, probably represent reduced forms. In structure the plants range from solitary, free-living or attached unicells to large coenobia (*Hydrodictyon*) up to 20 cm long. A characteristic feature of the order is the lack of vegetative cell division, though this has not precluded nuclear division, so that cells in some species become multinucleate. Many of the genera in this order possess the ability to form secondary carotenoids such as canthaxanthin, astaxanthin, echinenone. These are formed extra-plastidically in old resting cells to which they impart a distinct yellow to red hue. The unicellular, spherical, zoospore-producing genera have, in the past, been responsible for much taxonomic confusion but new microscopical techniques and the development of unialgal cultures have clarified much of this confusion. The three most important taxonomic characters for these particular genera are the type of zoospore, the type of chloroplast and the presence or absence of pyrenoids (Starr, 1953). More recent approaches have involved chemical taxonomy, especially with species of *Chlorococcum* and *Chlorella*.

In the various examples discussed it will be noted that motility has been suppressed more than once in the order, a feature which, taken in conjunction with the diverse morphology, suggests that the organisms may well be polyphyletic in origin. Some genera show indications of a relationship, but there are others whose relationships are extremely vague. Some forms (*Trebouxia*) have become associated with fungi and are an integral part of lichens (see below and p. 385). Sexual reproduction when present is usually by means of motile isogametes, but oogamy has been recorded for a species of *Dictyosphaerium*.

CHLOROCOCCACEAE: *Chlorococcum* (*chloro*, green; *coccum*, berry). Fig. 3.16

Much confusion has existed over this genus, and many of the species formerly described are now known to be phases in the life cycles of species from other genera. Bold and Parker (1962) consider that there are 18 valid species which are not always readily distinguished on morphological grounds, in which case changes in colony shape, cell size, colour or colony texture under specific culture conditions may be useful. The plants are represented by non-motile, spherical or ellipsoidal cells which vary in size, occurring singly or else forming a stratum on, or in the soil; *C. humicolum*, for example, is a very common soil species. There is no eye spot or contractile vacuole; the chloroplast is hollow and parietal with one or more pyrenoids. The characteristic features of this genus are first, the hollow, parietal chloroplast; second, the presence of one or more pyrenoids; and third, biflagellate zoospores that retain their oval shape for some days after losing their motility. The cell walls are two-layered with a thin inner layer and an outer gelatinous

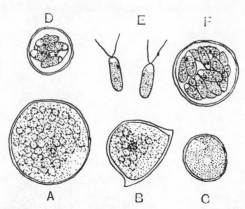

Fig. 3.16 *Chlorococcum humicolum*. A–C, vegetative cells; D–F, zoospore formation. (D–F × 800) (after Smith)

one which is sometimes lamellose and irregularly thickened. Old cells may be coloured with an excess of secondary carotenoids. The young cells with thin walls are uninucleate but the adult ones with thick walls are commonly multinucleate, and it is in this older condition that the protoplast divides and gives rise to 8, 16 or more biflagellate zoospores or isogametes which are liberated all together in a vesicle, usually in the early hours of the morning. After a short motile phase the flagella are withdrawn and a new vegetative phase commences.

Under dry conditions aplanospores are formed and when this occurs the

parent cell gelatinises and a 'palmella' stage results, the cells of which subsequently give rise to 2 or 4 biflagellate gametes. In culture solutions of low
nutrient concentration, reproduction takes place by means of zoospores, but
in highly concentrated solutions these are replaced by aplanospores. The
environment can therefore affect the mode of reproduction. The production
of aplanospores suggests how the genus *Chlorella* may have arisen, while
the multinucleate state indicates a transition to a coenocytic phase that is
further developed in other genera (*Protosiphon, Hydrodictyon*).

In *Trebouxia* (see p. 385), the cells differ from those of *Chlorococcum* in
lacking an irregularly thickened wall and the plastid is axile rather than
parietal. Certain species of the genus *Codiolum* (a little fleece), Fig. 3.17, have
now been shown to be stages in the life history of *Spongomorpha* and *Acrosiphonia* (see p. 94). *Codiolum kuckuckii* is restricted as an epiphyte to the apical
region of the southern brown alga *Splachnidium* (p. 207).

CHARACIACEAE: *Characium* (a slip or cutting). Fig. 3.17
Each plant is a solitary unicell with a single parietal chloroplast, and
motile only in the reproductive phase. The ellipsoidal cells occur singly, or in
aggregates on submerged plants or living aquatic larvae and they are borne
on a short stalk that emerges from a small basal disc. In most species the
mature cells contain more than one nucleus. Asexual reproduction is by

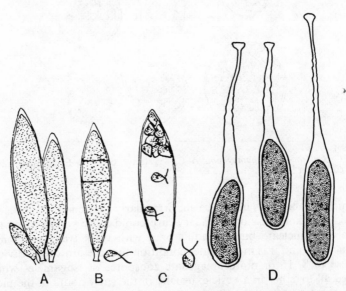

Fig. 3.17 *Characium angustatum.* A, vegetative cells; B, cell commencing zoospore formation; C, liberation of zoospores: the cell is probably broken accidentally; D, *Codiolum
gregarium* — a group of three plants. (A–C × 650) (A–C, after Smith; D, after Setchell and
Gardner)

means of biflagellate zoospores which are liberated through a terminal or lateral aperture. Sexual reproduction is generally isogamous but some species exhibit anisogamy. In *C. saccatum* the + and − gametes are produced from separate plants and fuse to give a zygote which, on germination, divides to produce a number of zoospores.

ENDOSPHAERACEAE: *Chlorochytrium* (*chloro*, green; *chytrium*, vessel). Fig. 3.18

The species grow endophytically in other plants such as mosses, angiosperms and red algae. The cells are spherical or ellipsoidal, the walls varying in thickness. Reproduction is by means of biflagellate gametes that normally fuse to give a motile zygote. This eventually settles on the leaf of a host plant and secretes a wall. Tubular prolongations grow out from these bodies and enter the host, either by the stomata or between two epidermal cells. Sub-

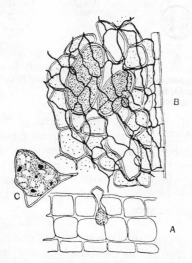

Fig. 3.18 *Chlorochytrium lemnae.* A, entrance of zygote into host; B, resting cells in leaf of *Lemna*; C, resting cell. (after Fritsch)

sequently the end of the tube swells out into an ellipsoidal or lobed structure into which the contents of the swarmer pass. In *Chlorochytrium lemnae*, which attacks duckweed (*Lemna*), the endoparasitic cells sink in the old *Lemna* fronds to the bottom of the pond or stream and remain dormant until the next spring. There is evidence that reduction division takes place at gamete formation so that the vegetative phase is therefore diploid.

CHLORELLACEAE: *Chlorella* (*chlor*, green; *ella*, diminutive). Fig. 3.19

The globular cells are non-motile, solitary or aggregated into groups, and usually lack pyrenoids. The species are not readily determined taxonomically, so cultures are often given strain numbers. Using morphological and physiological criteria some eight well-defined species can be recognized. They will grow in a wide range of media, including salt solution (0·2 M.). It seems that cells grown in the dark differ from those grown in the light. Quiescent dark cells are small and have a photosynthetic quotient (O_2/CO_2) of 1. They give rise to dividing dark cells which in daylight give rise to 'light' cells that are larger, poorer in chloroplast material and less active photosynthetically

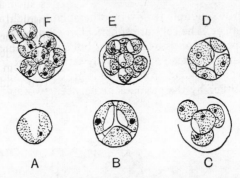

Fig. 3.19 *Chlorella vulgaris.* A, single cell; B, division into four; C, final stage of division into four daughter cells; D, first stage of division into eight; E, F, second and third stages of division into eight daughter cells. (after Grintzesco)

(photosynthetic quotient = 3). During the course of development the amount of DNA increases. 'Light' cells give rise to a new generation of 'dark' cells in the dark and reproduction is by means of division into 2, 4, 8 or 16 autospores. Several species form a symbiotic association with lower animals and they are then known as zoochlorella (cf. p. 385). Because the species grow well in uni-algal culture they are often used in physiological investigations; they are also used in sewage purification.

COELASTRACEAE: *Scenedesmus* (*scene*, rope; *desmus*, fetter). Fig. 3.20

The planktonic colonies are composed of four, eight, or, more rarely, sixteen cells attached to each other at one point by mucilage pads. This type of colony has probably originated from that in which cells are attached at one point indiscriminately, as in *Ankistrodesmus*. A large number of species are known to occur widely, and often in some abundance, in ponds and lakes. Taxonomically the two end cells are highly important. They may differ in shape from the others and often have processes that are elaborations of the cell envelope (e.g. *S. quadricauda*); these processes can probably be correlated with the planktonic mode of life. Other species (e.g. *S. acuminatus*) have tufts of bristles that perform the same function and are similar to those of *Pediastrum*.

The cell wall of *S. quadricauda* exhibits a complex structure under the electron microscope. On the outside there is a pectic, hexagonal net layer which encloses a complex prop system in pectin. There is a middle layer and an innermost cellulose layer.

Reproduction takes place when the contents of each cell in the colony divide to give daughter coenobia which are then liberated.

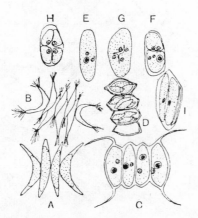

Fig. 3.20 *Scenedesmus.* A, *S. acuminatus*; B, *S. acuminatus* with mucilage bristles; C, *S. quadricauda*; D, *S. quadricauda* reproducing; E–I, stages in the formation of daughter coenobia in *S. quadricauda.* (after Fritsch)

HYDRODICTYACEAE: *Hydrodictyon* (*Hydro*, water; *dictyon*, net). Fig. 3.21

The algae included in the Hydrodictyaceae occur as coenobia and represent a more advanced stage of the simple condition seen in *Ankistrodesmus* and *Scenedesmus*. Species of *Pediastrum* are common components of freshwater plankton, with up to 128 cells united into flat, disc-like coenobia.

The most advanced condition is found in *Hydrodictyon* of which only three species are known, the commonest, *H. reticulatum*, being cosmopolitan in distribution. It is a hollow, free-floating, cylindrical network closed at either end and up to 20 cm in length. Further increase in the size of colony is probably impossible for purely mechanical reasons. The chloroplast is reticulate with numerous pyrenoids, though in the young uninucleate cells there is only a simple parietal chloroplast which later becomes spiral and then reticulate. *H. africanum* and *H. patenaeforme* develop into saucer-shaped nets, and in the former, the original young cylindrical cells later become spherical and up to 1 cm in diameter, then become detached and lie on the substratum. The latter species is composed of cells which may grow to be 4 cm long and 2 mm in diameter. Experiments have shown that for growth of *H. reticulatum* the actual growing period depends upon light intensity, e.g. it is 3 weeks at 600–700 lx, $5\frac{1}{2}$ weeks at 250 lx. Under normal conditions growth rate is periodic, increasing by day and decreasing at night and altering the length of the light period merely alters the degree of growth in the time. Under conditions of continuous light the periodicity is much reduced, though this effect is more marked with old cells.

Asexual reproduction in *Pediastrum* and *H. reticulatum* is by means of numerous uninucleate biflagellate zoospores, which swarm in the parent cell about day-break and then come together to form a new coenobium which is

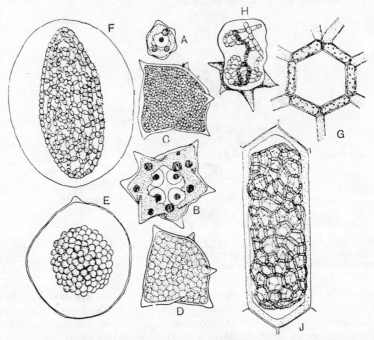

Fig. 3.21 *Hydrodictyon.* A-F, development of young net of *H. patenaeforme* from the zygote: A, young polyhedron; B, older polyhedron with four nuclei; C, protoplasm granular just before zoospore formation; D, 'pavement' stage; E, zoospores rounding off and wall of polyhedron expanding to form vesicle; F, fully formed net still enclosed in vesicle. G, portion of mature net of *H. reticulatum*; H, polyhedron and young net of *H. reticulatum*; J, *H. reticulatum*, formation of net in parent cell from zoospores. (A–E × 250, F × 175) (A–F, after Pocock; G, H, after Oltmanns; J, after Fritsch)

subsequently liberated, further growth being brought about by elongation of the coenocytic cells. It has been found that zoospore production in *H. reticulatum* depends not only upon light intensity but also upon its duration:

	Hours of illumination					
	2	4	6	8	10	
500 lx	0	0	6	88	100	% of cells producing
4000 lx	11	87	100	100	100	zoospores

Sometimes zoospores, instead of coming to rest to form a daughter net, settle down to form haploid hypnospores. These are not very common in *H. reticulatum* but are normal in *H. patanaeforme* and possibly also in *H. africanum*, where asexual reproduction of the coenocytes as described above is unknown. In the first two species the hypnospore germinates to give a large, rather slow-moving biflagellate zoospore that settles down to form a polyhedron (Fig. 3.21), whose contents later divide and after forming a pavement stage develop swarmers that are liberated into a vesicle to give a new 'germ net'. The duration of the 'germ net' in *H. reticulatum* is short because each coenocyte rapidly produces a daughter net. This does not happen in *H. patanaeforme.*

Sexual reproduction is isogamous in *H. reticulatum* and *H. patanaeforme,* but anisogamous in *H. africanum.* In the first species the biflagellate gametes are smaller than the zoospores, but in the second species there is no distinction in size. Both species are generally monoecious. The planozygote arising from conjunction is quadriflagellate, and after a motile period it comes to rest and forms a diploid hypnospore. At germination the zygote generally enlarges and germinates meiotically to give 4, or more rarely 8, slow-moving biflagellate swarmers that later pass into the polyhedron stage and develop as described above. Sexual reproduction in *Pediastrum* follows the same course. Gamete and zoospore reproduction can be induced in *H. reticulatum* by varying the external conditions artificially. If plants are cultured in weak maltose solution in bright light or in the dark and are then transferred to distilled water, zoospores will develop from the light plants and gametes from the dark ones.

Hydrodictyon is essentially a collection of a number of individual plants because it arises as the result of the fusion of a number of swarmers. *Volvox,* on the other hand, must be regarded as a single plant composed of many cells because it arises from a single zygote or gonidium.

PROTOSIPHONACEAE: *Protosiphon (proto,* first; *siphon,* tube). Fig. 3.22

There is probably but a single species, *P. botryoides,* but it exists in a number of different strains, one strain from the deserts of Egypt tolerating temperatures up to 91° C and salt concentrations up to 1 per cent. It commonly grows associated with a very similar alga, *Botrydium* (cf. p. 136) in the damp mud at the edges of ponds. The green aerial portion is more or less spherical, up to 100μ in diameter, grading into a colourless rhizoidal portion that may branch occasionally. The multinucleate cell contains one parietal reticulate chloroplast with many pyrenoids. The shape and colour of the thallus can be modified by varying the external conditions; bright light and low moisture, for example, causes an old thallus to turn brick red due to secondary carotenoid formation. During dry weather the contents of the vesicle encyst to give large aplanospores or 'coenocysts'. When conditions are once more favourable these cysts either grow directly into vegetative cells or else produce biflagellate gametes or zoospores.

Vegetative reproduction can take place by lateral buds which are cut off by cell walls. Flooding of plants on damp soil causes them to produce biflagellate swarmers which behave as isogametes though these swarmers can behave as zoospores. The zygote either germinates directly or may remain dormant for some time. The plant is probably haploid as there is evidence of meiosis at zygote germination. Morphologically it is of great interest as a possible source of origin for the Siphonales (see p. 331).

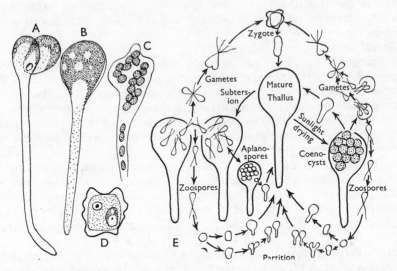

Fig. 3.22 *Protosiphon botryoides.* A, B, plants, one showing budding; C, cyst formation; D, zygote; E, schematic diagram of life cycle. (d × 1666) (A–C, after Fritsch; D, E, after Bold)

REFERENCES

GENERAL
Bourelly, P. (1968) *Les Algues d'eau Douce,* Tome **1** (Boubée et Cie, Paris).
Brown, Jr., R. M. and Bold, H. C. (1964). *Univ. Texas Publ.,* No. **6417**.
Fritsch, F. E. (1935). *Structure and Reproduction of the Algae.* Vol. 1, pp. 145–198, (Cambridge University Press).
Van Den, Hoek, C. (1965). *Nova Hedwigia,* **10**, 367.

Characium
Starr, R. C. (1953). *Bull. Torrey Bot. Club,* **80**, 308.

Chlorella
Kihara, W. and Krauss, R. W. (1963). *Chlorella — Physiology and Taxonomy of 41 Isolates* (University of Maryland).

Chlorococcum
Bold, H. C. and Parker, B. C. (1962). *Arch. Mikrobiol.,* **52**, 267.

Hydrodictyon
Neeb, O. (1952) *Flora,* **139**, 39.
Pirson, A. and Doring, H. (1952). *Flora,* **139**, 314.
Pocock, M. A. (1960). *Jl. S. Afr. Bot.,* **26**, 170.
Sorenson, I. (1951). *Oikos,* **2**, 197.

Pediastrum
Davis, J. S. (1967), *J. Phycol.,* **3**, 95.

Scenedesmus
Bisalputra, T. and Weier, T. E. (1963). *Amer. J. Bot.,* **50**, 1011.
Bisalputra, T., Weier, T. E., Risley, E. B. and Engelbrecht, A. H. P. (1964). *Amer. J. Bot.,* **51**, 548.

ULOTRICHALES

This order comprises filamentous and parenchymatous types which have clearly evolved from the filamentous form. There is little or no differentiation between base and apex in many of the forms though there is usually a special attachment cell or pad of cells. In most genera all the cells except those for attachment are capable of producing reproductive bodies, but in two genera from New Zealand (*Gemina*, *Lobata*) there appears to be a considerable degree of sterilization. Within the genus *Enteromorpha* there is also evidence of the development of heterotrichy, or the condition in which the thallus possesses a prostrate adherent portion and an erect aerial portion.

In the past the treatment of the genera in this order has varied considerably.

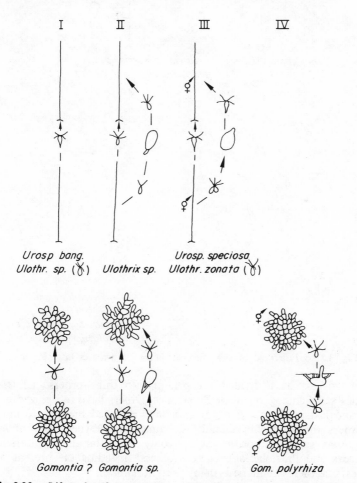

I   II   III   IV

Urosp bang.
Ulothr. sp. (☿)   *Ulothrix sp.*   *Urosp. speciosa*
Ulothr. zonata (☿)

Gomontia ? *Gomontia sp.*   *Gom. polyrhiza*

Fig. 3.23a   Life cycles of some members of the Ulotrichales. (after Kornmann)

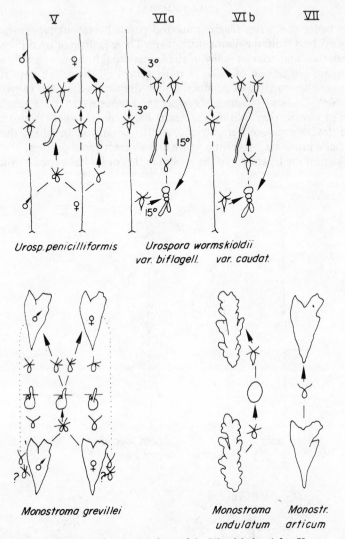

V   VI a   VI b   VII

Urosp. penicilliformis   *Urospora wormskioldii*
                            var. biflagell.   *var. caudat.*

Monostroma grevillei   *Monostroma*   *Monostr.*
                                          undulatum   *articum*

Fig. 3.23b Life cycles of some members of the Ulotrichales. (after Kornmann)

Some workers have divided the genera into three orders: Ulotrichales, Ulvales and Schizogoniales or Prasiolales. Others have recognized only the first and last of these orders, or else have regarded them as forming two suborders, the Ulotrichineae and Prasiolineae. Even when only one order is recognized some authorities accept only three families (Ulotrichaceae, Ulvaceae and Prasiolaceae) whereas others, including the present writers, suggest that there should be as many as eight families.

The Ulotrichaceae and Microsporaceae contain only filamentous unbranched forms which can be separated from one another by chloroplast and cell wall structure. The Cylindrocapsaceae are also unbranched but reproduction is oogamous as distinct from the isogamy or anisogamy in the first two families. The Sphaeropleaceae are also filamentous and oogamous but differ morphologically from the Cylindrocapsaceae. Some of the remaining four families consist of plants with a flattened or tubular parenchymatous thallus. Some members of the Monostromataceae are distinct because of the heteromorphic alternation and some of the remaining species (*Ulvaria*) could either be included in the Ulvaceae or treated as a separate family (see p. 65). The Prasiolaceae are distinguished by their thallus and plastid structure and the Capsosiphonaceae by thallus structure, habitat and reproduction.

Until very recently the taxonomic treatment of the order was based upon morphology. It was then found that some species of *Monostroma* (see p. 66) had a life history with a much reduced but persistent sporophyte stage, and so a separate family for *Monostroma* was proposed. Since then more attention has been given to life histories and Kornmann (1963) has suggested that the order should be restricted to species with an alternation of heteromorphic generations and to their derivative species. Such a radical re-alignment brings together species of *Ulothrix*, *Urospora*, *Gomontia* and *Monostroma* (Fig. 3.23). Morphologically these genera differ widely, and there are also differences in their early stages of development. If therefore we can accept that alternation of heteromorphic generations may have arisen more than once in the course of evolution the traditional taxonomic treatment would seem preferable. Despite the great interest of Kornmann's approach it would seem that it raises more problems than it solves. Another approach by Vinogradova (1969) that unites *Monostroma*, *Gomontia*, *Kornmannia* and *Enteromorpha* (*Blidingia*) *nana*, presents difficulties with respect to the last genus which is more closely related to *Enteromorpha* (see p. 69).

ULOTRICHACEAE: *Ulothrix* (*ulo*, shaggy; *thrix*, hair). Fig. 3.24
The unbranched filaments are attached to the substrate by means of a modified basal cell which frequently lacks chlorophyll, but even though attached at first, the plants may later become free-floating. Under unfavourable conditions, such as nutrient deficiency, rhizoids may grow out from the cells or the filaments become branched. This suggests one way at least in which the branched habit may have evolved from the simple filament, and in this case it probably represents an attempt to increase the absorbing surface in order to counteract the deficiency of salts. The cells of the different species vary considerably in size, shape and wall thickness. When the walls are thick they are often lamellate. The single chloroplast forms a circular band around the whole or most of the cell circumference and possesses two or more pyrenoids. When it does not encircle the cell completely the chloroplast margins are lobed. Vegetative reproduction can take place through fragmentation.

The life histories of the various species are not well known even though the genus has been recognized for a long time. There are two distinct groups of species: those in which there is an alternation of isomorphic generations and those in which only the haploid generation is known or in which there is a dwarf phase (see above).

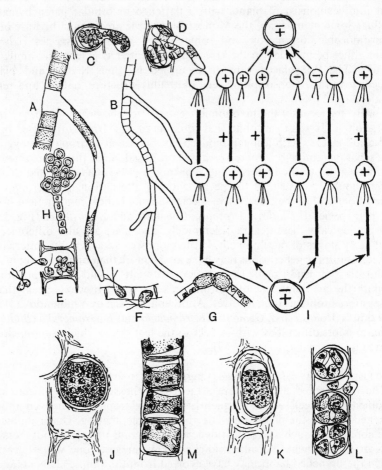

Fig. 3.24 *Ulothrix zonata.* A, B, rhizoid formation; C, liberation of swarmers into vesicle; D, germination of aplanospores in the cell; E, liberation of gametes; F, escape of zoospores; G, akinetes of *U. idiospora*; H, palmelloid condition; I, schema to illustrate the different types of filaments and swarmers; J, K, aplanospores; L, zoospore formation; M, banded chloroplasts in a portion of the vegetative filament. (A–C, E, F, × 375; D × 250; J–M × 400) (A, B, I, after Gross; C–F, J–M, after West; G, H, after Fritsch)

Swarmers are formed from all cells of the filament except the attachment cell, but they usually appear first in the apical cells and then in successive lower cells. They are liberated through a hole in the side of the cell into a delicate vesicle which subsequently bursts to free the swarmers. Asexual reproduction is normally by means of quadriflagellate zoospores, though in one species (II in Fig. 3.23) they are biflagellate. In the most recent revision of the genus Mattox and Bold (1962) transfer species with biflagellate zoospores to the genus *Hormidium* (see below). The numbers of zoospores produced per cell depend on the width of the filament: wide cells give 2, 4, 8, 16 and 32 zoospores per cell, narrow cells 1, 2 or 4 per cell. Some species, in the presence of nutrient deficiency (see above), or as a normal course, produce bi- or quadriflagellate microzoospores which only germinate at low temperatures. The place of these in the life history should be clarified when further species have been studied. Undischarged zoospores can turn into aplanospores that may even germinate *in situ* before liberation. Akinetes (*U. idiospora*) are another means of asexual reproduction.

Sexual reproduction is by means of biflagellate gametes of which 8, 16, 32 or 64 are produced per cell. The plants are dioecious with + or − strains. In *U. flacca*, gametes are produced and fuse under long-day conditions but under short-day conditions they germinate without fusing. Light periodicity is therefore significant and other species of the genus may also be affected (see also p. 437). On germination the zygote either gives rise to 4–16 bodies which function as aplanospores or zoospores, or develops into a 'Codiolum' phase (*U. zonata*). The species occur most abundantly in winter or spring and optimum conditions would seem to include low temperatures. The genus is well represented in both fresh and salt waters.

Genera closely allied to *Ulothrix* include *Schizomeris* in which some of the cells may divide longitudinally and in which sexual reproduction has not been observed. *Hormidium* species are often included in *Ulothrix* but the filaments have a tendency to fragment, the chloroplast only contains a single pyrenoid and the edges are not lobed, and the zoospores are always biflagellate. *H. flaccidum* is a widely distributed soil alga (see p. 382). In *Binuclearia* the cells are arranged in pairs in a gelatinous filament.

MICROSPORACEAE: *Microspora* (*micro*, small; *spora*, seed). Fig. 3.25.

The species, all free-floating, are restricted to fresh-water and consist of unbranched threads with walls of varying thickness, the thicker walls often showing stratification. Each cell wall consists of two overlapping halves held

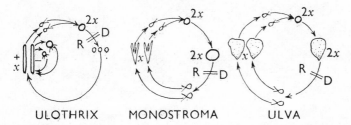

ULOTHRIX   MONOSTROMA   ULVA

Fig. 3.25 *Microspora amoena*. A, portion of thread; B, early cleavage in swarmer formation; C, two young cells; D, akinete formation; E, formation of aplanospores; F, G, stages in germination of aplanospores; H, liberation of zoospores; I, zoospores. (C, I × 745; D × 550) (C, D, I, after Meyer; A, B, E, F–H, after Fritsch)

in place by a delicate membrane. This structure arises from the fact that the filament is made up of a series of H pieces and as a result the filament readily fragments. At cell division a new H piece is introduced between the two daughter protoplasts. In young cells the parietal chloroplast is commonly reticulate but older cells are so filled with starch that the plastid structure is difficult to see. There are no pyrenoids. Asexual reproduction is by means of biflagellate zoospores (quadriflagellate in one species), 1–16 being produced per cell and liberated either when the thread fragments into H pieces or else by gelatinization of the cell walls. Any cell can produce single aplanospores instead of zoospores, and some species also produce multinucleate, thick-walled akinetes. These germinate directly to form a new filament or with later germination to give four protoplasts that then grow into new filaments.

The genus appears to be polyploid with 2n = 16 in. *M. stagnorum, M. humidulum*, 2n = 24 in. *M. aequabilis*, and 2n = 32 in., *M. amoena*, though hybridisation may account for the condition in *M. amoena* where there is a second race with 2n = 20.

Microspora with its cell walls in two pieces has an analogue with *Tribonema* in the Xanthophyceae and it seems that this type of construction has arisen more than once in the course of evolution.

CYLINDROCAPSACEAE: *Cylindrocapsa* (*cylindro*, cylinder; *capsa*, box). Fig. 3.26.

The single genus is found in freshwater and each plant consists of un-branched filaments attached at the base by means of a gelatinous holdfast. When young each thread is composed of a single row of elliptical cells with

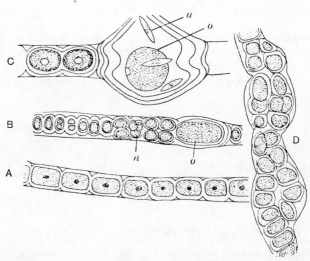

Fig. 3.26 *Cylindrocapsa*. A, vegetative filament; B, thread with young anthideridia (*n*) and young oogonium (*o*); C, fusion of gametes. (*a* = antherozoid, *o* = ovum); D, old mature filament. (after Fritsch)

thick, stratified walls, the whole being enclosed in a tubular sheath. Each cell contains a stellate chloroplast and a single pyrenoid.

In older filaments the cells divide longitudinally, usually in pairs, and this suggests how genera such as *Monostroma* may have evolved. Reproduction is more advanced because, apart from bi- or quadriflagellate zoospores, sexual reproduction is oogamous, the plants being monoecious or dioecious. Each antheridium gives rise to two biflagellate antherozoids and each oogonium to a single ovum.

MONOSTROMATACEAE: *Monostroma* (*mono*, single; *stroma*, layer). Figs. 3.27, 3.28.

The thallus originates as a flat, one-layered plate or a small sac, which in most species ruptures very early to give a plate of cells one layer in thickness, the cells often being arranged in groups of two or four. Because there is only one layer the thallus tends to be a paler green and more fragile than the distromatic thalli of *Ulva* (see below).

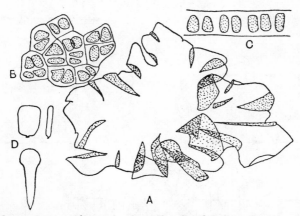

Fig. 3.27 *Monostroma crepidinum*. A, plant; B, cells of thallus; C, transverse section of thallus; D, *M. lindaueri*, plants. (A, D, × ⅔; B, C, × 200) (after Chapman)

The species of the original genus *Monostroma* are now best treated as falling into three genera: those with the macroscopic leafy thallus alternating with an enlarged zygote or cyst which after some months undergoes meiosis and gives rise to zoospores (*Monostroma*); those with the flattened thallus producing only one kind of swarmer (*M. oxyspermum*) or with alternation of two kinds of flattened thalli (*Ulvaria*); at least one species (*M. zostericola*) with a sporophyte thallus and a small gametophyte disc producing biflagellate gametes (Yamada and Tatewaki, 1965). Bliding (1968) places this species in the new genus *Kornmannia*, while Vinogradova (1969) removes *M. oxyspermum* from *Ulvaria* and places it in the new genus *Gayralia*, family Gayraliaceae. *M. latissimum*, *M. wittrockii* (sensu Moewus), *M. angicava* and *M. nitidum* are wholly gametophytic and the enlarged zygote produces quadriflagellate swarmers which give rise to new adults via a flat dwarf thallus (*M. angicava*) or an erect tube (*M. nitidum*) (Fig. 3.28). In some cases it

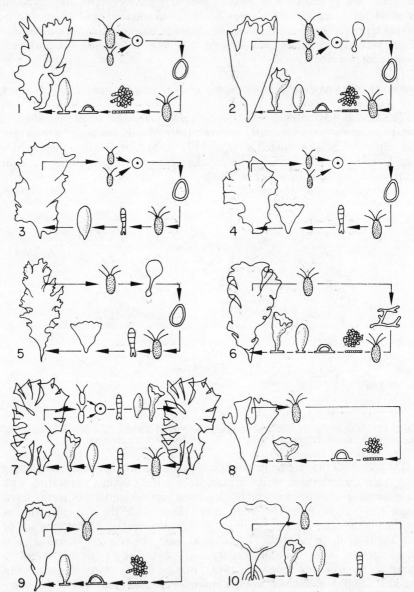

Fig. 3.28 Hitherto known life histories of the genus *Monostroma*: 1, *M. angicava*; 2, *M. grevillei*; 3, *M. nitidum*; 4, *M. latissimum*; 5, *M. undulatum* (= *M. pulchrum*); 6, *M. leptodermum*; 7, *M. fuscum var. splendens*; 8, *M. zostericola*; 9, *M. arcticum*; 10, *M. tubiforme*. (after Hirose and Yoshida, 1964)

would appear that not only is there an enlarged zygote but there may also be a dwarf shell-boring 'codiolum' sporophytic phase (Tokuda and Arasaki, 1967). In *Ulvaria obscura* vars. *blytii* and *wittrockii* (sensu Bliding) there is apparently a normal alternation of isomorphic gametophytic and sporophytic generations. *M. arcticum* and *M. tubiforme* only produce biciliate swarmers and so could be either wholly haploid or wholly diploid (Hirose and Yoshida, 1964). Two species, *M. groenlandicum* and *M. lindaueri* retain the saccate condition throughout their life and probably should be removed to a new genus. A complete understanding of the original genus will not be possible until the life history of every known species has been thoroughly studied. At the present time no two workers are agreed on the distribution of the various species originally included in the old genus *Monostroma*. It is, however, clear that both life histories and frond development must be considered in any final arrangement.

ULVACEAE: *Ulva* (marsh plant). Fig. 3.29.

The thallus, which characteristically is a flat, expanded, distromatic (two-layered) plate, originates from a single uniseriate filament that subsequently expands by lateral divisions. There is no hollow sac as in *Entero-morpha* (see below), but the two layers may separate towards the margins or

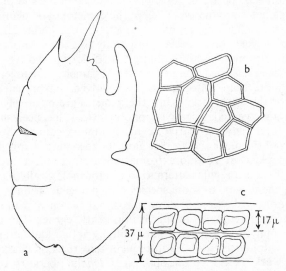

Fig. 3.29 *Ulva sorensenii.* A, plant; B, cells in surface view; C, transverse section of thallus. (after Chapman)

at the base as in *U. linza, Enteromorpha stipitata* and *E. flabellata* (Dangeard, 1958), so that it becomes difficult to know in what genus to place such transitional species. Typically the flat thallus is unbranched though it can be deeply divided. Branching or proliferating is characteristic of *Enteromorpha* and *Letterstedtia* (see below), but branching has been reported in *U. linearis*.

The young uniseriate filaments are attached by a single cell, but as expansion takes place multinucleate rhizoids grow down from the lower cells and form a basal attachment disc that may persist throughout the winter, new plants arising from it in the spring. Detached fragments are another means of forming new thalli especially in the free-floating forms of quiet waters. The form of the thallus of some species can vary considerably. Plants of *U. linza* found near high water mark in Japan are long and lanceolate and they gradually change to short and broad fronds further down the shore. Varying exposure times and water loss may be the factors involved here but culture experiments have shown that amino acids and kinetin may also be involved. The plasticity of the species has undoubtedly led in the past to taxonomic confusions, and many of these have yet to be resolved. The taxon *U. lactuca* has probably included four separate species, *U. lactuca* proper (*sensu stricto*) being anisogamous and with a high latitude distribution, while one of the new taxa, *U. thuretii* is isogamous and a short day plant with a S. European distribution. Similarly a study of *U. linza* from the Mediterranean has shown that the plants differ from those ascribed to the species elsewhere and they have been renamed *U. bertoloni*. It is clear that further work, accompanied by breeding experiments is vitally necessary before the species in this genus are finally established.

Reproduction is by means of zoospores produced in the sporophytic plants and 32 or 64 gametes produced per cell in the gametophytic plants. The latter are generally isogamous, the + and − gametes being liberated from separate plants, but in at least three species anisogamy is known to occur. In some species the gametes may fuse into clumps. In *U. lobata* two types of 'clump' reaction have been observed. In one type 10–12 gametes fuse together whereas in the other more than 100 gametes are involved. At present there is no explanation for this difference. While the gametes are positively phototactic before fusion, the zygote is negatively phototactic. It has been shown that in certain cases there may be relative sexuality among gametes from different plants, the sex of the older and weaker gametes becoming changed (see p. 35). Meiosis takes place at zoospore formation and, in most of the species investigated, there is a regular alternation of morphologically similar diploid and haploid generations. In some species only one generation is present, the swarmers invariably developing directly without fusion or without meiosis. *U. mutabilis* is more complex in that it apparently can exist in the haploid, diploid (gametophytic and sporophytic plants), triploid and tetraploid states. This kind of cytological study is infrequent and there is clearly scope for much further work. In at least one species, *U. lobata*, liberation of gametes is associated with the spring tidal cycles (see p. 197).

The plants occur in saline or fresh water and become particularly abundant when the waters are polluted by organic matter or sewage.

In the allied sub-antarctic genus *Gemina* the plants resemble either *Ulva* or *Enteromorpha*. The cells are normally in pairs and apparently only some cells are capable of reproduction whereas in *Ulva* all cells of the thallus can give rise to swarmers. Similarly, in the genus *Lobata*, the reproductive cells are restricted to the thickened central portion of the thallus.

Papenfuss (1960) has argued that *Gemina*, *Lobata* and *Letterstedtia* should all be merged into the genus *Ulva*. Segregation into somatic and germ cells in these genera would seem sufficient justification for retaining them as distinct from *Ulva*. Species of *Letterstedtia* occur in South Africa and New Zealand but Pocock (1959) maintains, probably correctly, that the New Zealand *L. petiolata* should be the type of a new genus. *L. insignis* in South Africa is a long strap-shaped thallus with leafy marginal frills and reproductive zones which are separated by purely somatic areas. *L. petiolata* characteristically has a broad rounded blade which arises from several branched, stipe-like portions.

In the genus *Enteromorpha* the plants also start life as uniseriate filaments but they soon become multiseriate and tubular. Like *Ulva* many of the species are attached to the substrate by rhizoids, but in *E. nana* there is a basal prostrate disc from which the erect plants arise. This feature, together with small cell size, has been used to create the independent genus *Blidingia* (*E. nana = E. minima = Blidingia minima*), but because basal discs are known to occur in *E. hendayensis* and *E. tubulosa*, and small cells are a feature of *E. bulbosa*, there seems no justification for retaining the genus *Blidingia* nor for removing it to the family Monostromataceae (Bliding, 1968). The existence of a basal thallus as well as an erect one forms what is known as *heterotrichy*, a feature characteristic of the Chaetophorales (see p. 81). It seems very likely that the phenomenon of heterotrichy has arisen on a number of occasions in the course of evolution. In addition to the many attached species there are also a few species, occurring especially on salt marshes (p. 372), which are unattached for the whole or part of their existence.

There is considerable plasticity and hence polymorphism in this genus, as in *Ulva*, so that taxonomists have either accepted a very large number of species or else have described few species with many forms. Further work, including culture and crossing experiments will be necessary before a final solution is reached. Taxonomically one can conveniently divide the species into groups which appear to have common features: the *compressa-intestinalis* group, the *clathrata*, the *nana*, the *procera* and the *prolifera* groups. Bliding (1963) also recognises *torta*, *jugoslavica*, *flexuosa* and *linza* groups, but he removes the *nana* group to *Blidingia*.

Growth in length of the thallus is primarily intercalary, though in the early stages there is also an apical cell that undergoes divisions. In some species the chloroplasts exhibit polarity, occurring normally in the apical portion of the cell. Such polarity is apparently absent in *Ulva* and *Monostroma*. When the tubular thallus is damaged by grazers or wave action near the apex the wounded cells put out papillate undergrowths but if the damage occurs near the base, then rhizoids are produced.

In many species there is a dioecious gametophytic generation which produces isogametes and this alternates with a morphologically identical sporophytic generation which produces zoospores. In *E. prolifera*, however, there is alternation between heteromorphic generations, one being simple and the other much branched. It is thought that a similar situation may exist in the case of *E. compressa* var. *australiensis*. Scagel (1960) reports that the sporophytic stage of *E. intestinalis* gives rise to tuberculate-like plants which may be identical with the plant known as *Collinsiella tuberculata*. Zygotes produced from gametes of this generation give rise to a disc from which the characteristically tubular *Enteromorpha* plant arises. This represents an important deviation and should be confirmed. Some species, including *E. nana*, *E. procera*, *E. aragoensis*, *E. adriatica*, are probably wholly diploid as only zoospores are known. In the first-named species meiosis has not been observed and the swarmers have been designated neutrospores. In the other species the swarmers can be either bi- or quadriflagellate and it is clear that cytological study is required to show whether the genus comprises a polyploid series or not.

Anisogamy occurs in *E. intestinalis* where the male gamete is small with only a rudimentary pyrenoid. Parthenogenesis has been recorded for some species, e.g. *E. clathrata*, and this presumably results in new sexual plants. In the reproductive areas some of the cells fail to divide. These are termed

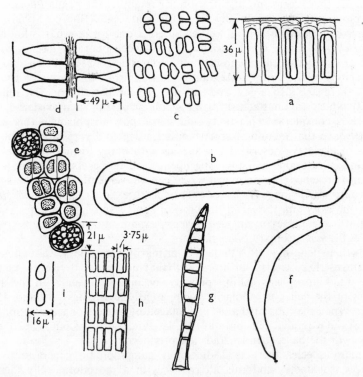

Fig. 3.30 *Gemina* and *Capsosiphon*: a, t.s. membrane of *Gemina enteromorphoidea*; b, t.s. thallus of *G. enteromorphoidea*; c, cells in surface view, *G. enteromorphoidea*; d, t.s. *G. linzoidea*; e, reproductive and vegetative cells of *G. linzoidea*; f, *Capsosiphon aurea*; g, *C. aurea* plantling; h, part of thallus of young plant of *C. aurea*; i, membrane of *C. aurea*. (after Chapman)

rest cells, and later they may germinate and give rise to sporelings which eventually separate from the parent thallus.

Within the genus there are several lines of development including two probable reduction lines, (see p. 334), and one such reduction series may be represented by the sequence *E. clathrata – E. 'ralfsii – E. salina – E. percursa*. The last named species consists usually of only two rows of cells and has been placed in a separate genus *Percursaria*. An isomorphic alternation of generations has been demonstrated, however, and the filaments are said to develop from basal pads (see above) so that there seems little justification for generic separation. Bliding (1963) would regard the series as proceeding in the opposite direction with *E. percursa* as the most primitive form, though in a later paper (1968) he removes *Percursaria* to a new family, Percursariaceae, which he considers may not even belong in the Ulotrichales. The semi-terrestrial habitat of the species would seem to be against such a view.

CAPSOSIPHONACEAE: *Capsosiphon* (*capso*, box; *siphon*, small tube). Fig. 3.30.

The family is founded upon the golden-green *Capsosiphon fulvescens* (formerly *Enteromorpha aurea*), but now another species, also golden-green is known in New Zealand. Some authorities (e.g. Papenfuss, 1959) do not consider the characters justify a separate family, but on the basis of morphology, reproduction and the unusual brackish water habitat, it can be argued that there is sufficient justification for the treatment adopted (Bliding, 1968; Vinogradova, 1969). The cells with gelatinous walls are arranged, often 2–4 enclosed in a common mother cell wall, in horizontal and longitudinal rows but do not separate off into longitudinal areolae as in *Prasiola*. Reproduction commonly takes place by means of akinetes or 4-flagellate zoospores and biflagellate gametes. After fusion a cyst is formed from which 16–32 zoospores emerge. Chihara (1967) considers that these facts justify a relationship with the Monostromataceae.

PRASIOLACEAE: *Prasiola* (*prasio*, green). Figs. 3.31, 3.32.

The young unbranched filament, which is known as the 'hormidium' stage, consists of a single row of cylindrical cells with thick walls which frequently exhibit striations. Later on the cells divide longitudinally and produce a thin expanded thallus, known as the 'schizogonium' stage, which tapers to the base. The cells of the mature expanded thallus are often arranged in fours and possess axile, stellate chloroplasts, while in some species another feature is the presence of short rhizoids that occur in the stalk-like portion or else are produced from the marginal cells. In the juvenile filament reproduction takes place by means of fragmentation as a result of the death of isolated cells, while in the older, more leafy thallus, 'buds' can arise from the margin. Sometimes the cells produce large, thick-walled akinetes that germinate to form aplanospores from which new plants arise. In *P. japonica*, sexual reproduction is brought about by non-motile macrogametes (16 per cell) and motile microgametes (64 or 128 per cell) that are both produced from the same plant, so that this species at least is monoecious and anisogamous.

Despite the existence of sexual reproduction, multiplication in this species is usually by asexual means. The plants are haploid (n = 3), but in view of discoveries about *P. stipitata* (Friedmann, 1959) it is desirable that *P. japonica* be restudied.

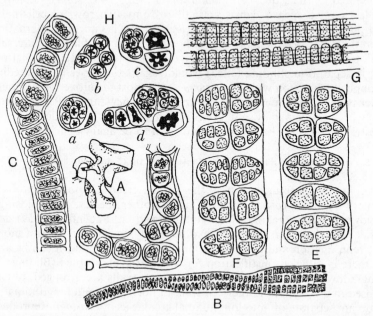

Fig. 3.31 *Prasiola*. A, plant of *P. crispa*; B, '*Schizogonium*' stage of *P. crispa* forma *muralis*; C, D, '*Hormidium*' stage of *P. crispa* f. *muralis* with akinetes; E, development of macrogametes in *P. japonica*; F, development of microgametes in *P. japonica*; G, *P. crispa*, membrane striations in '*Schizogonium*' stage; H (a-d), formation of aplanospores in akinetes and young plants. (E, F, × 665; G × 650) (A, B, after Fritsch; C, D, H, after Oltmanns; E–G, after Knebel)

In *P. stipitata* the adult plant is at least diploid (12–14 chromosomes), though in view of the count for *P. japonica* it may be tetraploid, and it reproduces by means of aplanospores. Subsequently the cells at the distal periphery (meiospores of Friedmann) undergo reduction division and give rise to cells which are haploid. These cells are monosexual and give rise to either male or female gametes that are anisogamous. The gametophyte segments are therefore attached to the diploid plant. A similar situation exists in *P. meridionalis* where the segments were formerly regarded as another alga, *Gayella constricta* (Fig. 3.32). In some species swarmers are produced from sporangia and Friedmann (1959) considers that these subsequently were retained and finally produced the so-called meiospores and meiosporophytes of *P. stipitata* and *P. meridionalis*, which must therefore be regarded as highly evolved forms. Ecologically, it has been observed that where *P. stipitata* forms a belt on the shore, wholly diploid plants occupy the upper levels and plants with gametophytic lobes occur lower down.

The shape of the thallus in *P. crispa*, and probably in other species also, varies considerably with the environment. The genus, which is generally absent from the tropics and sub-tropics, is represented by saline, freshwater or terrestrial species, the last named being tolerant towards considerable desiccation. This can be attributed to the lack of vacuoles in the cells and also to a high protoplasmic viscosity. Some authors believe the genus should be separated from the Ulotrichales (see p. 60) but this hardly seems warranted (see p. 331), especially in view of Friedmann's interpretation of the development of the *P. stipitata* condition.

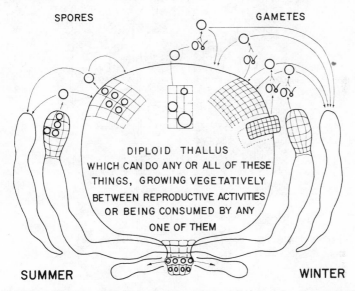

Fig. 3.32 Diagram of life history of *Prasiola meridionalis*. (after Bravo, 1965)

SPHAEROPLEACEAE: *Sphaeroplea* (*sphaero*, sphere; *plea*, full). Figs. 3.33, 3.34.

This genus is widely distributed but is only frequent locally, being most abundant on ground that is periodically flooded by freshwater. The long, free, unbranched filaments consist of elongated coenocytic cells containing up to seventy parietal chloroplasts with pyrenoids. The chloroplasts occupy the periphery of discs of cytoplasm, the discs being separated from each other by vacuoles, although occasionally they may come together to form a diffuse network. Each disc normally possesses one or two nuclei. In most of the species the septa develop as ingrowths, though in *S. africana* they are replaced by a series of strands which sometimes fail to meet at the centre so that the coenocytes are continuous.

Vegetative reproduction is secured by means of fragmentation, and there is apparently no asexual reproduction though zoospores have been reported in *S. wilmani*: these may, however, be ova developing by parthenogenesis.

In sexual reproduction, although the cells do not change in shape, both oogonia and antheridia are formed singly or in series, the species being either monoecious or dioecious. In the formation of oogonia the annular chloroplasts first become reticulate and then the ova are formed without any nuclear divisions. In *S. annulina* the ova are non-motile, but in *S. cambrica* the large female gametes are biflagellate and thus represent an advanced

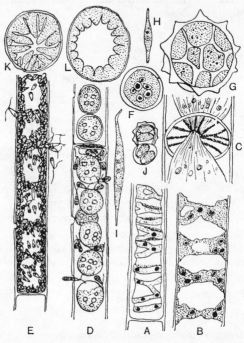

Fig. 3.33 *Sphaeroplea.* A, *S. annulina*, portion of thallus; B, *S. annulina* chloroplast; C. structure of septum in *S. africana*; D, female plant with ova and antherozoids; E, male plant; F, young zygote; G, zygote with thickened wall; H, I, young gametophytes; J, spores emerging from zygote; K, L, *S. africana*, transverse sections across the septa. (A–L × 375) (A–C, K, L, after Fritsch; D–J, after Oltmanns)

anisogamy. In the antheridia the nuclei do undergo division and numerous, elongated, narrow antherozoids are formed which are liberated through small pores, subsequently entering the oogonial cells through similar perforations. The fertilized ovum (oospore) becomes surrounded by a hyaline membrane inside which two new membranes are laid down and then the first one disappears. The new external membrane is ornamented and the contents become brick red.

Germination stages are only known for some species and the oospores lie dormant for several months or even years. On germination 1–4 biflagellate swarmers (8 occasionally) are produced and these come to rest and then grow into new plants. Sometimes the swarmers do not separate and a four- or eight-flagellate synzoospore may form depending on whether it comprises two or four zoids. These develop to a four-fold sporeling (i.e. with four claws). In some cases the swarmers from the oospore are completely suppressed and a new filament develops directly (cf. Fig. 3.34), this type of reproduction being known as azoosporic. In the most complete life cycle investigated the motile ova may also develop parthenogenetically after forming a short non-motile structure which divides to produce three new filaments. Various reduction stages from this life cycle occur among the different species (cf. Fig. 3.34). The adult plants are haploid since meiosis occurs at the segmentation of the oospore.

Primitive characters, which seem to be a feature of the genus, are the numerous ova and the entire lack of specialized organs for reproduction. The plant must probably be regarded as an Ulotrichaceous filament which has

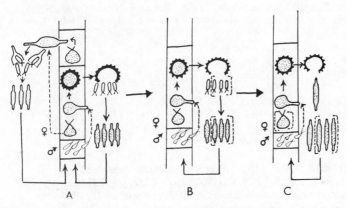

Fig. 3.34 Evolutionary development of life cycles in the genus *Sphaeroplea*. A, *S. Wilmani* type; B, *S. cambrica* type; C, *S. annulina* type. (after Rieth)

still retained some primitive features. In *S. annulina* cells are frequently found with only one or two plastids thus showing a gradation towards *Ulothrix*. There would seem to be very little justification for following those authors who would place the family in the Siphonocladales (p. 90), even though *S. africana* does have some features characteristic of that order.

REFERENCES

GENERAL

Bliding, C. (1963). *Op. bot. Soc. bot. Lund.*, **8**, 5.

Bliding, C. (1968). *Bot. Notiser*, **121**, 535.

Brown, R. M. (1967). *J. Phycol.*, **3**, Suppl. 5.

Chapman, V. J. (1964). *Oceanogr. Mar. Biol. Ann. Rev.*, **2**, 193.

Fritsch, F. E. (1945). *Structure and Reproduction of the Algae*, Vol. I pp. 220–6, (Cambridge University Press).

Kornmann, P. (1963). *Phycologia*, **3** (2), 60.

Van Den Hoek, C. (1968). *Nova Hedwigia*, **10**, 367.
Vinogradova, K. L. (1969). *Botan. Zhurn.*, **54**, 1347.
Ulotrichales
  Ramanathan, K. R. (1964). *Ulotrichales* (I.C.A.R. New Delhi).
Capsosiphon
  Chihara, M. (1967). *Bull. Nat. Sci. Mus.*, **10**, 163.
Enteromorpha
  Muller-Stoll, W. R. (1952). *Flora*, **139**, 148.
  Scagel, R. F. (1960). *Can. J. Bot.*, **38**, 969.
Monostroma and Ulvaria
  Dube, M. A. (1967). *J. Phycol.*, **3**, 64.
  Gayral, P. (1965). *Revue gen. Bot.*, **72**, 627.
  Hirose, H. and Yoshida, K. (1964). *Bull. Jap. Soc. Phyc.*, **12**, 19.
  Jonsson, S. (1968). *C.r. hebd. Séanc. Acad. Sci. Paris*, **267D**, 402.
  Kida, W. (1967). *Rep. Fac. Fish. Pref. Univ. Mie.*, **7**, 81.
  Tatewaki, M. (1969). *Sci. Pap. Inst. Alg. Res. Hokk. Univ.*, **6**, 1.
  Tokuda, H. and Arasaki, S. (1967). *Rec. Oceanog. Works Japan*, **9**, 139.
  Yamada, Y. and Tatewaki, (1965). *Sci. Pap. Inst. Alg. Res. Hokk. Univ.*, **5**, 105.
Prasiola
  Bravo, L. M. (1965). *J. Phycol.*, **4**, 177.
  Cole, K. and Akintobi, S. (1963). *Can. J. Bot.*, **41**, 661.
  Friedmann, I. (1959a). *Ann. Bot. N.S.*, **23**, 571.
  Friedmann, I. (1959b). *Nova Hedwigia*, **1**, 333.
  Friedmann, I. (1969). *Öst. Bot. Z.*, **116**, 203.
  Fujiyama, T. (1955). *Journ. Fac. Fish. Anim. Husb. Hiroshima Univ.*, **1**, 15.
Sphaeroplea
  Rieth, A. (1952). *Flora*, **139**, 28.
Ulothrix
  Kornmann, P. (1962). *Helgoländer wiss. Meeresunters.* **8**, 357.
  Mattox, R. K. and Bold, H. C. (1962). *Publ. 6222, Univ. Texas.*
  Perrot, V. (1968). *C.r. hebd Séanc. Acad. Sci., Paris*, **266D**: 1953.
Ulva
  Dangeard, P. (1958). *Botaniste*, **42**, 65.
  Dangeard, P. (1963). *Botaniste*, **46**, 184.
  Lovlie, A. (1964). *C.r. Trav. Lab. Carlsberg*, **34**, 77.
  Lovlie, A. and Braten, T. (1968). *Expt. Cell Res.*, **51**, 211.
  Papenfuss, G. F. (1960). *J. Linn. Soc. (Bot.)*, **56**, 303.
Letterstedtia
  Pocock, M. (1959). *Hydrobiol.*, **14**, 1.

4

CHLOROPHYTA

CHLOROPHYCEAE

OEDOGONIALES

Three genera, *Oedogonium, Oedocladium* and *Bulbochaete* are described for this order. They are characterized by the presence of zoospores and antherozoids with a ring of flagella around their anterior end. An advanced oogamy and specialized cell division are the unique features of this order. These characters are not found elsewhere and suggest that the order represents an evolutionary dead end. The species are all freshwater and their taxonomy is difficult because it depends on a knowledge of the sexual plants and on the type of reproduction.

Oedogonium (*oedo*, swelling; *gonium*, vessel). Figs. 4.1, 4.2

In *Oedogonium* the thallus consists of long unbranched filaments which are attached when young, though later they become free-floating; in the other genera the filaments are usually branched. Each cell possesses a single nucleus together with an elaborate reticulate chloroplast containing numerous pyrenoids. According to some workers the cell wall contains an outer layer of a chitinous material and there is an inner layer of pectin and cellulose.

Vegetative cell division is so peculiar and characteristic that many accounts of the process have appeared. A thickened three-layer transverse ring develops internally on the cell wall at the anterior end, enlarges and invaginates into the cell cavity. The nucleus migrates to the anterior end where nuclear

77

division commences. The site of the new transverse wall, between the two nuclei, is marked by a region of microtubules, and cross-wall formation occurs by the condensation of vesicles at this region. Concurrent with cross-wall formation is the splitting of the outer cell wall at the region of the

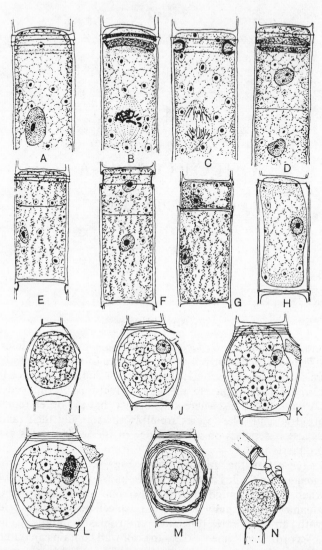

Fig. 4.1 *Oedogonium.* A–G, stages in cell division in *Oe. grande*: B, C, formation of ring; F, G, expansion of ring to form new cell; H, formation of aplanospore in *Oe. nebraskense*; I, *Oe. ciliatum*, position of antherozoid two hours after entering egg. J–M, stages in fertilization of ovum of *Oe. americanum*: K, entrance of sperm; L, fusion of gamete nuclei; M, zygote; N, *Oe. kurzii*, dwarf male. (A–G × 526; N × 175) (A–M, after Ohashi; N, after Pringsheim)

thickened ring. The ring stretches and expands upwards forming the lateral cuticle of the new cell. A new outer cell wall is formed on the inside of this cuticle, and fuses with the transverse wall at its periphery. The old outer wall forms a cap at one end and a bottom sheath at the other, and as successive divisions always occur at the same place, a number of caps develop there and give a characteristic striated appearance to some of the cells. The new upper cell is thus composed of the old anterior cap and a new cell wall derived from the invaginated ring. The bottom cell is comprised of the old cell wall. This method of growth in *Oedogonium* is intercalary, but in the other two genera, as each cell can only divide once, there is usually only a single cap.

Vegetative reproduction commonly occurs by means of fragmentation, while asexual reproduction depends on multiflagellate zoospores (one per cell) or akinetes. In *O. cardiacum* the zoospore liberation rhythm can be affected by temperature and day length. The flagella, which may have one or two rings of basal granules, form a circular ring around an anteriorly situated beak-like structure. This is the typical oedogonian swarmer, and two theories have been put forward to explain its origin: The first theory claims that the group arose independently from flagellate organisms which possessed a ring of flagella. If this is true then there could be very little connection with the other members of the Chlorophyceae. The second theory suggests that several divisions of the two original blepharoplasts and flagella took place, thus resulting in the ring structure. If this is correct then development might well have occurred from a Ulothrichalean type of swarmer.

When the zoospore is ripe the cell wall ruptures near the upper end and the swarmer is liberated into a delicate mucilaginous vesicle, but this soon disappears and the zoospore can then escape. After remaining motile for about an hour the anterior end becomes attached to the substrate and develops into a holdfast, or else the zoospore flattens to form an almost hemispherical basal cell. The type of holdfast depends on the species and the nature of the substrate, a smooth surface inducing a simple holdfast and a rough surface inducing the development of a branched holdfast. Development of the one-celled germling can proceed along one of two lines, depending on the species:

(a) The single cell divides near the apex by the normal method described above, in which case the basal daughter cell persists as the attachment organ and the upper cell goes on to form the new filament.

(b) The apex of the cell develops a cap and then a cylinder of protoplast grows out pushing it aside, and when the protoplast has reached a certain length a cross-wall is formed at the junction of the cylinder and the basal cell. The upper cell subsequently develops along the normal lines.

Sexual reproduction is by means of an advanced type of oogamy, the development of sex organs being promoted by an alkaline pH and some nitrogen deficiency. In some species the oogonia and antheridia are produced on the same plant (monoecious forms): in other species the oogonia and antheridia appear on different filaments which are morphologically alike (dioecious isomorphic forms). The species belonging to both these groups are termed macrandrous because the male filament is normal in size. There is a third group of species in which the male filament is much reduced and forms dwarf male plants. Such species are heteromorphic and form the nannandrous group. The dwarf males arise from motile androspores which are formed singly in flat discoid cells, the androsporangia, produced by repeated divisions of ordinary vegetative cells. The androsporangia are formed in the oogonial

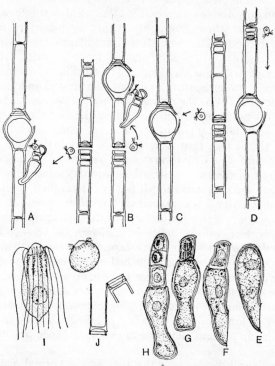

Fig. 4.2 *Oedogonium*. A, idioandrosporous nannandrous filament; B, gynandrosporous nannandrous filament; C, dioecious macrandrous filament; D, monoecious filament; E–H, stages in development of dwarf male plant; I, antherozoid; J, escape of zoospore (E–H × 480: I × 480; J × 138) (A–D, after Mainx; E–I, after Ohashi; J, after West)

filament for gynandrosporous species but in filaments that do not bear oogonia for idioandrosporous species (Fig. 4.2). In shape and structure the androspores are small editions of the zoospores, and after swimming about they settle on the wall of an adjacent cell and germinate into a small male plant, which is composed of a rhizoidal holdfast with one or two flat antheridial cells above, though in some cases only one antheridial cell without any rhizoidal portion is formed. Usually two antherozoids are freed from each antheridium into a delicate vesicle which later dissolves. The antherozoids are also like small zoospores, and if they fail to enter an ovum immediately they may remain motile for as long as thirteen hours. In the macrandrous mono-ecious species the antheridia are usually to be found immediately below the oogonia, where they arise by an ordinary vegetative division in which the upper cell subsequently continues to divide rapidly, thus producing a series of from two to forty antheridia. The antheridia frequently develop one day later than the oogonia, thus ensuring cross-fertilization.

The oogonia are enlarged spherical or ellipsoidal cells arising by one division in which the upper segment forms the oogonium and the lower a support cell, or else the latter subsequently divides to give antheridia. In some species the lower cell may also become an oogonium so that a series of oogonia can form on one filament. Each oogonium contains one ovum with a colourless receptive spot situated opposite to the opening in the oogonium wall from which a small quantity of mucilage is extruded. The opening is either a very small pore, formed by gelatinization of a tiny papilla, or else a slit, but in either case there is an internal membrane forming a conduit to the ovum. After fertilization the oospore often becomes reddish in colour and develops a thick membrane which is usually composed of three layers. The markings on the outer membrane are important taxonomically. At germina-tion the protoplast divides into four segments, which may each develop flagella and escape as zoospores, or else they function as aplanospores that later give rise to zoospores. Meiosis takes place at the germination of the zygote so that the adult filaments are haploid. In one species* it has been definitely established that two of the zygote segments ultimately develop into male plants and two into female plants. Zygote germination without meiosis is reported in which case large diploid swarmers are formed and these develop into abnormally large threads that are always female. Oogonia appear on these diploid filaments and can be fertilized, but the fate of the zygote is unknown.

It remains to discuss the possible origin of the androspores, and there are two hypotheses that may be considered:

First, the androspore may be equivalent to a second and smaller type of asexual zoospore, such as the microzoospores found in some species of *Ulothrix*, but in the Oedogoniales they can no longer give rise to normal filaments. On this view the nannandrous forms are the more primitive, the macrandrous having been derived by the androsporangium acquiring the capacity to produce antheridia immediately and hence never appearing.

Second, the androspore might be equivalent to a prematurely liberated an-theridial mother cell which subsequently undergoes further development. On this view the macrandrous species are the more primitive. At present there does not appear to be any very convincing evidence in support of either theory.

CHAETOPHORALES

In this order the fundamental structure is heterotrichous, that is, it possesses both a basal and erect system of branched threads. Some authorities do not

O. plagiostomum var. *gracilis*.

consider this character a sufficient distinction to warrant removal from the Ulotrichales. In some of the genera reduction has probably taken place and only the basal (e.g. *Ulvella*) or erect system (e.g. *Draparnaldia* spp.) is now represented. Many of these reduced forms are epiphytic or endophytic upon other algae or phanerogams. Another feature, by no means universal, of the Chaetophorales is the presence of setae (e.g. *Coleochaete*) or hairs (e.g. *Stigeoclonium*). There is considerable variation among the reproductive cells. Thus there may be quadriflagellate macro- and microzoospores and bi- or quadriflagellate gametes. Aplanospores and akinetes are also known. Sexual reproduction is primarily isogamous, occasionally anisogamous (e.g. *Aphano-chaete*) or oogamous (e.g. *Coleochaete*). The order is important because it is possible that its ancestral forms evolved into the first land plants. There are marked terrestrial tendencies and the genus *Fritschiella* (p. 83) is of special importance in this connection.

CHAETOPHORACEAE: *Stigeoclonium* (*stigeo*: sharp-pointed; *clonium*, branch)
    Fig. 4.3
    Many of the species and varieties are heterotrichous and branched and the plants are commonly surrounded by a layer of mucilage of varying thickness. They can carry aquatic fungi as parasites and this causes cell modification (Islam, 1963). The chloroplast is band-like and often does not fill the entire cell, especially in the older parts of the thallus. The aerial part bears branches that terminate in a colourless hair, the degree and nature of the branching depending upon illumination, nutrition and the rate of water flow. There is no localized area for cell division in the aerial portion, but in the creeping system only the terminal cells are meristematic. The prostrate system may be (*a*) loosely branched, (*b*) richly and compactly branched, or (*c*) a compact disc, but the more developed the basal portion the less elaborate is the aerial and *vice versa*. Vegetative reproduction is by means of fragmentation.
    Swarmers are formed in any cell, but mostly in those of the erect part. In addition to asexual macrozoospores, which are produced singly in cells, there are intermediate quadriflagellate microswarmers (asexual or sexual) and smaller biflagellate swarmers that often seem to develop by parthenogenesis into either hypnospores or new plants. Macrozoospores are mostly produced in summer and microswarmers throughout the year. The zygote, derived from fusion of microswarmers, is said to germinate either to a new filament or to zoospores, and these then give rise to the germlings in which the erect filament arises first and the prostrate portion subsequently or *vice versa*. By increasing the osmotic pressure or by adding toxic salts to the environment the thallus passes into a palmelloid state, while under other conditions akinetes can be formed. The plants are confined to well-aerated fresh water; they have also been found growing on fish living in stagnant water, but in these cases the movements of the fish presumably provide adequate aeration. Some species occur as epiphytes or endophytes and others seem able to tolerate a wide range of conditions.
    In all but one species so far studied the dominant phase in the life history is the haploid, reduction division taking place at germination of the orange or

bright-red zygote. In *S. subspinosum*, however Islam (1963) reports that the sexual plant, reproducing by gametes, gives rise after gamete fusion to an isomorphic diploid generation that reproduces by haploid, quadroflagellate swarmers.

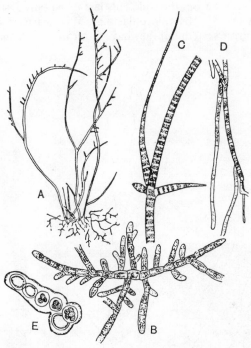

Fig. 4.3 *Stigeoclonium*. A, plant of *S. tenue*; B, basal portion of *S. lubricum*; C, aerial portion of *S. protensum*; D, rhizoids in *S. aestivale*; E, palmelloid state. (A–C, E, after Oltmanns; D, after Fritsch)

Other genera exhibit signs of reduction, being represented by only the basal portion of the thallus. Such genera are the epiphytic disc-like *Chaetopeltis* (with emergent setae), the plate-like *Pseudendoclonium* and *Ulvella*, and the endophytic and endozoic *Entocladia* (*Endoderma*) consisting of branched threads ramifying in the host. This process of reduction is also represented in the next example.

CHAETOPHORACEAE: *Fritschiella* (after F. E. Fritsch). Figs. 4.4, 4.5

This genus was first recorded (*F. tuberosa*) in India on land undergoing a drying process after the monsoon rains. It has now been recorded from a wide variety of soils, not all subject to water logging. It has also been recorded from Burma, the Sudan and Japan. A second species, *F. simplex*, has been recorded from East Pakistan. The mature plant, which forms dark–green clusters on the soil, possesses four different thallus systems. There is a septate rhizoidal

D

system (lacking in *F. simplex*) of elongated colourless cells; these arise from a
prostrate system of short congested branches, and since the cells are rich in
starch they can be regarded as a storage tissue. There is a primary projecting
system of upright, short-celled, simple or branched threads from which, in
F. tuberosa only, arises the secondary projecting system of elongate branches
with longer cells. In the Sudan plants this system is reduced because of the
high evaporation rate from the soil. This last is the only system that emerges
above the soil surface. The chloroplasts are not clearly recognizable in the
cells of the prostrate system but they have fewer pyrenoids than do the cells of

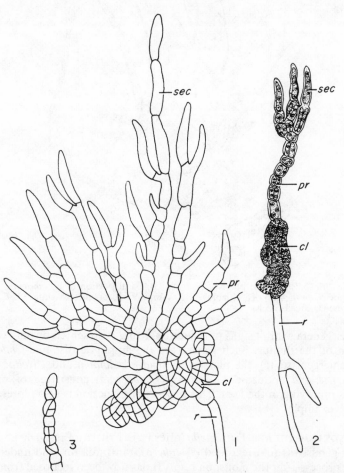

Fig. 4.4 *Fritschiella tuberosa* Iyeng.: 1, a small mature plant with a single rhizoid; 2, a
fifteen-days old plant with a single septate rhizoid; 3, a portion of the primary projecting
system longitudinal and diagonal divisions. (cl = cluster of cells of the prostrate system;
r = rhizoid; pr = primary projecting branches; sec = secondary projecting branches) (1–3 ×
900) (after Singh, 1941)

the erect systems where the plastid is a curved plate-shaped body with 2–8 pyrenoids. The cells are uninucleate.

Spore formation in *F. tuberosa* is restricted to the prostrate system, each cell producing one or more swarmers. Prior to swarmer formation the prostrate branches form club-shaped endings protected by a cuticle of dark-brown colour. These may be of three kinds: quadriflagellate macro-zoospores

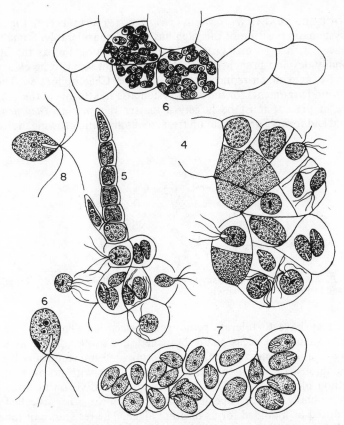

Fig. 4.5 *Fritschiella tuberosa* Iyeng.: 4, a portion of the prostrate system, showing the formation of macro- and microzoospores; 5, a portion of the plant, showing the liberation of zoospores in living condition; 6, a portion of the prostrate system, showing the gametes; 7, a portion of the prostrate system, showing a number of zoospores with dense contents, which have escaped liberation; 8, 9, microzoospores. (4, 6, 7, × 1900; 5, × 1250; 8, 9, × 3200) (after Singh, 1941)

(one per cell) formed from the primary erect cells in *F. simplex*; quadri- or biflagellate macrozoospores (2 or 4 per cell); numerous biflagellate gametes. The first two kinds are produced on plants other than those that produce gametes so that it is likely there is alternation of two similar generations. The sexual plants are also probably segregated because gametes will only fuse if they arise from separate plants (McBride, 1968).

During dry conditions the two erect systems disappear leaving the prostrate system as the perennating structure. The alga is of great interest in tracing the origin of a land flora. Bower's statement of 1935 — 'Its interest lies in the terrestrial habit, and in the fact that this amphibial plant possesses the essentials of a three-dimensional, photosynthetic sub-aerial thallus originating from a simple filament', still holds good today.

PLEUROCOCCACEAE: *Pleurococcus* (*pleuro*, box; *coccus*, berry). Fig. 4.6
The systematic position of this alga has varied considerably. Some authors have placed it in a special group, the Pleurococcales, but as the alga can occasionally develop branched threads there would seem to be evidence for regarding it as a much reduced member of the Chaetophorales. There are equally sound arguments for the other systematic treatments of the genus, and its place at present must be largely a matter of opinion. *Pleurococcus* is terrestrial and forms a green coat on trees, rocks and soil, growing in situations

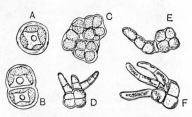

Fig. 4.6 *Pleurococcus naegelii*. A, single cell. B, single-celled colony. C, normal colony. D–F, thread formation. (after Fritsch)

where it may have to tolerate prolonged desiccation. The cells, which are globose in shape and occasionally branched, are single, or else as many as four may be united into a group. Under certain cultural conditions branching may be copious. Each cell contains one chloroplast with no pyrenoids. The sole method of reproduction is through vegetative division in three planes. There is probably only one species, *P. naegelii*, all the other so-called species being reduced or modified forms of other algae. The resistance of the cells to desiccation is aided by a highly concentrated cell sap and a capacity to imbibe water directly from the air.

TRENTEPOHLIACEAE: *Trentepohlia* (after J. F. Trentepohl). Fig. 4.7
The species grow as epiphytes or on stones in damp tropical and sub-tropical regions, but they will also grow under temperate conditions if there is an adequate supply of moisture. The threads have a characteristic orange-red colouring due to the excess of β-carotene. The cells contain discoid or band-shaped chloroplasts and lack pyrenoids. Usually both prostrate and erect threads are present, though the latter are reduced in some species.

Growth is apical, and the terminal cells often bear a pectose cap or series of caps which are periodically shed and replaced by new ones. The origin of the cap is not properly understood but is thought to be due to a secretion. The cellulose walls are frequently thickened by parallel or divergent stratifications,

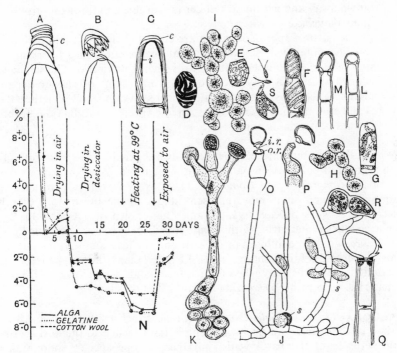

Fig. 4.7 *Trentepohlia*. A, B, *T. montis-tabulae* with pectin caps; C, *T. montis-tabulae*, cell structure (*c* = cap, *i* = innermost layer of cell wall); D–F, types of chloroplast; G, chloroplast in *T. iolithus*; H, I, *T. umbrina*, fragmentation of prostrate system; J, threads of *T. aurea* bearing sporangia (*s*); K, *T. umbrina*, sporangia; L, M, two stages in the development of the 'funnel' sporangium in *T. annulata*; N, graph showing decreasing water contents of *Trentepohlia*, gelatin and cotton-wool on drying; O, P, T, *umbrina*, detachment of stalked sporangium (*i.r.* = inner, *o.r.* = outer thickening of sporangial septum); Q, mature 'funnel' sporangium, *T. annulata*; R, S, gametangia of *T. umbrina*. (A–G, L, M, O–Q, after Fritsch; H, I, K, R, S, after Oltmanns; N, after Howland)

while each septum between the cells may also have a single large pit which is penetrated by a protoplasmic strand. The cells are uninucleate when young and multinucleate when old. Vegetative reproduction is through fragmentation, while other means of reproduction are to be found in three different types of sporangia:

   (*a*) Sessile sporangia that have become detached. These consist of enlarged cells which develop in almost any position and they produce bi-flagellate swarmers that may be isogametes.

(*b*) Stalked terminal or lateral sporangia that are cut off from an enlarged support cell which may give rise to several such bodies. The apical portion swells out to form the sporangium and underneath cuts off a stalk cell that frequently becomes bent. The dividing septum possesses two ring shaped cellulose thickenings which may play a part in the detachment of the mature sporangium. The detached sporangium is blown away and germinates under favourable conditions to give bi- or quadriflagellate swarmers.

(*c*) Funnel-shaped sporangia which are cut off at the apex of a cylindrical cell, the outer wall splitting later at the septum, thus liberating the sporangium, the subsequent fate of which is unknown. The sessile and stalked sporangia may occur on the same plant or else on separate plants. There has been no cytological work to show whether there is any alternation of generations. In one species reproduction is wholly by means of aplanospores.

Within the Trentepohliaceae there are several genera, the species of which form a green discolouration on and in stones and shells of gastropods. Such genera are *Gongrosira* and *Gomontia*. These are very difficult to determine taxonomically and some may represent stages in the life history of other algae. Recent work on *Gomontia polyrhiza* has shown that it has a life cycle of a large unicell alternating with a plate-like disc. Species of the allied genus *Cephaleuros* grow as epiphytes or parasites on the leaves of various phanerogams. *C. virescens* forms the red rust of the tea plant and can cause serious economic losses, especially when the tea bush is growing slowly. Bushes can be made less susceptible by treating the soil with potash.

COLEOCHAETACEAE: *Coleochaete* (*coleo*, sheath; *chaete*, hair). Fig. 4.8
Most of the species are freshwater epiphytes attached to the host by small outgrowths from the basal walls, but there is one species endophytic in *Nitella* (p. 117). Some of the species are truly heterotrichous while others only possess the prostrate basal portion, which is either composed of loosely branched threads or is a compact disc. The growth of the erect filaments is by means of an apical cell and the basal cushion possesses a marginal meristem. Each cell contains one chloroplast with one or two pyrenoids, and the characteristic sheathed bristle which arises from each cell may be broken off in the old plants. These bristles develop above a pore in the cell wall through which the protoplast extrudes, and at the same time a membrane is secreted over the protruding bare protoplast (McBride, 1968). Asexual reproduction takes place in spring and early summer by means of biflagellate zoospores which have no eye-spot and are produced singly. After a motile phase the zoospore settles down and divides either horizontally, when the upper segment develops into a hair and the lower forms the embryo disc, or vertically, when each segment grows out laterally; in either case hair formation takes place at a very early stage.

Sexual reproduction is by means of a specialized oogamy, some of the species being dioecious and the remainder monoecious. The female organs, or carpogonia, are borne on short lateral branches and are later displaced. Each carpogonium possesses a short neck or trichogyne (the long neck of *Coleochaete scutata* being an exception) the top of which bursts when the carpogonium is mature. In the disc forms, the carpogonia originate as terminal

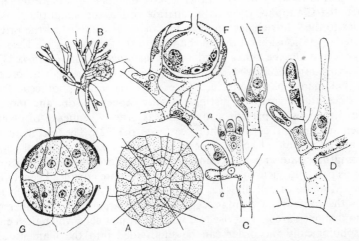

Fig. 4.8 *Coleochaete*. A, *C. scutata*, thallus with hairs; B, *C. pulvinata* with spermocarp; C, *C. pulvinata* with antheridia (*a*) and young carpogonium (*c*); D, *C. pulvinata*, almost mature carpogonium; E, *C. pulvinata*, fertilized carpogonium; F, *C. pulvinata* formation of envelope around fertilized carpogonium; G, *C. pulvinata*, mature spermocarp with carpospores (A × 150; B × 45) (A, B, after Smith; C–G, after Fritsch)

bodies on the outside of the disc, but as the neighbouring cells grow they eventually become surrounded and appear to be in the older part of the thallus. The antheridia develop in clusters at the end of branches (*C. pulvinata*) or from prostrate cells. They finally appear as small outgrowths cut off from a mother cell. Each antheridium produces one biflagellate colourless antherozoid.

After fertilization the neck of the carpogonium is cut off and the basal part enlarges; branches arise from the underlying cells and eventually surround the oospore where they form a red or reddish-brown wall, though in the disc forms this wall is only formed on the side away from the substrate. At the same time the enclosed oospore develops a thick brown wall and the cells of the outer envelope then die. The oospore hibernates until spring when it becomes green and divides into sixteen or thirty-two cells, and these, when the wall bursts, each give rise to a single swarmer which is a zoospore. Meiosis takes place at the segmentation of the zygote so that there is only the haploid generation. On the other hand, some observers have recorded the development of dwarf asexual plants before the reappearance of new sexual ones; if this is correct it may mean that there is an alternation of two unlike generations, an unusual phenomenon in the Chlorophyceae. Under certain conditions the cells will also produce aplanospores. The relation of this genus, with its advanced oogamy, to the other green algae is by no means clear.

SIPHONOCLADALES

Until 1935 this represented a well-established order, but in that year Fritsch placed most of the genera as septate members of the Siphonales but retained the Cladophoraceae as a separate order, the Cladophorales, with affinities to the Ulotrichales. In 1938 Feldmann re-established the order and suggested a relationship on the one hand with the Siphonales via *Valonia* and *Halicystis* and on the other hand with *Chaetophora* and *Ulothrix*. Egerod (1952) and, more recently, Nizamuddin (1964) have accepted the order but, unlike Feldmann, have continued to exclude the Cladophoraceae from it. There does not seem to be justification for this exclusion, and indeed the families in the order present a more or less orderly sequence of evolution, commencing from the Cladophoraceae (see p. 332). The members of the order are all septate at some stage of their existence, the cells or segments being multinucleate and possessing reticulate chloroplasts. Such plants are said to be partially coenocytic. A number of the more advanced genera have a specialized mode of cell division known as segregative cell division. In this process the contents of a single cell become separated by a number of cell walls so that one parent gives rise to a number of new cells which then enlarge.

Morphologically the order can be subdivided into the filamentous and vesicular forms. The former comprise genera such as *Lola*, *Rhizoclonium*, *Chaetomorpha*, *Cladophora* and *Microdictyon*, and the latter contain *Siphono-cladus*, *Valonia*, *Boodlea* and *Dictyosphaeria*. It is the vesicular forms that mostly possess segregative cell division.

CLADOPHORACEAE: *Cladophora* (*clado*, branch; *phora*, bearing). Figs. 4.9, 4.10

In one of the simplest genera, *Chaetomorpha*, the plants consist of un-branched filaments which are either attached or free-floating.

The genus *Rama*, with species mainly in the South Pacific, possesses very long branches and can be mistaken for *Chaetomorpha*. It also has short rhizoids at the base and because of these characters the genus appears to occupy a significant phylogenetic position (see p. 332). A genus that is often mistaken for *Chaetomorpha* is *Urospora* (*Hormiscia*). The most important feature is the presence of a 'Codiolum' stage in the life history, and in this respect it is possibly important in the evolution of *Spongomorpha* (see p. 94).

The genus *Cladophora* occurs in both fresh* and saline waters. Most species are attached by branched septate rhizoids, but some of them (e.g. *C. fracta*) may become free-floating later, and there is one section (*Aegagropila*) which is commonly free-living, the species forming ball-like growths. The attachment rhizoids in certain species form a prostrate thallus (*C. kosteri*) while in a few (*C. basiramosa*, *C. pygmaea*) the rhizoids are replaced by a single, disci-

* The optimum temperature for fresh-water species is 15–25° C. Growth increases with increasing water movement. They are not common in waters with low NO_3^- and $PO_4^=$ content (Whitton, 1970).

form holdfast. The colour of the frond varies from whitish-yellow to dark-, almost black-green. This feature is used taxonomically, but Van den Hoek (1963) has pointed out that colour varies with light intensity.

Growth is of three types: wholly apical, apical and intercalary, or wholly intercalary. Branches can arise either laterally or terminally. In the latter case the branch often grows into an almost horizontal position, a process termed

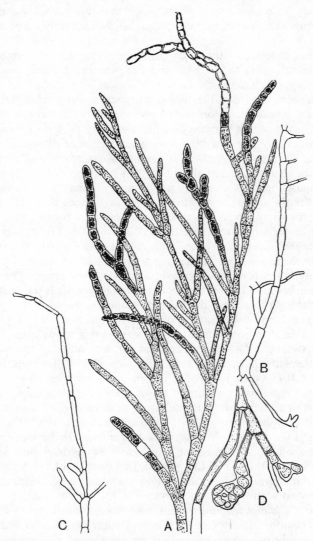

Fig. 4.9 *Cladophora.* A, plant with sporangia; B, shoot of *Aegagropila holsatica* bearing rhizoids; C, stolon of *Ae. holsatica*; D, rhizoids of *Spongomorpha vernalis* developing storage cells at the apices. (A, after Oltmanns; B, C, after Acton; D, after Fritsch)

D2

evection, which gives the appearance of a dichotomy. The cells are multi-nucleate, the number of nuclei possibly being related to the cell volume. The chloroplasts generally appear as a complex of disc-like structures united into a reticulum containing numerous pyrenoids. The plastids are parietal and bear processes projecting into the central vacuole. The wall in *Cladophora* consists of alternate layers of longitudinal and horizontal microfibrils of cellulose. In young cells these fibrils are arranged in spirals but as the cell elongates they become pulled out to lie almost transversely and longitudinally. There is very little production of mucilage, and this probably accounts for the dense epiphytic flora of minute species that is frequently found associated with plants of this genus. The genus is divided into a number of sections based upon various criteria. In the most primitive (sect. *Affines* and *Basicladia*) branching is restricted to the base and the branches are very long. In this respect they approach the genus *Rama* (see above).

In the *Aegagropila* group the species can exist as (*a*) threads, (*b*) cushions and (*c*) balls. The destruction of the old threads in the centre of the ball results in a cavity which may become filled with water, gas or mud. In Lake Söro the water in April and May is sufficiently free of diatoms for light to penetrate to such an extent that photosynthesis increases and so much gas collects in the centre of these balls that they float to the surface. Their characteristic shape is brought about by a continual rolling motion over the soil surface under the influence of wave action, and hence the 'ball' forms are found near the shore whilst the 'thread' and 'cushion' forms are to be found farther out in deeper water where there is less motion. The harder the floor the more regular is the shape of the balls, but even so the ball structure would also appear to be inherent in the *Aegagropila* because samples have been kept in a laboratory for eight years without losing their shape. The following types of branches have been recognized in the *Aegagropila* forms: (*a*) rhizoids, (*b*) cirrhoids, both these and the rhizoids being neutral or non-reproductive branches, (*c*) stolons or vegetative reproductive branches.

In the section *Aegagropila* most of the species reproduce vegetatively, but biflagellate swarmers have been reported for one species, *Ae. sauteri*, and these are interesting in that they may germinate while still within the sporangium (Fig. 4.10). Asexual reproduction in the other species, excluding the section *Aegagropila*, is by means of bi- or quadriflagellate zoospores which escape through a small pore in the cell wall. In some of the freshwater species certain cells may become swollen to form akinetes in which the walls are thickened and food is stored. Biflagellate isogametes are the means of sexual reproduction, all the species so far investigated being dioecious. The zygote develops at once without a resting period. In a number of species alternation of two morphologically identical haploid and diploid generations has now been established with meiosis taking place at zoospore formation. In one or two cases, e.g. *Cladophora flavescens*, the zoospores sometimes fuse, and this irregular behaviour is very comparable to similar phenomena found in the more primitive brown algae (cf. p. 186). In one species, *C. parriaudii*, reproduction is by 4-flagellate zoospores only.

In a few species there can be an odd or hetero-chromosome, and in a cell the number of zoospores with the odd chromosome are equal to the number lacking it. Haploid plants of *C. suhriana* have six or seven chromosomes, but in *C. repens* the cells contain either four or five. In a freshwater species, *C. glomerata*, a wholly different type of life cycle has been reported, and this difference may perhaps be compared with the various cycles found for *Ectocarpus siliculosus* under different conditions (cf. p. 188). Gametes and zoospores are both formed on diploid plants and meiosis takes place at gamete formation so that there is no haploid generation. Although zoospore

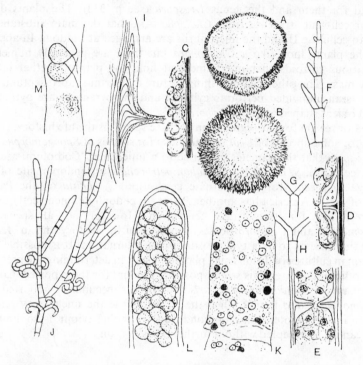

Fig. 4.10 *Cladophora.* A, ball of *Aegagropila holsatica* cut through and the dirt removed; B, same before cutting; C, *C. glomerata*, commencement of septum formation; D, *C. glomerata*, second stage in septum formation; E, *C. glomerata*, septum almost complete; F, diagram illustrating evection; G, H, types of branching; I, *C. glomerata*, structure of wall at a septum; J, *Spongomorpha coalita* with hook branches; K, *C. callicoma*, structure of chloroplast with nuclei and pyrenoids; L, *Ae. Sauteri*, zoospores in zoosporangium; M, *Ae. Sauteri*, zoospores. (A, B, × ¾) (A, B, after Acton; C–K, after Fritsch; L, M, after Nishimura and Kanno)

formation takes place all the year round gametes only appear in the spring, and the reason for this seasonal restriction is not understood. Van den Hoek (1963) considers that the evidence for gametes in *C. glomerata* is not wholly convincing and the life history therefore needs substantiation. Parthenogenetic development of gametes has also been recorded in a number of species. Of the species so far investigated the chromosomes appear to be present in multiples of 4 though Schussnig (1954) argues for 12, and this probably indicates polyploidy. The following diploid chromosome numbers have been recorded: *C. repens* 8 + 1, *C. suhriana* 12 + 1, *C. flavescens* 24, *C. flaccida*, 24, *C. pellucida* 32, *C. glomerata* 48, 72, 92 + 4 (octoploid).

ACROSIPHONIACEAE: *Spongomorpha* (*spongo*, sponge; *morpha*, construction). Fig. 4.10 J

The species of *Spongomorpha* and *Acrosiphonia* were originally regarded as species of *Cladophora*. Recent work indicates that a separate family is justified for them and the genus *Urospora* (see p. 333). The plants differ in their cell-wall structure from *Cladophora* in that the main substance is allied to cellulose II and the microfibrils are arranged at random. In appearance the plants look like a *Cladophora* but there are generally numerous adventitious rhizoids that intertwine and hold the plants together into a spongy mass. The other distinguishing feature is the rim-shaped septum. The plastid is an undivided perforate cylinder and the pyrenoids are pyramidal instead of lens shaped as in *Cladophora*.

A major reason for the segregation of these species is the life history. It was shown for some species of *Urospora* and also for species of *Spongomorpha* and *Acrosiphonia* that the gametes gave rise to a unicellular 'Codiolum' asexual stage (see p. 52). *S. lanosa* has *Codiolum petrocelidis*, an endoparasite of the red alga *Petrocelis*, for the alternate generation. *S. coalita* on the Pacific Coast of North America has another *Petrocelis* endophyte (incorrectly called *C. petrocelidis*) for its alternate generation. However, not all species of *Spongomorpha* and *Urospora* possess this type of life history and in *Acrosiphonia* there seems to be a typical isomorphic alternation. It is possible that variation in culture conditions may play some part in determining the presence of a 'Codiolum' phase. This would possibly account for the reported 'Codiolum' phase in *Cladophora rupestris*. A further peculiarity reported for *Acrosiphonia incurva* is that when the gametes fuse the nuclei may remain separate in the zygote (dikaryon condition). This behaviour is so unusual that it cannot be accepted without further confirmation.

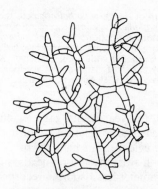

Fig. 4.11 Portion of net of *Microdictyon mutabile* (× 24). (after Dellow)

ANADYOMENACEAE: *Microdictyon* (*micro*, small; *dictyon*, net). Fig. 4.11

Branching in these plants is usually multiple in one plane, and sooner or later cells of the branches become attached to those of other branches by means of special pads or tenaculae, so that a flat net-like thallus eventually results.

M. mutabile from New Zealand can exist in a spongiose form with very few anastomoses so that it can easily be mistaken for a coarse species of *Clado-phora*, though it also gives rise to the typical net thallus. In those species that have been investigated, sexual reproduction is isogamous and there is alternation of isomorphic generations. The same is also true of *Anadyomene stellata* and *A. wrightii* (Enomoto and Hirose, 1970).

VALONIACEAE: *Valonia* (after the Valoni, an Italian race). 'Sea-bottle'. Fig. 4.12

In this genus, which is restricted to warm waters, the young coenocyte consists of one or more large vesicles which, as they mature, become divided up into a number of multinucleate segments. In some respects, therefore, the

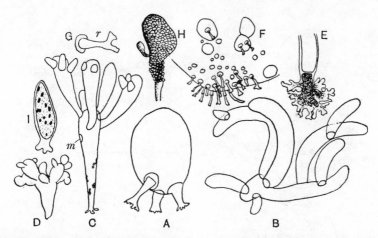

Fig. 4.12 *Valonia*. A, young plant of *V. ventricosa*; B, young plant of *V. utricularis*; C, adult plant of same (*m* = marginal cell); D, plant of *V. macrophysa*; E, rhizoid of *V. utricularis*; F, rhizoids from marginal cells at base of vesicle of *V. ventricosa*; G, single marginal cell and rhizoid (*r*) of *V. ventricosa*; H, *V. utricularis* fruiting; I, *V. utricularis*, germinating swarmer. (B × 1·4; D × 0·8) (B, D, after Taylor; A, C, E–I, after Fritsch)

genus provides a link with the Siphonales. The macroscopic club-shaped vesicle is attached to the substrate by rhizoids of various types. There is a lobed chloroplast that congregates with the cytoplasm at certain points in the older plants and then each group is cut off by a membrane to produce a number of marginal cells (segregative cell division). The cells do not neces-sarily form a continuous layer and are frequently restricted to the basal region where they may develop rhizoids, while in other species they are nearer to the apex where they may give rise to proliferations. The lower cells can form short creeping branches, and as these bear more of the erect vesicles a tuft of plants is produced.

The wall is generally composed of two sets of micro-fibrils oriented at an angle of rather less than 90° to each other, and in places there may be a third layer in between. Because of the large size of the cells, especially in *V. ventricosa*, they have been much used in studies of cytoplasm and cell-wall permeability.

Reproduction takes place by means of bi- or quadriflagellate swarmers, which are liberated from the cells through several pores, and although no sexual fusion has been observed as yet, meiosis occurs in *V. utricularis* at swarmer formation. The plants are therefore presumably diploid, a condition that is also characteristic of most of the Siphonales. The reproductive cells may encyst themselves, and it has been suggested on this evidence that the plant is a colonial aggregate of coenocytic individuals resulting from the retention of cysts that have developed *in situ*. The correctness or otherwise of this interpretation can only be obtained through a better knowledge of their phylogenetic history and the reproductive processes of other members of the group.

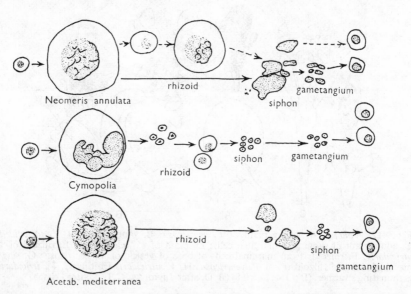

Fig. 4.13 Behaviour of the nucleus in certain members of the Dasycladales. Break-up of the large nucleus occurs in the rhizoid and the fragments, which may change shape, move up the siphon into the gametangia. (after Dao)

DASYCLADALES

Formerly members of this order were included in the Siphonocladales, but because they possess distinct characteristics they are best treated as a separate order. The thallus body is first characterized by the whorled (verticillate) arrangement of the lateral branches (in *Acetabularia* they are fused to form a disc), and secondly by the fact that the plastids are independent, disc-like, and not reticulate as in the Siphonocladales. In the species that have been investigated the vegetative thallus appears to be uninucleate. Before the gametes are formed the single primary nucleus fragments into smaller portions which may subsequently change shape either in the rhizoids or when moving up the main axis because the small daughter nuclei eventually migrate into the gametangia. This peculiar behaviour has been established for *Acetabularia*, *Cymopolia*, and *Neomeris* (Fig. 4.13). It is thought that all adult plants are diploid. In some genera the gametes are produced from special cysts (cf. *Acetabularia*).

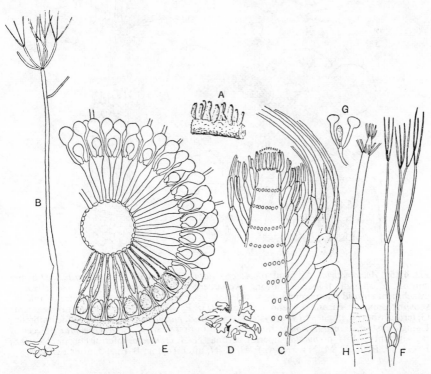

Fig. 4.14 *Neomeris*. A, plants of *N. annulata*; B, young plant of *N. dumetosa*; C, longitudinal section through apex of *N. dumetosa*; D, rhizoid in *N. dumetosa*; E, transverse section of thallus of *N. dumetosa* in middle of calcified area; F, *N. dumetosa*, assimilating filaments with sporangium; G, *N. annulata*, sporangium; H, regeneration of an injured axis. (A × ½; B, D, H × ¼; C, E, F × ⅓; G × 33) (A, G, after Taylor, B–F, H, after Church)

The order is very ancient and has apparently not changed greatly as a large number of fossil forms are known (cf. p. 318) and the living species of *Neomeris* (Fig. 4.14) and *Bornetella* are very like some of the extinct genera.

The genera are primarily confined to warm waters and in many there is extensive calcification.

DASYCLADACEAE: *Acetabularia* (*acetabula*, little cup; *aria*, derived from). Fig. 4.15.

This is a lime-encrusted genus which is confined to warm waters, extending as far north as the Mediterranean. The plants consist of an erect, elongate axis bearing one or more whorls of branched sterile laterals with a single

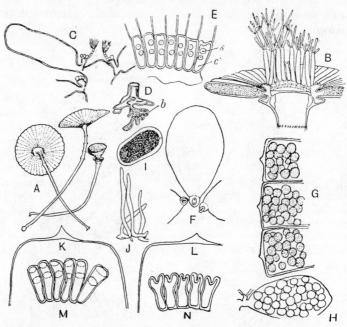

Fig. 4.15 *Acetabularia.* A, plant of *A. crenulata*; B, apex of *A. mediterranea* showing corona; C, apex of *A. moebii* showing two superposed fertile whorls; D, *A. mediterranea*, attachment rhizoid and perennating vesicle (*b*); E, *A. crenulata*, cells near centre of thallus, showing superior corona (*c*) and leaf scars (*s*); F, *A. pusilla*, vegetative ray segment; G, fertile lobes of *A. schenkii* with cysts; H, cysts in *A. pusilla* in a single lobe of the umbrella; I, single cyst of *A. mediterranea*; J, young plant in first year; K, L, *A. crenulata*, apieces of ray segments; M, *A. crenulata*, superior corona; N, *A. crenulata*, inferior corona. (A ×0·8, F. G ×44, H, K, L, M, N ×37) (A, F–H, K–N, after Taylor; B–E, I, J, after Fritsch)

fertile whorl at the apex. The sterile whorl or whorls are frequently shed in the adult plant leaving a mark or annulus on the stem where they were formerly attached. The fertile whorl is composed of a series of long sac-like gametangia which are commonly fused, though they are sometimes separate; these are borne on short basal segments which are morphologically equivalent to the primary branches. The basal segments also bear small projections on their upper surface, and these are with or without hairs which form the *corona*; in one section of the genus there is also an inferior corona on the lower surface. In *A. mediterranea* two or three years elapse before the plant attains maturity. In the first year the branched holdfast produces an upright umbilical thread, together with a thin-walled, lobed outgrowth that penetrates the substrate to act as the perennating organ. The aerial part dies, and in the next year or years a new cylinder arises that bears one or more sterile whorls of branches, until in the third or even a later year, a shoot develops which produces one deciduous sterile whorl and a single fertile whorl or umbrella. Each sac-like gametangial umbrella lobe produces a number of multi-nucleate cysts which are eventually set free through disintegration of the anterior end of the gametangium. In the spring biflagellate isogametes are liberated from these cysts and fuse in pairs. In *A. moebii* the cysts give rise to anisogametes. Use has been made of the uninucleate condition to determine the effect on morphology of varying ratios of nuclei of different species (cf. p. 475). This is possible because part of one species can be grafted onto another.

REFERENCES

GENERAL

Bourelly, P. (1966). *Les Algues d'eau douce*, Tome 1, (Boubée et Cie, Paris).
Fritsch, F. E. (1945). *Structure and Reproduction of the Algae*, Vol. 1, 293–309, 368–427 (Cambridge University Press).

CLADOPHORACEAE

Jónsson, S. (1965). *Annls. Sci. nat. (Bot.)*, *Ser.* 12, 3, 25.

Dasycladales

Valet, G. (1969). *Nova Hedwigia*, 16, 21.

Siphonocladales

Chapman, V. J. (1954). *Bull. Torrey Bot. Club*, 81, 76.
Egerod, L. E. (1952). *Univ. Calif. Publs. Bot.*, 25, 325.
Feldmann, J. (1938). *Rev. gén. Bot.*, 50, 57.
Nizamuddin, M. (1964). *Trans. Am. Microscop. Soc.*, 83, 282.

Acetabularia

Puiseaux-Dao, S. (1963). *Ann. Biol.*, 2, 99.
Puiseaux-Dao, S. (1970). *Acetabularia and Cell Biology*, Lagos Press.
Van Gansen, P. (1966). *J. Microscopie*, 4, 347.
Werz, G. (1965). *Brookhaven Symp. Biol.*, 18, 185.

Anadyomene

Enomoto, S. and Hirose, H. (1970). *Bot. Mag. Tokyo*, 83, 270.

Cladophora

Hoek, C. van Den. (1963). *Revision of the European Species of Cladophora* (E. J. Brill, Leiden).
Sakai, Y. (1964). *Scient. Pap. Inst. Algol. Res. Hokkaido Univ.*, 5, 1.
Soderstrom, J. (1963). *Bot. Gotheborg*, 1, 1.
Schussnig, B. (1954). *Arch. Protistenk.*, 100, 287.
Whitton, B. A. (1970). *Water Res.*, 4, 457.

Coleochaete

McBride, G. E. (1968). Ph.D. Thesis, (University of California).

Fritschiella

McBride, G. E. (1968). Ph.D. Thesis, (University of California).
Varma, A. K. and Mitra, A. K. (1964). *Bull. Jap. Soc. Phycol.*, 12, 44.

Neomeris
Dao, S. (1958). *Revue Algol.*, **3**, 192.
Oedogonium
Hill, G. J. C. and Machlis, L. (1968). *J. Phycol.*, **4**, 261.
Hoffman, L. (1964). *Am. J. Bot.*, **52**, 173.
Hoffman, L. (1967). *J. Phycol.*, **3**, 212.
Hoffman, L. and Manton, I. (1962). *J. Exp. Bot.*, **13**, 443.
Spongomorpha-Acrosiphonia
Fan, K-C. (1959). *Bull. Torrey Bot. Club*, **86**, 1.
Jónsson, S. (1968). *C.r. hebd. Séanc. Acad. Sci. Paris*, **267**D, 53.
Kornmann, P. (1962). *Helgoländer wiss. Meeresunters.*, **8**, 219.
Stigeoclonium
Islam, A. K. M. N. (1963). *Nova Hedwigia*, **10**, 1.

5

CHLOROPHYTA

CHLOROPHYCEAE III

CHAROPHYCEAE

CHLOROPHYCEAE

DERBESIALES

The members of this order were formerly placed in the Siphonales, but the discovery of the remarkable life history, with an alternation of two unlike generations has resulted in the establishment of a separate order by Feldmann.

HALICYSTIDACEAE: *Halicystis* (*hali*, salt; *cystis*, bladder) and *Derbesia* (after A. Derbes). Fig. 5.1.

In some classifications the family Derbesiaceae is used to accomodate the *Derbesia-Halicystis* complex.

The gametophytic plants consist of an oval vesicle, up to 3 cm in diameter, arising from a slender branched tuberous rhizoid embedded in calcareous *Lithothamnion* (cf. p. 281) growing at or below low-tide mark. There are only a few species all containing numerous nuclei in the peripheral cytoplasm. There does not appear to be any cellulose in the cell wall, though the wall is polylamellar as in *Valonia*. Growth of the *Halicystis* vesicles is very slow and they are shed at the end of the growing season by abscission, new vesicles arising later from the perennating rhizoid, and in this manner regeneration may go on for several years. In *H. boergesenii* from India as many as 6–12 annual scars can be found. Swarmers develop in the cytoplasm at the apex of the vesicle in an area which becomes cut off by a thin cytoplasmic membrane, the area thus cut off representing a gametangium. Macro- and micro-gametes are formed and in the early hours of the morning they are forcibly discharged through one or more pores. Gamete formation appears to be under the control of an endogenous rhythm with a basic period of about four days.

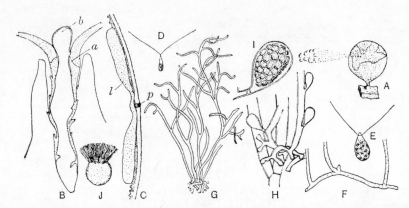

Fig. 5.1 *Halicystis ovalis* (and *Derbesia marina*). A, plant of *Halicystis* liberating gametes; B, rooting portion of *Halicystis* showing old rhizome and line of abscission (*a*) and new vesicle (*b*); C, gathering of protoplasm to form gametes (*l* = lining cytoplasm, *p* = pore of dehiscence); D, male gamete; E, female gamete; F, protonemal germling of *Halicystis*; G, *Derbesia* plant; H, *Derbesia*, with zoosporangia, growing on *Cladophora*; I, *Derbesia*, zoospore. (D, E × 600) (A–C, F–J, after Fritsch; D, E, after Kuckuck)

Several crops of these swarmers are produced as cytoplasm moves into the apical areas at bi-weekly intervals coincident with the spring tidal cycles. The zygote in *H. ovalis* germinates into a branched protonemal thread that in three months develops into a typical '*Derbesia*' plant with the erect aerial filaments arising from the basal rhizoidal portion.

It has been demonstrated that *Halicystis ovalis* and *Derbesia marina* are simply stages in the life cycle of one alga. The mature *Derbesia* threads produce zoospores that germinate into prostrate filaments; these later give rise to slender branched rhizoids which, after eight months, produce the characteristic *Halicystis* bladder. Although the cytology of the two plants has not yet been worked out, the *Derbesia* generation is presumably diploid and the *Halicystis* haploid. It also remains to be ascertained whether all the other species of *Halicystis* have a similar life cycle. It is possible that the same kind of life cycle holds for *Halicystis parvula* and *Derbesia tenuissima*. A spherical *Halicystis* has recently been reported from New Zealand and this probably has as its alternate generation *Derbesia novae-zelandiae* which has been known for a number of years.

SIPHONALES

The main characteristic of this order is its coenocytic structure in which true septa are rare or absent. The coenocyte normally has a cytoplasmic lining surrounding a central vacuole and containing numerous discoid chloroplasts. The pigment siphonoxanthin is characteristic of the order. There are sub-orders: the Eusiphoneae (Bryopsidaceae, Codiaceae) where the membrane contains mannan and there is only one type of plastid, and the Caulerpeae (Caulerpaceae, Udoteaceae) where the cellulose of the membrane is replaced by xylan and where amylogenic leucoplasts are present as well as chloroplasts. Some workers suggest that the difference is sufficient to justify the establish-ment of separate orders. In all cases so far studied the thallus is diploid. Except for some species of *Halimeda*, sexual reproduction is anisogamous. In some of the tropical genera, e.g. *Halimeda*, there is intensive lime deposition. Representatives of the order are known to have existed from very early times (see p. 315).

BRYOPSIDACEAE: *Bryopsis* (*bryo*, moss; *opsis*, an appearance). Fig. 5.2.

Most of the species of this genus are restricted to warmer seas, though a few, of which *B. plumosa* is the most common, occur in colder waters. The principal axis, which is often naked in its lower part, arises from an in-conspicuous, filamentous, branched rhizome that creeps along the substrate

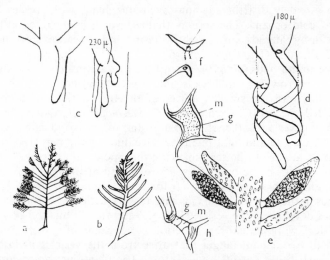

Fig. 5.2 *Bryopsis*. a, plant of *B. plumosa*; b, portion of same; c, *B. corticulans*, rhizoid formation from lower branches; d, the same, *B. scarfei*; e, *B. plumosa*, female gemetangia; f–h, stages in septum formation at base of gametangium (g = gelatinized material, m = membrane, r = ring of thickening initiating septum). (a × 0·6; b × 7) (a, b, after Taylor; c, d, after Chapman; e–g, after Fritsch)

to which it is attached by rhizoids. The bi- or tripinnate fronds either have branching confined to one plane or are radially branched, the branches being constricted at the point of origin, and here the cell membrane is also thickened. In *B. pennata* the microfibrils forming the wall are arranged transversely and if this holds true for the genus it differs from the Siphonocladales (p. 90). The cytoplasm in the main axis and branches frequently exhibits streaming movements. The function of the rhizome, especially in warmer waters, is probably that of a perennating organ, although vegetative multiplication can also occur through abstriction of the pinnae, which then develop rhizoids at their lower end. The only other known method of reproduction is sexual. The plants are dioecious and produce biflagellate anisogametes which develop in gametangia that are cut off from the parent thallus by means of septa. The microgametes differ from the macrogametes in that they lack pyrenoids. Gamete formation is hastened by moving the plants from light to dark or by changing the concentration of the nutrient solution. The gametes are liberated through gelatinization of the apex of the gametangium and, after fusion has taken place, the zygote germinates at once into a new plant. The plants are diploid and meiosis takes place at gamete formation. Inversion of the thallus takes place under conditions of dull light or when it is planted upside down, and under these circumstances the apices of the pinnae develop rhizoids. This exhibition of polarity indicates clearly that the thallus is differentiated internally, but it is still a matter for speculation how such differentiation can occur in an organism which is to all intents and purposes one unit.

CODIACEAE: *Codium* (fleece). Fig. 5.3.

This is a widely distributed, non-calcareous genus with species living in warm and cold oceans. The spongy thallus, which is anchored either by a basal disc or by rhizoids, varies greatly in form and appears as erect, branched, worm-like threads (*C. tomentosum* group), flat cushions (*C. adhaerens* group), or as large round balls (*C. bursa* group). In *C. tomentosum* there is a central medulla of narrow forked threads and a peripheral cortex of club-shaped utricles which are the swollen apices of the forked threads. Some of the large-utricled species (*C. cranwelliae*,* *C. megalophysum, C. papenfussii*) are restricted to caves and other shady places and this may indicate relationship between utricle size and light intensity. Deciduous hairs may develop on the vesicle and scars are to be seen marking their point of attachment; annular thickenings occur at the base of each vesicle and at the bases of the lateral branches, although a fine pore is left for intercommunication. The width of these pores in the case of *C. bursa* is said to vary with the season. Detachable propagules develop on the vesicle (Fig. 5.3) and form a method of vegetative reproduction, while sexual reproduction is by means of gametes, which are produced in ovoid gametangia that arise from the vesicles as lateral outgrowths, each being cut off by a septum. The plants are anisogamous, the macrogamete mass being green and the microgamete mass yellow. Some of the species are dioecious and others are monoecious, and in two of the latter the male and female gametangia are borne on the same utricles. In *C. elongatum* it would appear that the determination of the sex may be a seasonal phenomenon, females appearing first, then hermaphrodites and finally males. The gametes fuse or else develop parthenogenetically, but in either case a single thread-like protonema develops and this has a lobed basal portion, from which the adult develops through the growth of numerous ramifications of the one or more primary filaments. Meiosis occurs at gametogenesis.

* This species also occurs in deep water (100′) (F. I. Dromgoole, pers. comm.).

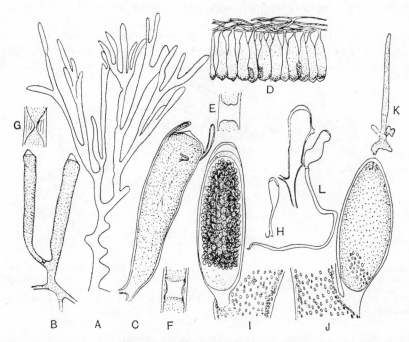

Fig. 5.3 *Codium*. A, plant of *C. tomentosum*; B, *C. fragile*, utricles; C, *C. tomentosum*, single utricle with hairs; D, *C. tomentosum*, portion of thallus with medulla and cortical utricles; E–G, stages in formation of constriction at base of utricle; H, propagule of *C. isthmocladum*; I, *C. tomentosum*. female gametangium; J, *C. tomentosum*, male gametangium; K, *C. tomentosum*, juvenile thread; L, *C. isthmocladum*, utrical with propagule. (A, after Taylor; B, C, E–G, J, after Tilden; D, H, K, L, after Fritsch; I, after Oltmanns)

CODIACEAE: *Halimeda* (daughter of Halimedon, King of the Sea). Fig. 5.4.

This wholly tropical genus has existed from Tertiary time, and it has played a considerable part in the formation of coral reefs and sands where the species are very abundant. The plants are borne on a short basal stalk that arises from a prostrate system of creeping rhizoids which can also function for vegetative reproduction (Kamura, 1966). The branched aerial thallus is composed of flat, cordate or reniform segments which are strongly calcified on the outside, the segments being separated from each other by non-calcified constrictions. Branching in some species is restricted to one plane and the size of the segments varies greatly from species to species and with water depth. The behaviour of the medullary filaments at the nodes, whether remaining separate, fusing into small units or into a single one, is a basic taxonomic character. The segments are composed of interwoven threads with lateral branches that develop perpendicularly and produce a surface of hexagonal facets through fusion of the swollen ends. Reproductive organs develop at the ends of forked threads which vary greatly in their mode of branching: these threads, which are cut off from the parent thallus by basal plugs, arise from the surface of the segments or, more frequently, are confined

105

to the edges.* Isomorphic biflagellate swarmers are liberated from male and female gametangia on separate plants, but in *H. incrassata* the gametes are anisogamous (Kamura, 1966). It is likely that reproduction usually occurs vegetatively by means of underground rhizoidal filaments (Colinvaux, 1968).

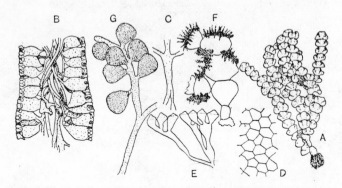

Fig. 5.4 *Halimeda.* A, plant of *H. simulans*; B, *H. discoidea*, longitudinal section showing structure; C, central filament: two fuse and subsequently divide into three; D, cuticle of *H. opuntia*; E, *H. scabra*, termination of filaments; F, fruiting plant; G, sporangia; (A × 33; B, C × 20; D × 132·5; E × 100) (A, D, E, after Taylor; B, C, after Howe; F, G, after Oltmanns)

CAULERPACEAE: *Caulerpa* (*caul*, stem; *erpa*, creep). Fig. 5.5.

Most species frequent the quiet shallow waters of the tropics where they are often rooted in sand or mud. The prostrate rhizome is attached by means of colourless rhizoids and gives rise to numerous, erect, assimilatory shoots with apical growth, the form and arrangement of which varies very considerably (Fig. 5.5). Radial branching is regarded as primitive, and the more evolved forms which are found in quieter waters possess a bilateral branching system. The genus has been divided by Börgesen into three groups:

(*a*) The species of this group, which grow where there is plenty of mud, possess rhizomes that are vertical or oblique to enable them to reach the surface even when covered successively by mud (e.g. *C. verticillata*).

(*b*) The rhizome in these species branches first at some distance from its point of origin; it possesses a pointed apex which aids in boring through sand or mud (e.g. *C. cupressoides*).

(*c*) The rhizome is richly branched immediately from its point of origin and the various species are principally found attached to rocks and coral reefs (e.g. *C. racemosa*).

It has also been shown that the form of the thallus in some of the species depends upon the conditions of the habitat, and this is particularly well illustrated by the plastic *C. cupressoides* and *C. racemosa*: In exposed situations the plants are small and stoutly built; in more sheltered habitats

* Similar bodies have been found in *Udotea* (Moorjani, 1969).

the shoots are longer and more branched; in deep water the plants are very large with richly branched flagellate shoots.

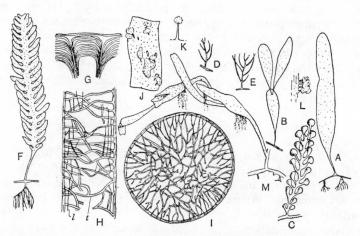

Fig. 5.5 *Caulerpa.* A, B, *C. prolifera*; C, *C. racemosa* f. *macrophysa*; D, *C. sertularioides*, side branches; F, *C. crassifolia* f. *mexicana*; G, structure of wall and two skeletal strands; H, longitudinal section of aerial portion showing longitudinal (*l*) and transverse (*t*) support strands; I, transverse section of rhizome with skeletal strands; J, K, L, *C. prolifera*, reproductive papillae; M, *C. prolifera* with gametes being liberated. (A–F × ½) (A–F, after Taylor; G–I, after Fritsch; J–M, after Dostal)

The coenocyte is traversed by numerous cylindrical skeletal strands, or *trabeculae*, which are more or less perpendicular to the surface and which are most highly developed in the rhizomes. The function of the trabeculae, which increase in thickness at the same time as the walls by successive deposition of callose, is uncertain. They may be for mechanical support or they may be concerned with diffusion, because movement of mineral salts is more rapid through these strands than through the cytoplasm; it is possible that they serve no useful purpose.

In addition to the trabeculae there are internal peg-like projections. Vegetative reproduction occurs through fragmentation of the old rhizome thus leaving a number of separate plants. In those species in which reproduction has been studied (e.g. *C. prolifera*) the anisogametes are formed in the aerial portions and are liberated through special papillae that develop on the frond. The sexual reproductive fronds have a variegated appearance caused by the massing of the biflagellate gametes at the different points. In certain species the whole plant can produce swarmers, while in others the reproductive area is limited, and in such cases the morphological identity and differentiation of the frond becomes of great interest. The thallus can be regarded as being composed of a number of individual cells which only become evident at gametogenesis. Fusion between the swarmers has been observed in *C. racemosa*, and it is probable that in all the species the motile bodies are functional gametes and that the adult plants are diploid.

DICHOTOMOSIPHONALES

The sole member of this order differs from the Siphonales in the following respects: first and foremost sexual reproduction is by a distinct oogamy, secondly the chloroplasts lack pyrenoids and starch is deposited between the lens-shaped plastids instead of on them. The characteristic siphonalean pigment siphonoxanthin is present, but the cell wall differs in that it lacks cellulose (Siddiqui and Nizamuddin, 1965). After fertilization the zygote rests for a period before germination whereas in the Siphonales it develops directly into a new plant.

DICHOTOMOSIPHONACEAE: *Dichotomosiphon* (*dichotomo*, forked; *siphon*, tube). Fig. 5.6.

This is a monotypic genus, the sole species, *D. tuberosus*, being found in fresh waters of Europe, the U.S.A., India, Burma and Pakistan. The thallus looks like *Derbesia* (p. 102) or *Vaucheria* (p. 137) with di-, tri- or quadrichotomous branching. Where branching occurs the filaments bear constrictions with an ingrowth of the wall material as well. Asexual reproduction is by means of club-shaped akinetes formed usually at the ends of lateral

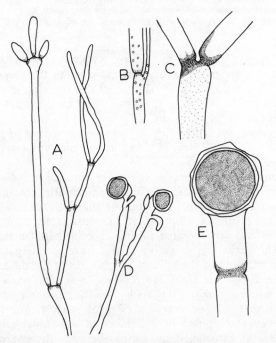

Fig. 5.6 *Dichotomosiphon tuberosus*. A, part of the filament; B, rhizoidal branch; C, filament showing constriction; D, oogonia and antheridia; E, oogonium. (after Siddiqi and Nizamuddin)

rhizoidal branches which are most abundant towards the base. These akinetes are densely packed with starch. The plants are monoecious and both sex organs are borne terminally on the ultimate branchlets. After fertilization the zygote, which develops a two-layered wall, remains quiescent within the oogonium for some time.

ZYGNEMATALES

The members of this order are distinct from the other members of the Chlorophyceae as a whole. The plants are either filamentous or single celled with a wall of one or two parts. There is considerable variation in form of the chloroplasts. The species are all characterised by a peculiar mode of sexual reproduction in which fusion takes place between masses of protoplasm which do not possess flagella and in this process cells come together in pairs and conjugate. Some authorities have suggested that the conjugate Chlorophyceae should be assigned to a separate class, Conjugatophyceae, because of the unusual reproduction, and this may well be justified.

The order has been divided into two suborders, Euconjugatae and Desmidioideae, or into two families, Zygnemataceae and Desmidiaceae. A recent revision creates three families, Mesotaeniaceae, Zygnemataceae and Desmidiaceae. The Mesotaeniaceae are single-celled species and are sometimes known as the saccoderm desmids. Most of the Desmidiaceae are also single-celled with a wall in two parts and are often known as the placoderm desmids. There are two views concerning the inter-relationships of the families: one is that both desmids are primitive to the filamentous Zygnemataceae, and the other that they are derived from the filamentous Zygnemataceae (see p. 333). Some authorities recognise two orders:

(a) Desmidiales, whose species are predominantly unicellular with cell wall and contents usually in two parts.
(b) Zygnematales, whose species are filamentous with no division of wall or cell contents into two parts.

Within the Zygnemataceae there are six types of gamete and zygospore formation.

(a) *Spirogyra* type. Vegetative cells become gametangial cells and the entire protoplasm is transformed to a gamete.
(b) *Mougeotiella* type. Vegetative cells become gametangial cells but a portion of the protoplasm persists as a cytoplasmic residue.
(c) *Zygnemopsis* type. Vegetative cells become gametangial cells but some of the contents develop into a bluish-white gelatinous substance. In this and the preceding type the conjugation tubes enlarge considerably.
(d) *Mougeotia* type. These resemble the *Mougeotiella* type but differ from it in that the zygospore is cut off from the gametangial cells by special walls.

    (*e*) *Temnogyra* type. The progametangial cells divide first into a shorter gametangial cell and a longer sterile cell. The entire protoplasm is used in the gametangial cells and there is a distinct conjugation tube.

    (*f*) *Sirogonium* type. Resembles the *Temnogyra* type but differs in that no conjugation tube is formed. When the gametangial cells make contact, a gelatinous substance is secreted and forms a disc.

The development and inter-relationships of these six types is illustrated in Fig. 5.7. It should be borne in mind, however, that the main series could be read in the reverse direction (dotted arrows).

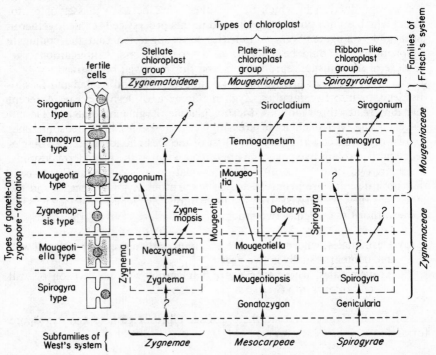

Fig. 5.7 Relationship between the types of chloroplast and the types of gamete- and zygospore-formation in the Zygnemataceae. (after Yamagishi)

The Zygnemataceae are conveniently placed into three sub-families based upon chloroplast type: the Zygnematoideae with stellate chloroplasts, the Mougeotioideae with flat plate-like chloroplasts and the Spirogyroideae with ribbon-like chloroplasts. Within each of these subfamilies most of the six gamete and zygospore formation types just described are to be found (Fig. 5.7). In some classifications *Gonatozygon* and *Genicularia* are placed in a separate family but Yamagishi (1963) has argued that this is not justified.

ZYGNEMATACEAE: *Spirogyra* (*spiro*, cell; *gyra*, curved). Figs. 5.8, 5.9.

The unbranched filaments are normally free-living although attached forms are known, e.g. *S. adnata*, and they form slimy threads known as 'water-silk' or 'mermaid's tresses'. They grow in stagnant or slowly moving water and are most abundant in either spring or autumn, the latter phase being due to the germination of a percentage of the spring zygospores. Each cell contains one or more chloroplasts possessing either a smooth or serrate margin and arranged in a characteristic parietal spiral band. The single nucleus is suspended in the middle of the large central vacuole by means of protoplasmic threads that radiate out to the parietal protoplasm. The chloroplasts, which may occasionally be branched, are T- or U-shaped in cross-section and contain numerous pyrenoids that project into the vacuole on the inner side, the majority of the pyrenoids arising *de novo* at cell division. The cell wall is thin and, according to some investigators, composed of two cellulose layers, while others maintain that there is only an inner cellulose layer with an outer layer of some other material. The whole filament is enclosed in a mucilage sheath

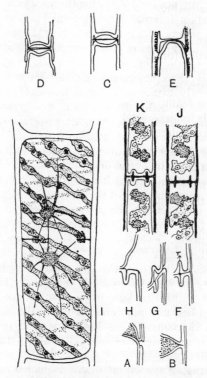

Fig. 5.8 *Spirogyra.* A, B, cell disjunction (diagrammatic); C–E, cell disjunction in *S. colligata*; F–H, *S. weberi*, cell disjunction by replicate fragmentation (*r* = replication of septum); I, vegetative structure and cell division, *S. nitida*; J, K, cell disjunction and development of replicate septa. (I × 266) (A–H, J, K, after Fritsch; I, after Scott)

of pectose. Any cell is capable of division, and vegetative reproduction by fragmentation is exceedingly common, three methods having been described:

(a) The septum between two cells splits and a mucilaginous jelly develops in between, so that when one cell subsequently develops a high turgor pressure the cells become forced apart.

(b) Ring-like projections develop on both sides of a septum and the middle lamella dissolves. Then the rings of one cell evaginate and force the cells apart while the rings of the other cell evaginate after separation (replicate fragmentation) (cf. Fig. 5.8).

(c) The septum develops an I-piece and then when the wall inverts, due to increased turgor, the I-piece is slipped off and the two cells come apart (cf. Fig. 5.8).

When two filaments touch they may form joints or geniculations and adhesion is brought about by a mucilaginous secretion produced by the stimulation of the contact. The formation of such geniculations, however, has no connection with reproduction.

Sexual reproduction occurs by conjugation, which is stimulated by a combination of certain internal physiological factors and the pH of the external medium. It commonly takes place during the spring when the threads come together in pairs, but more than two filaments may be involved. The threads first come together by slow movements, probably related to the secretion of mucilage; then they become glued together by their mucilage and later young and recently formed cells in both filaments put out papillae. These papillae meet almost immediately, elongate, and push the threads apart. Normally one of the threads produces male gametes and the other female, but occasionally the filaments may contain mixed cells. The papillae from male cells are usually longer and thinner than those from the female cells. The conjugating cells accumulate a great deal of starch, the nuclei decrease in size and the wall separating the papillae breaks to form a conjugation tube. The whole process so far described forms the maturation phase which is followed by the phase of gametic union. The protoplasm of the male cells contracts from the walls and the cytoplasmic mass then migrates through the conjugation tube into the female cell where fusion takes place and this is then followed by contraction of the female cytoplasm, though in the larger species it may contract before fusion. Fusion of the two nuclei may be delayed for some time, but in any case the male chloroplasts degenerate. The process described above is known as scalariform conjugation. In monoecious species lateral conjugation occurs between adjoining cells on the same filament. More recently a species (S. jogensis) has been described in which contents of adjacent cells fuse, the contents of one cell passing through a pore in the intervening cell wall.

The last phase is that of zygotic contraction which is brought on by further action of the contractile vacuoles, after which a thick three-layered wall develops around the zygote, the middle layer or mesospore frequently being highly sculptured. The zygospore occasionally germinates almost at once,

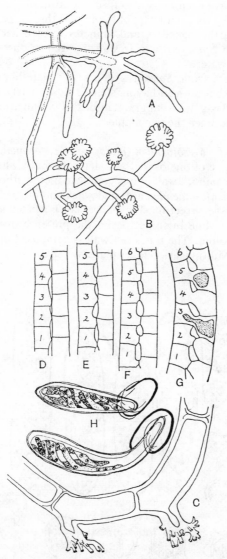

Fig. 5.9 *Spirogyra.* A, B, rhizoid formation in *S. fluviatilis*; C, rhizoids and haptophores of *S. adnata*; D–G, stages in conjugation, *S. varians*; H, germination of zygospore in *S. neglecta*. (A, B, after Czurda; C, after Delf; D–G, after Saunders; H, after Fritsch)

thus producing plants that account for the autumn maximum, but it is usually dormant until the following spring. Meiosis takes place very soon after fusion and four nuclei are formed and of these, three abort so that only the haploid generation exists. A two-celled germling is formed, the lower cell being relatively colourless and rhizoidal in character. Filaments of two different species have been known to fuse, the form of the hybrid zygospore being determined by the characters of the female thread. Azygospores, which have arisen by parthenogenesis, and akinetes form other means of reproduction. Although *Spirogyra* is a fresh-water genus, one species, *S. salina*, grows in brackish water.

Mougeotia differs from *Spirogyra* in its flat plate-like chloroplast, and *Zygnema* differs from both in having stellate chloroplasts. In *Mougeotia* the amoeboid protoplasts both migrate into the conjugation tube, and the zygospore forms between the two filaments.

MESOTAENIACEAE : *Mesotaenium* (*meso*, middle; *taenium*, band). Fig. 5.10.

This is an example of one of the saccoderm desmids, which as a group are characterized by a smooth wall in one complete piece without any pores. The rod-shaped cells of *Mesotaenium* are single, have no median constriction, and are circular in transverse section. The chloroplast is a flat axile plate containing several pyrenoids, while in some species the presence of a tannin pigment imparts a violet colour. The inner cell wall is composed of cellulose and the outer of pectose. Multiplication takes place by cell division, the daughter cells

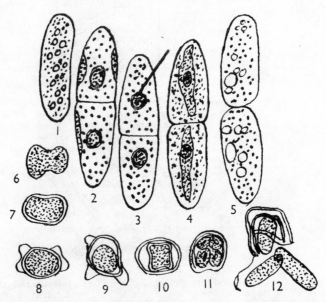

Fig. 5.10   *Mesotaenium*. 1–5, cell division. 6–12, conjugation of *Cylindrocystis brebissonii* and germination of zygospore. (after Fritsch)

being liberated by dissolution of the middle lamella after a constriction has been formed, though in some cases this may not occur until a number of cells have been enclosed in a common mucilaginous envelope. Sexual reproduction

is by means of conjugation, two processes being put out just as in the filamentous forms. Meiotic division occurs upon germination, giving rise to four
nuclei. The protoplast then divides giving rise to four daughter protoplasts
and hence new cells. Occasionally two nuclei degenerate, resulting in only
two new cells. The species are to be found in upland pools, peat bogs or on
the soil.

DESMIDIACEAE. Figs. 5.11, 5.12.

This family comprises the placoderm desmids, in which the wall is highly
ornamented and the wall and cell contents are in two sections. Each cell thus
comprises two half cells which may be delineated by a narrow (e.g. *Desmidium*)
or deep (e.g. *Staurastrum*) fissure or constriction called the sinus. The narrow
section between the two half-cells is called the isthmus. In a few genera (e.g.
Closterium) there is no half-cell arrangement but the cells contents are,

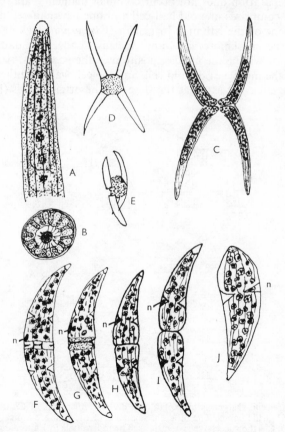

Fig. 5.11 *Closterium*. A, *C. lanceolatum*, chloroplast structure; B, *Closterium* sp., structure
in transverse section; C, *C. lineatum*, first stage in conjugation; D, *C. rostratum* var. *brevi-
rostratum*, zygospore formation, second stage; E, *C. calosporum*, mature zygospore; F–J,
C. ehrenbergii, stages in cell division (n = nucleus). (A–D, E–J, after Fritsch; E, after Smith)

however in two sections. The cell wall may be smooth (*Closterium*) or heavily
ornamented with spiny or granular projections (*Staurastrum*). The cell wall,
which is composed of cellulose and an outer pectin layer, is characterized by
numerous, regularly arranged pores. In *Closterium* the pores are in grooves
and mucilage secretion through the larger apical pores is responsible for the
jerky cell movement. The half-cells possess one, two or four axial chloroplasts.
In *Closterium* the two chloroplasts are conical and curved with several
pyrenoids. Some genera (e.g. *Closterium*) possess polar vacuoles in which
crystals of gypsum are located.

Cell division is peculiar and takes place in a variety of ways. In those
genera possessing a sinus, the isthmus elongates and mitotic nuclear division
follows. At the centre of the isthmus a new cellulose-pectin wall is laid down
and the isthmus enlarges (Fig. 5.11) to form two new semi-cells. A new sinus
forms in the plane of the old one. The cells now separate at the apical (new)
cell wall. If separation does not occur colonial 'filaments' are formed. Each
new cell thus comprises one old half-cell and one new half-cell derived from
the enlargement of the isthmus. In genera (*Closterium*) lacking the median
constriction the cell chloroplasts and nucleus divide. The daughter nuclei
migrate to what will be the median position of the new cells. A cross wall is
laid down in the middle of the old cell, and the cell wall on either side of the
new wall elongates and assumes the shape of the original cells (Fig. 5.12).

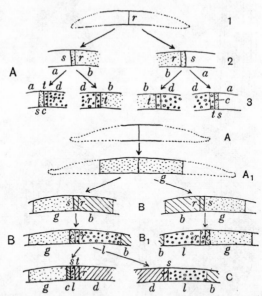

Fig. 5.12 *Closterium*. Diagrams to explain cell division in species of *Closterium* with (B)
and without (A) girdle bands. The different segments of the wall are indicated by shading.
1, 2, 3 and A, B, C = the successive generations. The individuals in 1 and A have each arisen
from a zygote and have not undergone division. (a, b, d = semi-cells of various ages; c = the
connecting band demarcated by the two sutures *s*, of the previous generation, and *t*, of the
present; *g* = girdle bands developed before and *l* after division; *s* = suture between young
and older semi-cells; *r* = the line of the next division) (after Fritsch)

In sexual reproduction papillae from the two cells meet or else the naked amoeboid gametes fuse immediately outside the cells. After the gametes have fused two of the chloroplasts degenerate and the zygospore nucleus divides meiotically. Zygospore division gives rise to two daughter cells, each containing one chloroplast and two nuclei, one of which subsequently degenerates. Under unfavourable conditions asexually produced chlamydospores have been observed in *C. moniliferum*.

The genus is fresh water and many species are planktonic. The group is extremely widespread but is scarce in waters containing much lime, the individual species thriving best in soft or peat bog waters. The most favourable seasons for their development are the late spring and early summer, and their resistance in the vegetative state to adverse conditions would seem to be very great. Recent cytological work on desmids has revealed great variation within clones of the same species. Thus, *Cosmarium cucumis* gave chromosome counts of 44 or 52, and *Netrium digitus* gave counts of 122, 172–82, and 592.

CHAROPHYCEAE

The class comprises but a single order, the Charales, with more than 300 species. The plants represent a highly specialized group that must have diverged very early in the course of evolution from the rest of the green algae, the intermediate forms subsequently being lost. The young plants develop from a protonemal stage and the erect plants have a structure which is more elaborate than any type so far described while the thallus is sometimes encrusted with lime (*Chara* spp.). The group is known to be very ancient because fossil members are found from almost the earliest strata. There is no asexual reproduction but the process of oogamous sexual reproduction is very complex. The family is divided into two subfamilies, the Charoideae, in which the characteristic oogonium has a single ring of five coronal cells, and the Nitelloideae where the oogonium is surmounted by two rows of five cells. There are five genera in the former and two in the latter and it is generally agreed that the Charoideae are more advanced than the Nitelloideae. The living forms are widely distributed in quiet waters, fresh or brackish, where they may descend to considerable depths (12 m) so long as the bottom is either sandy or muddy.

The greatest number of species is found in Asia but there is a high degree of endemism in Australia, North America and Asia.

Nitella: (a little star). Figs. 5.13–5.16.
The plants bear whorls of lateral branches arising from the nodes. The nodes are formed by a transverse layer of cells in contradistinction to the internodes, which consist of one large cell whose individual length may be as much as 25 cm in *Nitella cernua*. The height of the different species varies and in the largest may extend up to 1 m, growth being brought about by an apical cell which cuts off successive segments parallel to the base. Each new segment divides transversely into two halves, the upper developing into a node and the

lower into an internode. (Fig. 5.13). The nodal cell divides longitudinally and each of the daughter cells divides longitudinally several times so that each node has two central cells and a number of peripheral cells. Each peripheral cell divides again longitudinally. The outer cells give rise to branches by protruding to form new apical cells, but these soon cease to grow after the

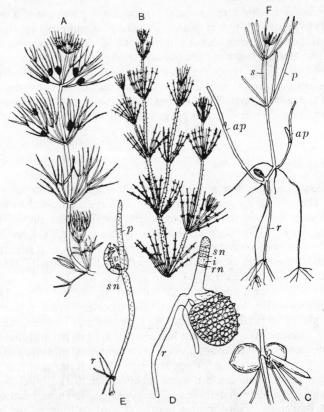

Fig. 5.13 *Charales.* A, *Nitella batrachosperma*; B, *Chara hispida*; C, underground bulbils of *C. aspera*; D, germinating oospore; E, protonema of *C. fragilis*; F, young plant of *C. crinita.* (*ap* =accessory protonema, *i* =internode, *p* =protonema, *r* =rhizoids, *rn* =rhizoid node, *s* =shoot, *sn* =stem node, *v* =initial of young plant) (after Fritsch)

branch has reached a short length. At the basal node of the main plant branches of unlimited growth are produced: these arise on the inner side of the oldest lateral in the whorl, thus producing a fictitious appearance of axillary branching.

Experimental work with *Nitella cristata* using varying conditions of light, temperature and day length has shown that the length, form and degree of branching of sterile and fertile whorls are influenced by these factors.

Multicellular branched rhizoids with oblique septa function as absorption organs and also serve for anchorage. The rhizoids develop from the lowest node of the main axis, but every node is potentially capable of producing them; normally the presence of the stem apex inhibits their appearance, but if this is cut off the rhizoids will then develop. This behaviour suggests an

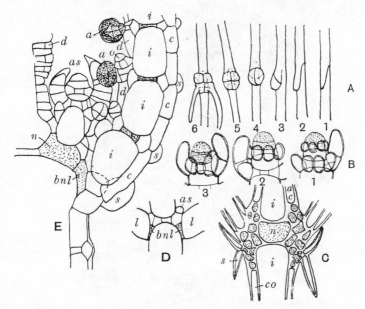

Fig. 5.14 *Charales*. A, 1–6, successive stages in development of root node of *Chara aspera*: 1, double foot joint; 2, dilation of toe of upper foot; 3, toe portion cut off; 4, 5, subdivision of toe cell; 6, rhizoids growing out; B, 1–3, successive growth stages of apex of *Nitella*: in 1 apical cell is undivided, in 2 it has divided, in 3 the lower cell has divided into an upper node and a lower internode; C, *C. hispida*, node with stipules; D, *N. gracilis*, longitudinal section of node; E, *C. fragilis*, branch at node with axillary bud. (*a* =antheridium, *ac* = ascending corticating cells, *as* =apex of side branch, *bnl* =basal node of branch (*1*) and *co* =cortical cells, *d* =descending cortical cells, *i* =internodal cell, *n* =nodal cell, *o* = oogonium initial, *s* =stipule) (A, B, after Grove; C–E, after Fritsch)

auxin control similar to that found in the higher plants. The cells, which have a cellulose membrane, contain discoid chloroplasts and a nucleus but no pyrenoids. In *N. opaca* the microfibrils of the cell wall are usually oriented in broad helices. Cytoplasmic streaming is very readily observed, especially in the internodal cells. Sexual reproduction is by a characteristic oogamy where light intensity plays a part in determining the production of the sex organs. The species are either dioecious or monoecious; in the latter case the oogonia and antheridia are juxtaposed, the oogonia being directed upwards and the antheridia downwards, both organs usually appearing on secondary lateral branches of limited growth.

ANTHERIDIA (Fig. 5.15)

The apical cell of the lateral branch cuts off one or two discoid cells at the base and then becomes spherical. The upper spherical cell divides into

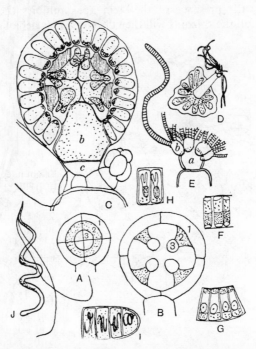

Fig. 5.15 *Charales*. A, B, stages in development of antheridium of *Chara*: 1–3, segments and cells to which they give rise; C, section of almost mature antheridium of *Nitella flexilis* (*b* = flask cell, *c* = extra basal cell); D, *C. tomentosa*, single plate with manubrium with spermatogenous threads; E, *C. tomentosa*, apex of manubrium with spermatogenous threads (*a* = primary head cell, *b* = secondary head cell); F–I, *C. foetida*, stages in formation of antherozoids in spermatogenous threads; J, mature antherozoid. (A, B, after Goebel; C–E, J, after Grove; F–I, after Frisch)

octants and this is followed by two periclinal divisions, after which the whole enlarges and the eight peripheral cells develop carved plates (shields), which give the wall a pseudo-cellular appearance. At maturity these peripheral cells acquire brilliant orange contents. The uppermost discoid basal cell protrudes into a hollow structure whose formation has already been described. The middle segment of each primary diagonal cell now develops into a rod-shaped structure, the manubrium, which bears at its distal end one or more small cells, the capitula; every one of these produces six secondary capitula from each of which arises a forked spermatogenous thread containing 100–200 cells. These antheridial cells each produce one antherozoid, an elongate body with two flagella situated just behind the apex. The complete structure has been regarded as one antheridium, though another view, probably more correct, regards the octants as secondary laterals, the manubrium as an internode, the capitula as a basal node and the spermatogenous threads as modified laterals. On this basis the antheridia are one-celled and conform to

120

the normal structure of the antheridia in the green algae. This second interpretation, if it is correct, helps considerably in understanding this peculiar
group.

OOGONIA (Fig. 5.16)

The apical cell of the lateral branch divides twice to produce a row of three
cells, the uppermost cell developing into the oogonium while the lowest forms
a short stalk. The middle cell cuts off five peripheral cells which grow up in a
spiral fashion and invest the oogonium; each one finally cuts off two small
coronal cells at the apex. The oogonial cell cuts off three cells at its base and
it is maintained that these, together with the oogonium, represent four

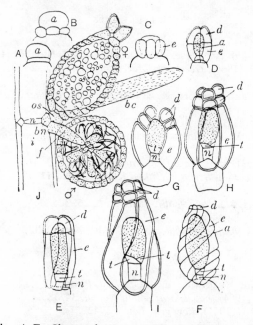

Fig. 5.16 *Charales*. A–F, *Chara vulgaris*, stages in formation of oogonium; A, first
division; B, C, division of periphery to form envelope cells; D, coronal cells cut off; F,
mature oogonium; G I, *Nitella flexilis*, stages in formation of oogonium; J, fertile branch
of *C. fragilis*. (a = oogonium, bc = bract cell, bn = branch nodal cell, d = coronal cells,
e = envelope cells, f = flask cell, i = internode, n = nodal cell, t = turning cell, os = oogonium
stalk cell) (after Grove)

quadrants, only one of which develops to maturity. When mature, the
investing threads part slightly to form a neck, and the apex of the oogonium
gelatinizes to allow the antherozoids to enter. After fertilization the zygote
nucleus travels to the apex of the oospore and a coloured cellulose membrane
is excreted around it, while the oogonium wall, together with inner walls of
the investing threads, thicken and silicify. Four nuclei are formed by two
successive divisions of the zygote nucleus, meiosis taking place during this

process. One of these nuclei is cut off by a cell wall and the other three degenerate. The small cell so formed then divides and two threads grow out in opposite directions, one a rhizoid, the other a protonema. The cell next to the basal cell of the protonema divides into three cells, the upper and lower forming nodes which become separated by elongation of the middle cell (Fig. 5.16). The lower node develops rhizoids while the upper produces a whorl of laterals from all the peripheral cells except the oldest, which instead forms the apex of the new plant. The mature plant is therefore morphologically a branch of the protonema. Vegetative reproduction can take place from secondary protonemata which develop from the primary rhizoid ring or from dormant apices.

REFERENCES

GENERAL

Bourelly, P. (1966). *Les Algues d'eau douce*, Tome **1**, (Boubée et Cie, Paris).

Chapman, V. J. (1964). *Oceanogr. Mar. Biol. Ann. Rev.*, **2**, 193.

CHAROPHYCEAE

Wood, R. D. and Imahori, K. (1965). *Revision of the Characeae*, (Cramer, Weinheim).

Wood, R. D. (1952). *Bot. Rev.*, **18**, 317.

Desikachary, T. V. and Sundaralingam, V. S. (1962). *Phycologia*, **2**, 9.

Conjugales

Okada, Y. (1953). *Mem. Fac. Fish. Kagoshima Univ.*, **3**, 165.

Yamagishi, T. (1963). *Sci. Rep. Tokyo Kyoiku Daig.*, **11**, 191.

Siphonales

Egerod, L. E. (1952). *Univ. Calif. Publs. Bot.*, **25**, 325.

Zygnemataceae

Randhawa, S. (1962). *Zygnemaceae*, (I.C.A.R., New Delhi).

Bryopsis

Green, P. B. (1960). *Am. J. Bot.*, **47**, 476.

Caulerpa

Dawes, C. J. and Rhamstine, E. L. (1967). *J. Phycol.*, **3**, 117.

Jacobs, W. P. (1964). *Pubbl. Staz. Zool. Napoli*, **34**, 185.

Feldmann, M. J. (1955). *Rev. gén. Bot.*, **62**, 422.

Closterium

Chardard, R. (1965). *Revue Cytol. Biol. Vég.*, **28**, 15.

Cook, P. W. (1963). *Phycologia*, **3**, 1.

Kies, L. (1964). *Arch. Protistenk.*, **107**, 331.

Lippert, B. E. 1967. *J. Phycol.*, **3**, 182.

Codium

Jacobi, G. (1961). *Arch. Protistenk.*, **105**, 345.

Silva, P. C. (1960). *Nova Hedwigia*, **1**, 497.

Dichotomosiphon

Siddiqi, M. Y. and Nizamuddin, M. (1965). *Sind. Univ. Sci. Res. Journ.*, **1**, 27.

Halicystis-Derbesia

Feldmann, J. (1954). *C.r. hebd. Séanc. Acad. Sci., Paris*, **222**, 752.

Hollenberg, G. J. (1935). *Am. J. Bot.*, **22**, 783.

Ziegler, J. R. and Kingsbury, J. M. (1964). *Phycologia*, **4**, 105.

Ziegler, J. R. and Kingsbury, J. M. (1968). *Am. J. Bot.*, **55**, 1.
Ziegler-Page, J. R. and Sweeney, B. M. (1968). *J. Phycol.*, **4**, 253.
Halimeda
Colinvaux, L. W. (1968). *J. Phycol.*, **4** (Suppl.), 4.
Hillis, L. W. (1959). *Publs. Inst. mar. Sci. Univ. Texas*, **6**, 321.
Kamura, S. (1966). *Bull. Univ. Ryukyus, Maths. Nat. Sci.*, **9**, 302.
Mougeotia
Puiseux-Dao, S. and Levain, N. (1963). *J. Microscopie.*, **2**, 461.
Pediastrum
Davis, J. S. (1967). *J. Phycol.*, **3**, 95.
Spirogyra
Butterfass, T. (1957). *Protoplasma*, **48**, 368.
Dawes, C. J. (1965). *J. Phycol.*, **1**, 121.
Staurastrum
Brook, A. J. (1959). *Trans. Roy. Soc. Edin.*, **63**, 589.
Lind, E. M. and Croasdale, H. (1966). *J. Phycol.*, **2**, 111.
Udotea
Moorjani, S. A. (1969). *J. E. Afr. Nat. hist. Soc. Nat. Mus.*, **27**, 227.
Zygnema
Chardard, R. (1967). *Revue Cytol. Biol. Vég.*, **30**, 191.

6

EUGLENOPHYTA, XANTHOPHYTA, CHLOROMONADOPHYTA

EUGLENOPHYTA

EUGLENOPHYCEAE

This class comprises mainly motile unicells, although there is one small group of palmelloid, non-flagellated species. The organisms lack a cellulose cell wall but there is an external semi-rigid pellicle or periplast. This appears under the electron microscope as a well-differentiated layer of proteinaceous strips with helical symmetry, which in many species is modified by warts and ridges. When the pellicle is rigid it is often markedly striated, but when it is flexible metabolic changes of shape are possible and permit locomotion. Some cells may also show a peculiar twisted shape, e.g. *Phacus*.

The motile cells possess two, rarely three, flagella. The long flagellum, which bears a lateral row of hairs ('stichonematic type'), emerges from an anterior reservoir and canal (called gullet and cytopharynx in the older literature). The short one does not emerge from the reservoir, around which are one or more contractile vacuoles. The single nucleus is generally conspicuous and located near the centre of the cell.

The majority of the euglenoids are colourless. The pigmented species contain numerous chloroplasts, which are disc-like, rod-like, ribbon-like or stellate, and distributed throughout the cytoplasm. The pigments themselves are chemically similar to those of the Chromophyta (see p. 5). Food reserve is deposited as granules of paramylum, a carbohydrate very similar to laminarin (see p. 184) of the Phaeophyceae. Most Euglenoids possess a red eye spot, which differs from that of *Chlamydomonas* (see p. 36) in being a collection of carotenoid — containing granules, located outside the chloroplast. Nutrition may be either photoauxotrophic, saprophytic or holozoic (phagotrophic). The Euglenophyceae possess an unusual form of nuclear division. At metaphase the nucleolus persists, the chromosomes arrange themselves parallel to the axis of division and there is no evidence for a mitotic spindle. Although meiosis has been reported there is apparently no sexual reproduction and it is assumed that meiosis occurs at the division of a zygote nucleus derived from nuclear fusion.

The predominant phase is the motile unicell, although non-motile cysts are known and reproduction is often by longitudinal cell division.

Euglenoids occur in both salt and fresh water where they are found in a variety of habitats, sometimes tolerating a very low pH, although they are more abundant in pools rich in organic matter. The saprophytic forms grow best in the presence of plenty of decaying matter, while colourless forms are endozoic. One member has been described as the cause of green snow.

The most recent treatment has divided the class into six orders. The Rhabdomonadales, Sphenomonadales and Heteronematales comprise only colourless species. Nutrition is either saprophytic (Rhabdomonadales: *Rhabdomonas*) phagotrophic (Heteronematales: *Peranema*) or both (Sphenomonadales: *Petalomonas*). The second flagellum is emerged in most species but not in the Rhabdomonadales. These three orders are further characterized by the absence of eye spots and flagellar swellings; with the exception of the Heteronematales they do not show euglenoid movement.

The remaining Euglenoids, which may be colourless or possess chloroplasts, are divided into the Euglenomorphales, Euglenales, Eutreptiales. With rare exceptions, nutrition is saprophytic or phototrophic. The saprophytes are a small group of Euglenoids living endosymbiotically in the digestive tract of tadpoles.

The Eutreptiales are characterised by two emergent flagella, one of which extends laterally, but the cells show typical euglenoid movement.

EUGLENALES

This is by far the largest group of Euglenoids, and includes both coloured (e.g. *Euglena*, *Phacus*, *Trachelomonas*, *Colacium*) and colourless forms (e.g *Astasia*). Nutrition is typically phototrophic in the coloured forms or holozoic or saprophytic in the colourless forms.

EUGLENACEAE: *Euglena* (*Euglena*, bright eye). Fig. 6.1

This family comprises coloured Euglenoids containing chloroplasts which are usually twisted and elongate. In *E. spirogyra* the pellicle, which underlies a plasmalemma membrane, is composed of a single unit wound in an interlocking helix around the cell. In cross section the unit is composed of grooves and teeth. Arranged parallel with it are a series of mucus bodies, whose function is not exactly known.

Contrary to earlier belief there are two flagella in most of the pigmented forms. The locomotory one (with lateral hairs down one side) is emergent and usually as long as the cell; the other does not emerge from the reservoir. At the base of the locomotory flagellum is a swelling which may be concerned with the photoreception mechanism. The distinctive red eye spot is situated on the edge of the reservoir immediately opposite the flagellar swelling. Both eye spot and flagellar swelling are absent in *Astasia*. The contractile vacuole (usually only one) is adjacent to the reservoir on the side opposite to the eye spot and discharges into the reservoir. Numerous lenticular chloroplasts are scattered throughout the cell, and pyrenoids may (*E. gracilis*) or may not (*E. spirogyra*) be present. Chromosome counts in algae are not always easy but in *E. spirogyra* the chromosome complement appears to be about 86.

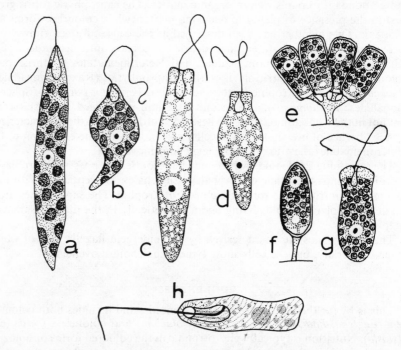

Fig. 6.1 a, b, *Euglena intermedia*; c, d, *Astasia fustis*; e, *Colacium caluum*, colony attached; f, *C. caluum*, single cell attached; g, *C. caluum*, euglenoid zoospore attached; h, *Peranema*. (after Leedale)

Asexual reproduction is by longitudinal cell division, which may occur either while the cells are motile or resting. Division commences at the anterior end, and where there are non-motile stages a gelatinous sheath may encircle the dividing cell. In some species thick-walled cysts have been recorded, and these germinate to give a single motile cell. Sexual reproduction has not been demonstrated with certainty.

COLACACEAE: *Colacium* (*colacium*, limb). Fig. 6.1

The genus comprises dendroid or palmelloid colonies of immotile cells enclosed in a wall. The species are epizoic and colonies are attached to the host by a gelatinous stalk. ·

Reproduction is by cell division which differs from that of other Euglenoids by occurring at the non-motile stage. Amoeboid stages are known and sometimes the protoplast of a colonial cell may escape and form a free-swimming flagellated individual that will eventually come to rest, lose its flagella and start a new colony.

Phacus is characterised by a flat leaf-like appearance while *Trachelomonas* possesses a distinctive brown cellular sheath.

REFERENCES

GENERAL

Buetow, D. E. (1969). *Biology of Euglena* (Edit.). Vol. 1, (Academic Press, New York and London).

Leedale, G. F. (1967). *Euglenoid Flagellates*, (Prentice Hall, New York).

Leedale, G. F. (1967). *A. Rev. Microbiol.*, **19**, 17.

Mignot, J-P. (1965). *Protistologica*, 3, 22.

Wolken, J. J. (1967). *Euglena, An Experimental Organism for Biochemical and Biophysical Studies*, 2nd Ed., (Rutgers University Press).

Colourless forms

Christen, H. R. (1963). *Nova Hedwiga*, **4**, 437.

Pringsheim, E. G. (1963). *Farblose Algen*, (Gustav Fischer Verlag, Stuttgart).

Colacium

Johnson, E. F. (1965). *Arch. Protistenk.*, **83**, 241.

Euglena

Gibbs, S. P. (1960). *J. Ultrastruct. Res.*, **4**, 127.

Leedale, G. F., Meeuse, B. J. D. and Pringsheim, E. G. (1965). *Arch. Mikrobiol.*, **50**, 68, 133.

Pringsheim, E. G. (1965). *Nova Acta Acad. Caesar. Leop. Carol.*, **18**, 1.

XANTHOPHYTA

XANTHOPHYCEAE

These algae were originally classified with the Bacillariophyceae and Chrysophyceae in the division Chrysophyta. Similarities to the Chrysophyceae include the presence of chrysolaminarin or oil as food reserves, some aspects of the flagella types, the presence in certain species of a distinct type of spore (the statospore), and also the frequent occurrence of cell walls with two overlapping halves. The principal difference is in pigmentation, as the cells lack a secondary chlorophyll and fucoxanthin and possess specific carotenoids. Although there is undoubtedly some relationship between Xanthophyceae and Chrysophyceae it is probably not close enough to warrant their inclusion in the same division.

Within this group there is an evolutionary trend from the flagellated unicell to the palmelloid state and also through the filamentous to the siphonaceous habit. Morphological development is therefore analogous to that of the Chlorophyceae.

In certain genera in which a cell wall is present, it may be of two over-lapping halves. In the filamentous forms the filaments are composed of articulated H pieces, two such pieces comprising the surrounding wall of an individual cell.

The flagellated cells of the Xanthophyceae usually possess two unequal flagella, the short one being of the simple type, the long one being pleurone-matic (tinsel type, 'flimmergeissel'). It is from this heteromorphic arrangement that the class acquired its early name 'Heterokontae'. The chloroplasts, of which there may be one or more, usually lack pyrenoids, and their yellow-green colour is derived from carotenoids which tend to mask the chlorophyll *a* (the chlorophyll *e*, recorded from *Tribonema bombycinum* and *Vaucheria haemata* is very probably an *in vitro* product). Asexual reproduction is by cell division, production of motile zoospores, non-motile aplanospores, akinetes, or in some unicellular forms by internal cysts (=statospores). The statospore or cyst germinates to give two or four zoospores or amoeboid bodies.

All flagellated cells are naked and possess a contractile vacuole system. Sexual reproduction has been recorded for only a few genera, but it is a very distinct feature of the siphonaceous species.

The majority of Xanthophyceae are fresh water, sub-aerial (growing on trees, walls etc.) or soil inhabitants and a few are marine.

Recent work has shown that certain species of Xanthophyceae differ in certain important aspects. In the true Xanthophyceae the eyespot is part of the chloroplast. In the following the eyespot is independent of the chloroplast: *Pleurochloris commuta*, *P. magna* (Pleurochloridaceae), *Polyedriella*, *Vischeria stellata*, *V. punctata*, *Chlorobotrys* (Chlorobotrydaceae) and *Ellipsoidion acuminatum*. Additionally these species differ in that the zoospores possess only one emergent flagellum, and the chloroplast lacks a girdle lamellae. Golgi bodies and the polygonal pyrenoids are not found in motile cells but are present in vegetative cells. Hibberd and Leedale (1970) have suggested that these algae should be transferred to a new class, the Eustigmatophyceae, though further work is needed before the validity of this class can be esta-blished. It does raise however the important differences within the Xantho-phyceae. Other coccoid forms, e.g. *Pleurochloris meiringensis*, *Botrydiopsis*, *Bumilleriopsis* are however 'true' Xanthophytes.

CHLORAMOEBALES (HETEROCHLORIDALES)

The organisms are all naked, motile, flagellated unicells, although tem-porary conversion to an amoeboid form is not uncommon. The ovoid or pear-shaped cells are uninucleate with one or more discoid chloroplasts. The two flagella are apically inserted, with one or more anterior contractile vacuoles. Reproduction is by cell division of either motile or amoeboid phases or by means of statospores.

Heterochloris (*hetero*, unlike; *chloris*, green). Fig. 6.2

This genus possesses two parietal chloroplasts and numerous oil droplets. Reversion to the amoeboid form and metabolic movements are quite common. Reproduction is by cell division, although statospores have been reported and in *Chloromeson* they are formed endogenously. The typical Xanthophycean heterokont condition of the flagella is quite noticeable, the difference ranging from 2–3 times the length of the shorter one in *Heterochloris* and *Chloramoeba* to 6–7 times in *Chloromeson*.

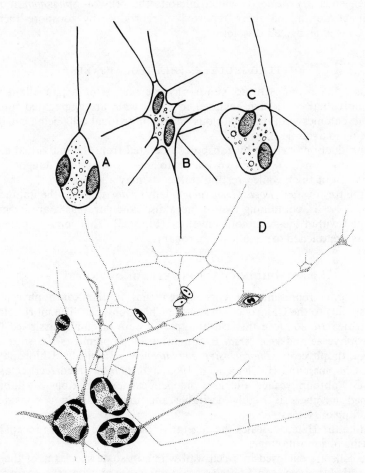

Fig. 6.2 A, *Heterochloris mutabilis*, flagellate stages; B, *H. mutabilis*, rhizopodal stages; C, *H. mutabilis*, amoeboid stages; D, *Chlorarachnion reptans*. (A–C, after Fritsch; D, after Pascher)

RHIZOCHLORIDALES

This order differs from the preceding one in that the vegetative phase is permanently amoeboid, and motile flagellates occur only in the zoospore stage. The majority of the members are uninucleate with one or more chloroplasts. Reproduction is by cell division. Often the daughter protoplasts do not completely separate but remain connected by their protoplasmic strands.

Chlorarachnion, which contains the biggest species, has uninucleate cells with many chloroplasts and pyrenoids. Nutrition, besides phototrophy, is also holozoic. Cyst formation occurs by rounding off of the amoeboid protoplast and formation of a membrane (Fig. 6.2).

The genus *Myxochloris*, which inhabits the cells of *Sphagnum* moss, is multinucleate and has been reported to reproduce by zoospore-liberating cysts as well as by cell division.

HETEROGLOEALES (HETEROCAPSALES)

This order includes those Xanthophyceae possessing a palmelloid habit. The individual cells, each enclosed in a cell wall, are aggregated into palmelloid colonies within a gelatinous matrix, individual cells being capable of reverting to a motile stage.

Reproduction is either by zoospores liberated from the individual cells or by cell division in one or two planes to give two or four daughter cells. Formation of thick-walled resting statocysts may also occur.

In the fresh water *Gloeochloris* (*gloeo*, jelly; *chloris*, green) the uninucleate, ellipsoidal cells, containing two or more discoid chloroplasts, are irregularly arranged within the gelatinous envelope (Fig. 6.3). This genus is either free floating or attached to aquatic phanerogams.

MISCHOCOCCALES (HETEROCOCCALES)

This order represents the coccoid tendency in the Xanthophyceae and corresponds to the Chlorococcales in the Chlorophyceae. The morphological similarities are so close that certain species were originally assigned to the Chlorophyceae and *vice versa*. Examination of its pigments has shown that the Xanthophycean *Chlorocloster engadinensis* is typically Chlorophycean (D. J. Chapman and H. Clemençon, 1965) and therefore incorrectly classified in the Xanthophyceae. Further biochemical examination of dubiously assigned members is highly desirable and may result in a considerable reshuffling of species.

Unlike the Heterogloeales the vegetative non-motile cells are incapable of reverting to a motile stage.

The cells are enclosed in a cell wall which in some species is of the overlapping halves type. Most species are uninucleate and possess one or more chloroplasts within the cell.

Reproduction is either by means of zoospores or aplanospores; strict vegetative cell division appears to be unknown. The majority of species are free floating, or tree, soil or rock dwellers, but epiphytic species are also known. In terms of genera and species, this is numerically, by far the largest order in the class, but, with few exceptions, the species, which are mainly freshwater inhabitants, are of rare occurrence.

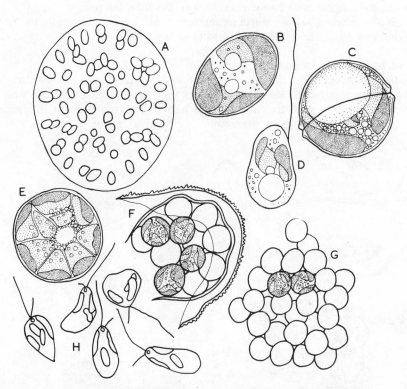

Fig. 6.3 Xanthophyceae. A–D, *Gloeochloris planctonica*: A, vegetative colony; B, colonial vegetative cells; C, resting cyst; D, swarmer; E–H, *Botrydiopsis arrhiza*: E, vegetative cell; F, liberation of swarmers; G, aplanospores; H, swarmers. (after Thienemann)

PLEUROCHLORIDACEAE

The family Pleurochloridaceae represents the free-living stage which is an unattached single cell.

Botrydiopsis (like *Botrydium*), is a widely distributed soil alga (Fig. 6.3). The cells tend towards a spherical shape, and the cell wall is apparently in two sections. The cells are uninucleate with numerous chloroplasts (frequently only a few in growing cells), usually without pyrenoids. Reproduction is by means of four, eight or more zoospores of a typical Xanthophyceae form.

Vegetative cells have been known to form aplanospores. These usually germinate to form other vegetative cells, but may pass to a thick-walled resting stage before germinating to zoospores.

CHLOROBOTRYDACEAE: *Chlorobotrys* (*chloro*, green; *botryo*, cluster). Fig. 6.4

The representatives of this family are single cells in free-living colonies enclosed in a gelatinous sheath. *Chlorobotrys* is an acid or alkaline swamp genus. The individual cells are spherical with a few chromatophores. Reproduction is commonly by autospores or thick-walled cysts and zoospore production appears to be the exception rather than the rule.

Botryochloris (Botryochloridaceae) represents a non-mucilaginous, free-living colony of many spherical cells. (Fig. 6.4).

CENTRITRACTACEAE

Bumilleriopsis, (like *Bumilleria*) (Fig. 6.4) is a typical member with single, free-living, unattached cells that are elongated or ellipsoidal. The cell wall is characteristically and distinctly of two halves. Reproduction in this genus (and family) appears to be predominantly by aplanospores.

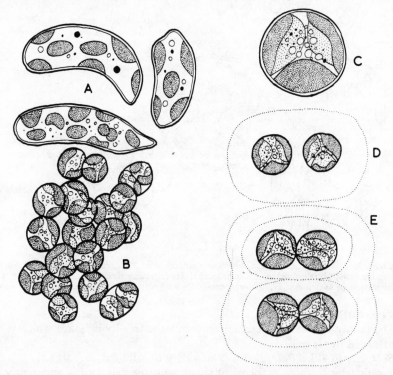

Fig. 6.4 Xanthophyceae. A, *Bumilleriopsis breve*; B, *Botryochloris commutata*; C–E, *Chlorobotrys regularis*: C, vegetative cell; D, two-cell colony; E, four-cell colony. (A, after Fritsch; B, after Fott; C–E, after Thienemann)

MISCHOCOCCACEAE: *Mischococcus* (*mischo*, stalk; *coccus*, sphere) Fig. 6.5

The members of this family usually form dichotomous dendroid colonies from cells supported on mucilage stalks arranged in a tree-like pattern.

Mischococcus is a freshwater epiphyte. Each mucilage stalk bears at its end a single spherical cell enclosing two chloroplasts. The zoospores are typically Xanthophycean, although those of some species have been reported to possess only one flagellum. When the zoospores settle they give rise to a single spherical cell resting on a mucilage cushion.

This cell then forms a new cell membrane and subsequent secretion of mucilage forms the initial stalk, pushing the cell ahead of it. The original site of the cell and its old cell membrane are marked by a basal swelling. Transverse division of the terminal cell gives two daughter cells that are together supported on a freshly secreted stalk. In some cases only the upper daughter cell secretes the mucilage, in which case the lower cell remains attached to the old stalk, while the other becomes separated by the freshly secreted intervening mucilage.

If longitudinal division occurs the two daughter cells are borne aloft on a common stalk, but at the next cell division the resultant two sets of daughter cells form separate stalks and thus produce a dichotomy. Division into four will give rise to occasional tetrachotomy. Continual repetition of these division processes eventually gives rise to the dendroid colonies. Reproduction also takes place by aplanospores or by vegetative propagation of cells detached from old colonies.

CHARACIOPSIDACEAE: *Characiopsis* (like *Characium*). Fig. 6.5

The cells, which are epiphytic, solitary or gregarious, vary greatly in shape, even in pure culture; they develop from a short stalk with a basal mucila-

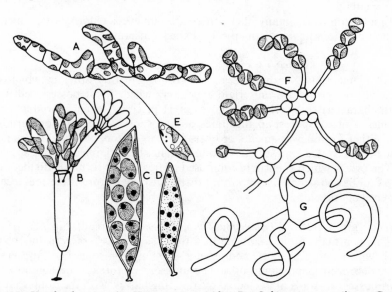

Fig. 6.5 Xanthophyceae. A, *Heterococcus viridis*; B, *Ophiocytium gracilipes*; C, D, *Characiopsis saccata*, swarmer formation; E, swarmer of *Heterococcus viridis*; F, *Mischococcus confervicola*; G, *Ophiocytium lagerheimii*. (A, after Smith; B, G, after Fott; C, D, after Fritsch after Carter; E, after Gernock; F, after Thienemann)

ginous cushion. The wall, composed of cellulose and pectins, is in two unequal portions, the smaller upper part forming a lid which is detached at swarmer formation, while in one species the lower part bears internal processes. Although the young cells are uninucleate and contain one or more chloroplasts, the adult cells are multinucleate and contain eight to sixty-four nuclei as well as several chloroplasts. Reproduction is either by means of zoospores (eight to sixty-four per cell) or by means of thick-walled aplanospores, which in one species are said to give rise to motile gametes, although this is a feature that requires further investigation.

CHLOROTHECIACEAE: *Ophiocytium* (*ophio*, snake; *cytium*, vessel). Fig. 6.5

The species of the freshwater genus *Ophiocytium* are generally elongated or cylindrical, curved, solitary, multinucleate cells of various shapes (e.g. *O. variabile*, Fig. 6.5), or attached colonies of cells. Dendroid colonies are formed in *O. gracilipes* by liberated zoospores attaching themselves to the rim of the parent cell.

There are many similarities between the Characiopsidaceae and the Chlorotheciaceae, the latter probably representing a specialization of the former. Unlike *Characiopsis* the cells are permanently multinucleate; like that genus, the cell wall of *Ophiocytium* is composed of two very unequal segments.

The lower and larger section of *Ophiocytium* is like an open-ended cylinder with a spine-like projection at the closed end, which serves as the attachment device for the epiphytic species. The open, upper end, through which aplanospores or zoospores are liberated, is covered by the other section, which acts as a small lid.

In addition to the many nuclei there are numerous chloroplasts, which in certain positions have a distinctive H-shaped appearance.

TRIBONEMATALES (HETEROTRICHALES)

All filamentous Xanthophyceae are grouped in this order in which the majority of species are uniseriate and unbranched. Most possess cell walls with characteristic overlapping halves or H pieces. Asexual reproduction is by means of zoospores, aplanospores or statospores. The latter on germination either form new filaments or liberate a pair of zoospores. Akinetes have also been recorded, and sexual reproduction has been claimed for one species.

The order, which is entirely composed of freshwater forms, is divided into two families, in which the important distinguishing feature is the presence or absence of branching.

TRIBONEMATACEAE: *Tribonema* (*tribo*, thin; *nema*, thread). Fig. 6.6

This is a filamentous analogue to a form such as *Microspora* (cf. p. 63) with which it is frequently confused. *T. bombycinum* sometimes appears in sheets covering ponds and pools. The unbranched threads are composed of cells possessing walls of two equal overlapping halves, with the result that the filaments are open-ended and tend to dissociate into H pieces. At cell division a new H piece arises in the centre and the two halves of the cell separate. Each cell contains one nucleus, although *Tribonema bombycinum* may have two, together with two or more parietal chloroplasts. Asexual reproduction is by means of heterokont zoospores (2–4 per cell) which are liberated by separation of the cell wall. On coming to rest the zoospore elongates and puts out an attachment process, and in this state it resembles *Characiopsis*. Aplanospores (1–2 per cell) and akinetes, which are formed in chains, also act as a means of propagation. Sexual reproduction is apparently very rare; in this some of the motile bodies come to rest first and are surrounded by other motile gametes. Iron bacteria sometimes live within the alga which becomes yellow or brown from ferric or manganese salts.

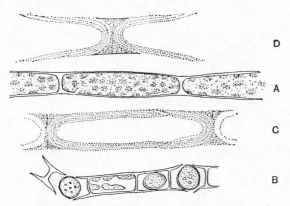

Fig. 6.6 *Tribonema*. A, *T. bombycinum*; B, *T. minus*, hypnospores; C, D, construction of H piece in *T. bombycinum* as shown after treatment with KOH. (A × 450, C, D, × 675) (A, C, D, after Smith; B, after Fritsch)

The branched habit in this order is represented by the family Heterodendraceae, of which *Heterococcus* (*hetero*, different; *coccus*, sphere), is an example (Fig. 6.5). This occurs as a discoid filamentous expanse or as single isolated filaments or as non-filamentous cell aggregates. The filaments are each composed of a few short squat cells with one or more parietal chloroplasts. The cell wall appears to be composed of one piece, rather than two H pieces. Reproduction can occur by zoospores, akinetes or by aplanospores. In view of some of its anomalous features, it is possible that this family is not properly included in the Tribonematales.

VAUCHERIALES (HETEROSIPHONALES)

This order includes the Xanthophyceae which are siphonaceous and coenocytic. The plant body is either filamentous or saccate and contains a large number of nuclei and chloroplasts.

Asexual reproduction is by zoospores or aplanospores. Unlike the other Xanthophyceae sexual reproduction (isogamy or oogamy) is quite common in members of this order. There are three or four genera, the majority of species being freshwater or terrestrial, with only a few marine. Two of the the most widely distributed Xanthophyceae genera, *Botrydium* and *Vaucheria*

are assigned to this order. They are placed in separate families distinguished
by their type of thallus and method of reproduction.

BOTRYDIACEAE: *Botrydium* (a small cluster). Fig. 6.7

This family is distinguished by a saccate plant body, which reproduces by
means of aplanospores or zoospores. The mud-dwelling *Botrydium* is analogous to a form such as *Protosiphon*, the commonest species, *Botrydium
granulatum*, being frequently confused with it, especially as these two plants
are often associated on areas of drying mud. *B. granulatum* makes its appearance during the warmer part of the year when the green, pear-shaped vesicles
are rooted by means of colourless, dichotomously branched rhizoids. The

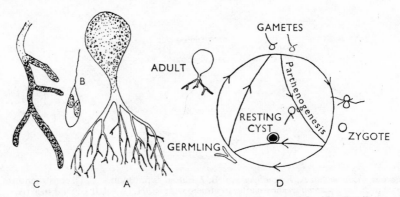

Fig. 6.7 *Botrydium granulatum.* A, plant; B, swarmer; C, cyst formation; D, diagram of
life cycle. (A–C, after Fritsch; D, after Miller)

wall is composed of cellulose and the lining cytoplasm contains numerous
nuclei, while the chloroplasts, containing pyrenoids, are confined to the
aerial saccate part. The shape of vesicle is influenced by the environment, the
shade forms being elongate or club-shaped. In *B. wallrothii* and *B. granulatum*
the unbranched vesicle is covered with lime. When the plants are
submerged, reproduction takes place by means of numerous zoospores which
are set free by gelatinization of the vesicle apex, but when the plants are only
wet but not submerged, aplanospores are produced instead. These are formed
by the cytoplasm rounding off and secreting a wall, and they may be either
uni- or multinucleate. They germinate directly to give a new plant. Under dry
or drought conditions, the cytoplasm may round off to form a single thick
walled cyst (macrocyst), or alternatively several multinucleate thick walled
sporocysts or hypnospores. Sometimes the protoplast migrates into the
rhizoid and there forms multinucleate hypnospores or rhizocysts. Upon
return to favourable conditions these hypnospores germinate to uninucleate
zoospores, or if the hypnospore is uninucleate, germination gives rise directly
to a new plant.

Sexual reproduction has been claimed for *B. granulatum*. In this species the contents of the cell round off into uninucleate isogametes with typical hetero-kont flagellation. Nuclear fusion takes place after gamete fusion and is followed by the formation of four or eight zoospores. Meiosis is said to occur at division of the zygote nucleus. In *B. granulatum* it is estimated that about 40,000 isogametes are formed in each vesicle, but as the plant is monoecious many fuse either in pairs or threes, rarely fours, before they are liberated. Those that do not fuse, develop parthenogenetically. The life cycle can be tentatively represented as in Fig. 6.7.

VAUCHERIACEAE: *Vaucheria* (after J. P. Vaucher). Figs. 6.8, 6.9

In the past this genus has been classified with the siphonaceous Chloro-phyceae in the order Siphonales (Codiales and Caulerpales). Recent work has shown that there are many fundamental differences between *Vaucheria* and the siphonaceous Chlorophyceae and that it is correctly classified in the Xanthophyceae.

The cell walls do not contain native cellulose; chlorophyll *b* is lacking in *Vaucheria* as in all other Xanthophyceae. The discoid chloroplasts, which lack pyrenoids, contain the typical Xanthophycean carotenoids, and lack siphonaxanthin so characteristic of the siphonaceous Chlorophyceae. The

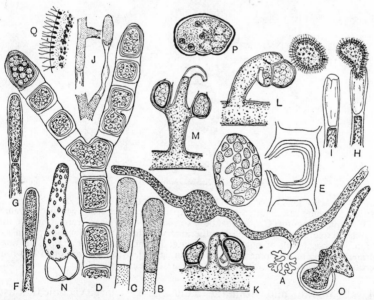

Fig. 6.8 *Vaucheria.* A, *V. sessilis*, germinating zoospore; B, *V. piloboloides*, developing aplanospore; C, *V. piloboloides*, escape of aplanospore; D, *V. geminata*, thread with cysts; E, escape of amoeboid protoplast from cyst; F–I, *V. repens*, development and escape of compound zoospore; J, regeneration and formation of septa in injured thalli; K, sex organs of *V. sessilis*; L, sex organs of *V. terrestris*; M, sex organs of *V. geminata*; N, *V. geminata*, germinating aplanospore; O, germinating zygote; P, zygote with four haploid nuclei; Q, portion of compound zoospore, much magnified. (K, L, M × 100) (A, D, E, N, O, after Oltmanns; B, C, F–J, Q, after Fritsch; K–M, after Hoppaugh; P, after Hanatschek)

storage reserve is oil. The two flagella present on motile cells are typically Xanthophycean; one is pleuronematic (flimmergeissel or tinsel type) while the other is simple (peitschengeissel or whip-lash type).

Vaucheria is essentially a temperate genus, inhabiting well aerated streams, soil or marine mudflats, where it sometimes forms a greenish covering on the surface (e.g. *V. haemata*). Some species (e.g. *V. de baryana*) may be slightly lime-encrusted.

The plant consists of a colourless, basal rhizoidal portion from which arise green, erect, aerial coenocytic filaments with apical growth and monopodial branching.

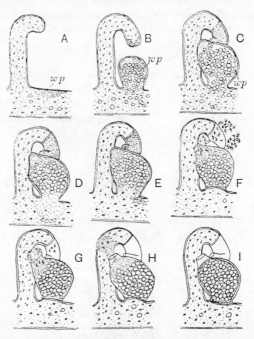

Fig. 6.9   *Vaucheria sessilis*. Stages in development and fertilization of oogonium. April 1–6, 1930. A, young antheridium and 'wanderplasm' in place from which oogonium will arise; B, young oogonium; C, oogonial beak formed; 'wanderplasm' retreating into thread; oil globules passing into oogonium; antheridial wall forming; D, 'wanderplasm' out of oogonium; E, basal wall of oogonium forming; F, antherozoids emerging; G, oogonial membrane forming at tip, some antherozoids in egg; H, cytoplasm extruded and rounded off; fertilization occurring; I, ripe egg. (*wp* =wanderplasm) (after Couch)

All walls are thin, and within the thallus there is a large vacuole that usually runs the length of the cell. The chloroplasts and tiny nuclei are confined to the cell periphery and septa are formed only with reproductive structures or after wounding. Vegetative reproduction is secured through fragmentation, and sexual reproduction is brought about by the well-known, compound, multi-flagellate zoospores (synzoospores) which are produced singly in club-shaped sporangia cut off from the ends of the erect aerial branches. The chloroplasts and nuclei congregate in the apex of a filament before the septum is laid down and the nuclei then arrange themselves peripherally. Finally, two slightly unequal flagella develop opposite each nucleus and then the zoospore is ready for liberation, a process which is achieved by gelatinization of the sporangium tip. This compound structure may be regarded as representing a group of biflagellate zoospores which have failed to separate. The syn-zoospore is motile for about fifteen minutes; then it comes to rest and germinates, the first thread often being more or less colourless. Zoospore formation can often be induced by transferring the plants from light to darkness, or from a nutrient solution to distilled water.

Under dry conditions in terrestrial species, thin-walled aplanospores may be formed in sporangia at the ends of short lateral or terminal branches, and if exposed to greater desiccation the threads of the terrestrial forms become septate and rows of cysts are formed to give the 'Gongrosira' stage. When conditions become more favourable these cysts germinate either into new filaments or divide into thin-walled bodies whose contents escape as amoeboid masses, which on becoming sessile germinate into new filaments.

Sexual reproduction is distinctly oogamous, the different species being either monoecious or dioecious. The oogonia are sessile or stalked. Development begins with a bulging in the main thread, and a large mass of colourless cytoplasm followed by many nuclei and chloroplasts moves into the bulge which gradually increases in size. It eventually becomes an oogonium with a single uninucleate egg, which is cut off from the main thread by a transverse system. Some authors maintain that the extra nuclei which passed into the oogonia are potential gametes which degenerate, while others maintain that the surplus nuclei travel back into the main thread just before formation of the septum. This latter view is considered to be the more likely explanation, although both possibilities may occur dependent upon the species. In the mature oogonium there is a beak, the apex of which gelatinizes, or apical pores through which the antherozoids can pass.

The antheridia, which are usually stalked, commonly arise close to the oogonia, and in most species their development commences before oogonia formation. The branch bearing the antheridium becomes densely packed with nuclei and a few chloroplasts, so that often the antheridia appear colourless.

When the antheridium, which is often hook-shaped, separates from the main thread by a transverse septum, the contents divide into uninucleate fragments which form pear-shaped antherozooids that possess apically inserted, unequal, heterokont flagella pointing in opposite directions, and are unusual in that the anterior, shorter flagella is of the flimmergeissel type. The antherozooids escape through apical pores and enter the oogonium, although only one antherozoid enters the egg. After fertilization, the zygote secretes a thick wall and remains dormant for some months before it germinates to give rise to a new filament.

The latest evidence indicates that meiosis occurs at germination of the zygote, thus indicating that the adult plant is haploid (cf. the siphonaceous Chlorophyceae in which the adult is diploid).

REFERENCES

GENERAL

Bourelly, P. (1968). *Les algues d'eau douce*, Tome **2**. p. 155–238, (Boubée et Cie, Paris).

Chapman, D. J. and Clemençon, H. (1965). Unpublished observations.

Ettl, H. (1956). *Bot. Notiser*, **109**, 411.

Fott, B. (1959). *Algenkunde*, (Jena, Stuttgart).

Fritsch, F. E. (1935). *Structure and reproduction of the algae*. Vol. **1**, p. 470–506 (Cambridge University Press).

Hibberd, D. J. and Leedale, G. F. (1970). *Nature, Lond.*, **225**, 758.

Smith, G. M. (1955). *Cryptogamic Botany*, Vol. **1**, 2nd Ed., p. 165–184, (McGraw-Hill, New York).

Venkataraman, G. S. (1961). *Vaucheriaceae*, (I.C.A.R., New Delhi).

Botrydium

Falk, H. (1967). *Arch. Mikrobiol.*, **58**, 212.

Tribonema

Falk, H. and Kleinig, H. (1968). *Arch. Mikrobiol.*, **61**, 347.

Vaucheria

Deschomps, S. (1963). *C.r. hebd. Séanc. Acad. Sci., Paris*, **257**, 727.

CHLOROMONADOPHYTA

CHLOROMONADOPHYCEAE

This is a very small division of little-known flagellates, with no recorded non-flagellated members. Apart from one marine species, they are of infrequent occurrence in freshwater bogs and ponds.

The cells, characteristically ovoid or pear-shaped, are naked and surrounded by a pliable periplast that permits cytoplasmic movement and changes in cell shape, sometimes referred to as 'metabolic movements' (c.f. Euglenophyceae). There is a gullet associated with a slight furrow, and trichocyst-like rods are distributed in the peripheral cytoplasm. These may be discharged either as slime threads or as club-shaped bodies. There are two flagella, one of which usually trails behind. The pigmented members possess numerous, discoid, yellow-green chloroplasts scattered throughout the cytoplasm. Only chlorophyll *a* is present together with the same carotenoids as in the Xanthophyceae, and recent work on the pigments suggests that the class should be absorbed in the Xanthophyta (Chapman, D. J. and Haxo, 1966). The organisms do not appear to store starch, but rather oils.

There is but a single order, the Chloromonadales, with one family.

CHLOROMONADACEAE: *Gonyostomum* (*gonu*, knee; *stomum*, mouth). Figs. 6.10, 6.11

There are three species, the one most widely distributed being the pigmented *G. semen*. The cell is typically ovoid with two flagella, a triangular gullet and a shallow longitudinal furrow. There is a distinct nucleus and contractile vacuoles.

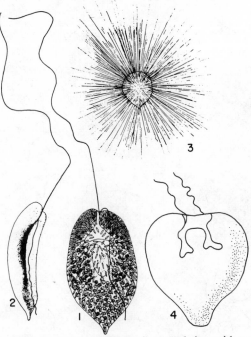

Fig. 6.10 *Gonoyostomum semen*. 1, vegetative cell, ventral view, with two flagella, chromatophores, triangular cavity, nucleus and trichocyst-like rods; 2, side and ventral view showing ventral groove and positions of flagella; 3, cell in 1 per cent acid fuchsin showing slime threads; 4, cell dividing and constricting at anterior end. (1, 2 × 630; 3 × 252; 4 × 1400) (after Drouet and Cohen)

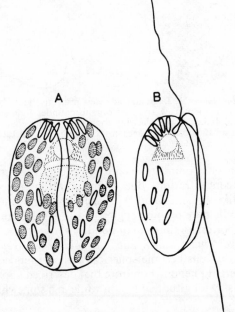

Fig. 6.11 *Gonyostomum ovatum*. A, surface view; B, lateral view.

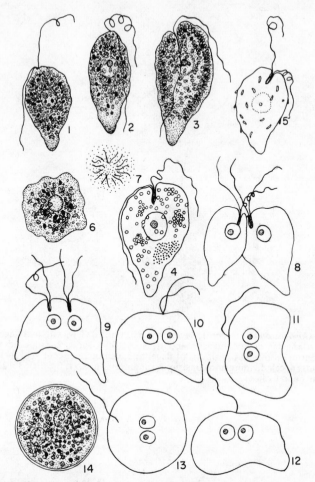

Fig. 6.12 *Hornellia marina*. 1, 2, flat and side view respectively; 3, flat view with chromato-phores oriented more to narrow sides of organism under influence of strong light; 4, flagellate with fat globules; 5, cell showing trichocysts; 6, flagellate bursting; 7, flagellate ejecting thread-like substance; 8–16, stages in sexual reproduction 1.25 pm — 3.40 pm. (after Subrahmanyan)

Under certain environmental conditions (e.g. viscous agar medium or increasing salt concentrations due to evaporation) the cells may become spherical, cyst-like or amoeboid with metabolic movements. Under other conditions the cells may become aggregated in a gelatinous matrix and then they sink to the bottom of the pool or pond.

In both *Gonyostomum semen* and *Vacuolaria virescens* encystment can occur, whereby the cells become spherical and surrounded by a sheath: this represents a different phenomenon from that described above.

Cell division is by longitudinal fission while in a somewhat amoeboid form.

Sexual reproduction has been recorded for *Vacuolaria virescens* and *Hornellia marina* (possibly *Chattonella subsalsa*). In the latter case two flagellated individuals fuse and lose their flagella. Germination of the resulting resting zygote has not been described.

Hornellia marina is the sole monotypic marine genus and occurs off the west coast of India, often in sufficient quantity to cause a green discolouration in the sea (Fig. 6.12).

REFERENCES

GENERAL

Chapman, D. J. and Haxo, F. T. (1966). *J. Phycol.*, **2**, 89.

Hollande, A. (1942). *Arch. Zool. exp. gén.*, **83**, 1.

Mignot, J-P. (1965). *Protistologica*, **3**, 5, 25.

Gonyostomum

Drouet, F. and Cohen, A. (1935). *Biol. Bull. Mar. Lab., Woods Hole*, **68**, 422.

Hovasse, R. (1945). *Arch. Zool. exp. gén.*, **84**, 239.

Hornellia

Subrahmanyan, R. (1954). *Indian J. Fish.*, **1**, 182.

Vacuolaria

Fott, B. (1943). *Arch. Protistenk.*, **84**, 282.

Poisson, R. and Hollande, A. (1943). *Annls. Sci. nat. (Bot. Zool.), Ser. II*, **5**, 147.

Schnepf, E. and Koch, W. (1966, 1967). *Arch. Mikrobiol.*, **54**, 229; **57**, 196.

7

CRYPTOPHYTA
PYRRHOPHYTA

CRYPTOPHYTA

CRYPTOPHYCEAE

The cryptomonads are characterized by a peculiar, normally naked, asymmetrical cell (ovoid or bean-shaped), which possesses two slightly unequal flagella of the flimmergeissel type.

At least one contractile vacuole is usually present. There is an antero-ventral gullet or an oblique ventral furrow, which is lined with a row of trichocysts in special sub-surface pits which shoot out the threads when stimulated.

The coloured species usually possess one or two lobed parietal chloroplasts with pyrenoids, around which a starch-like reserve is deposited. The colour varies from reddish-brown through olive green to blue green due to the presence of chlorophylls *a* and *c* and characteristic carotenoids and biliproteins. Instead of a wall the cells have a periplast, which is usually firm, thereby restricting 'metabolic movements' such as are found in the Chloromonadophyceae and Euglenophyceae. Reproduction is solely by longitudinal fission. Both marine and freshwater forms are known.

The colourless species (e.g. *Chilomonas paramecium*) possess the typical cryptomonad form.

Originally these algae were classified with the dinoflagellates in the Pyrrhophyta. Recently, however, they have been removed to a separate division.

There are two orders, the Cryptomonadales and the Cryptococcales.

CRYPTOMONADALES

CRYPTOMONADACEAE: *Cryptomonas* (*crypto*, hidden; *monas*, unit). Fig. 7.1

The species of this genus possess the typical naked cryptomonad cell. *Cryptomonas ovata* var. *palustris* has an olive-greenish hue whereas other genera may be red (e.g. *Rhodomonas*) or blue-green (e.g. *Cyanomonas*). Apart

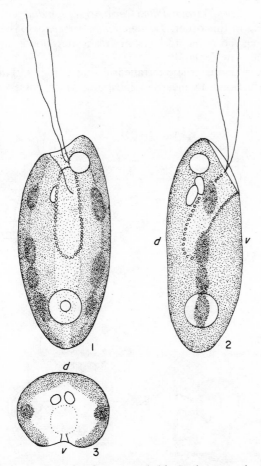

Fig. 7.1 *Cryptomonas* cell. 1, view from ventral side. In the centre the trichocysts line the gullet; on both sides are the chloroplasts with pyrenoids which are arranged in a row; nucleus at base, contractile vaciole at top. 2, lateral side view (d = dorsal, v = ventral side); on the dorsal side, to the left of the gullet, lie two light receptive bodies; 3, t.s. through a cell (d = dorsal, v = ventral side). (after Fott)

from the usual longitudinal division of the motile cell, some species have thick-walled, resting-stage cysts. *Hemiselmis*, (*hemi*, half; *selmis*, deck) (Fig. 7.2) a representative of the Hemiselmidaceae, is characterized by a bean-shaped cell, with laterally inserted flagella. The gullet is replaced by an oblique furrow with trichocysts. There are one or two parietal chloroplasts, which in *H. rufescens* are red but are blue-green in *H. virescens*.

Initially there is only a single chloroplast but during maturation this divides into two so that when division occurs the two chloroplasts distribute themselves between the two daughter cells.

CRYPTOCOCCALES

CRYPTOCOCCACEAE: *Tetragonidium* (*tetra*, four; *gonidia*, cell) Fig. 7.2

The one species in this order, *Tetragonidium verrucatum*, is a rarely found organism that has been recorded from freshwater ponds in Europe, the United States and New Zealand.

The cells have a characteristic tetrahedral shape (Fig. 7.2) and a cellulose cell wall. Reproduction is by means of zoospores which possess the character-

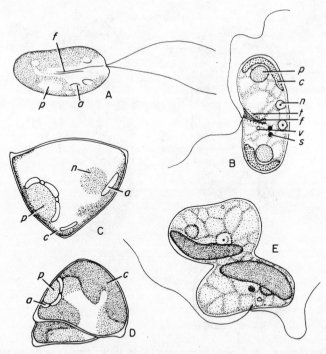

Fig. 7.2 A, *Tetragonidium verrucatum*, swarmer; C, D, *T. verrucatum* cells; B, *Hemiselmis rufescens*. Early stage in division; E, *H. rufescens*. Late division stage. (a = starch grain; c = chloroplast; f = furrow; n = nucleus; p = pyrenoid; s = stigma; t = trichocysts; v = contractile vacuole) (A, C, D, after Fritsch; B, E, after Parke)

istic features of the flagellate Cryptomonads. On coming to rest these zoo-spores develop into the typical coccoid parent.

A new genus *Bjornbergiella* has recently been described. It exists in the form of filaments of a few cells each and together with *Tetragonidium* is the only indication of evolution towards an algal organization.

REFERENCES

GENERAL

Butcher, R. W. (1967). *Introductory Account of the Smaller Algae of British Coastal Waters, Part IV. Fishery Invest., Lond.*, (*Ser* 4).

Chapman, D. J. (1966). *Phytochem.*, **5**, 1331.

Fritsch, F. E. (1935). *The Structure and Reproduction of the Algae*, Vol. 1, p. 652–663.

Hollande, A. (1942). *Arch. Zool. exp. gén.*, **83**, 24.

Hovasse, R. (1965). *C.r. hebd. Séanc. Acad. Sci. Paris*, **261**, 2947.

Joyon, L. (1963). *Ann. Fac. Sc. Clermont-Ferrand*, **22**, 1.

Mignot, J-P., Joyon, L. and Pringsheim, E. G. (1968). *Protistologica*, **4**, 493.

Pringsheim, E. G. (1935). *New Phytol.*, **43**, 143.

Colourless Forms

Pringsheim, E. G. (1962). *Farblose Algen*. (Gustav Fischer Verlag, Stuttgart).

Chilomonas

Anderson, E. (1962). *J. Protozool.*, **9**, 380.

Cryptomonas

Fott, B. (1959). *Algenkunde*, (Jena, Stuttgart).

Hollande, A. (1942). *Arch. Zool. exp. gén.*, **83**, 24.

Hemiselmis

Parke, M. (1949). *J. mar. biol. Assn. U.K.*, **28**, 255.

PYRRHOPHYTA

This division comprises the dinoflagellates, whose name describes the characteristic spiral motion of the motile cells. Although evolution towards a filamentous organization has occurred, the majority possess flagella.

The organisms are characterized by the structure of certain cellular organelles. The nucleus is large and distinct and the chromosomes appear as condensed threads (with helical coiling of the chromonema). In contrast to most other organisms, the chromosomes remain condensed during interphase, and at metaphase are arranged parallel to each other across the equator. A mitotic spindle appears to be absent. Like the procaryota no basic protein has been detected; there is continuous DNA synthesis and the chromosomes are fibrillar. Because of this, Dodge (1965) has suggested that the dinoflagellates might be assigned to a third group, the Mesocaryota.

The cells are biflagellate. Both flagella have the 9 + 2 fibril arrangement but are unusual in other respects (see later).

A distinctive, non-contractile vacuole or pusule is also present. This consists of a cluster of vesicles that open into the inner side of the flagellar canal.

F

Like some other flagellated groups, the dinoflagellates possess trichocysts. These consist of long membrane-bound crystalline cores connected by fibrils to their internal retainer sac. The trichocysts, which have a banded appearance, originate in vesicles of golgi bodies and are ejected through a fine pore at the cell surface. Their exact function is unknown but they are probably a mechanical sensing or defensive device.

The brownish colour of the chloroplast is due to the predominance of the group-specific carotenoid peridinin,* which masks the chlorophyll *a* and *c*. Species without chloroplasts often possess cytoplasmic pigments, of unknown nature, which impart various colours to the cell. Some dinoflagellates are also characterized by the ability to luminesce and by the presence of potent neurotoxins. The storage reserve is starch.

The Desmophyceae and Peridiniales are predominantly holophytic (and nearly all are obligate photoauxotrophs), although holozoic nutrition is known to occur amongst phototrophic species. The Gymnodiniales are predominantly holozoic, as are the Dinophysiales and Rhizodiniales, while the Blastodiniaceae are parasitic. The non-flagellated orders are typically holophytic.

In those dinoflagellates with holozoic nutrition ingestion takes place typically, but not always, while the alga is in a temporarily assumed amoeboid form.

Although the majority are free-swimming or attached, marine or fresh-water forms, they are also quite common as sand dwellers and parasites in fish and invertebrates. The endosymbiotic algae (Zooxanthellae) of corals, sea anemones and certain other coelenterates and radiolarians are dinoflagellates (see Ch. 7). There are two classes within the division.

DESMOPHYCEAE (= Desmokontae)

The genera, composed entirely of motile cells as the principal life stage, characteristically possess the two flagella apically inserted instead of laterally. The cells are surrounded by a periplast or a cellulose envelope (theca) consisting of two valves separated by a longitudinal suture.

PROROCENTRALES

PROROCENTRACEAE: *Prorocentrum* (*proro*, bow; *centrum*, spike). Fig. 7.3
The cell shape varies from spherical to ovoid with noticeable flattening, except in some species of *Exuviaella*. The flagella project through the envelope via the individual pores at the edge of which there may be teeth-like projections.

One flagellum projects straight forward to provide the forward motion while the other, causing the rotatory motion, extends at right angles. There are usually two yellow-brown chloroplasts commonly with associated pyrenoids, and numerous trichocysts are often present. Reproduction is by

---

* Recently *Glenodinium foliaceum* has been shown to possess fucoxanthin instead (Mandelli, 1968).

longitudinal cell division, where one valve of the envelope is retained for each daughter cell, but cysts have been recorded in older cultures. *Prorocentrum* is a common marine genus and *P. micans* is often a component of red tides (see p. 408). *Exuviaella marina* is a sand dweller.

DINOPHYCEAE (Dinokontae)

This class represents the main line of dinoflagellate development, and differs from the previous class, which is considered to be the more primitive, in certain important respects. The cell wall or envelope when present is not divided by a single vertical suture. The motile flagellates demonstrate the characteristic 'dinoflagellate orientation', because they are circled by a transverse or spiral groove, the girdle, which divides the outer wall into a forward portion, the epicone or epitheca and a hypocone or hypotheca (posterior part). Parallel to the axis of locomotion is a longitudinal furrow, the sulcus. Both the sulcus and girdle may be greatly modified, but the basic pattern remains the same. The two flagella arise on the ventral side, usually through separate pores at the junction of the girdle and sulcus. The flagellum in the transverse groove is a helical ribbon possessing a unilateral row of long fine hairs and a single striated longitudinal strand, while the longitudinal flagellum is pleuronematic with short fine lateral hairs, though these are absent in *Amphidinium*. Some workers have argued that the flagella are not latero-ventrally inserted, but still apically inserted and that the dinoflagellate motion is consequently a unidirectional sideways spiralling.

Apart from a few genera that have only a firm periplast, the Dinophyceae possess a cellulose wall. In a large majority of the flagellated genera, the 'armoured' dinoflagellates, the walls possess regularly arranged sculptured plates, whose number and arrangement are taxonomically important. Those genera without sculptured plates or without the cellulose wall are referred to as 'unarmoured' dinoflagellates. A very few rare forms of these latter have an internal skeleton, which in *Actiniscus* is siliceous.

One or more variously shaped peripheral chloroplasts are present (except in colourless members) with or without pyrenoids, and showing the typical yellow-green to golden-brown colour, and possess a triple outer membrane, rather than the usual double membrane. In addition to photoauxotrophy both holozoic and saprophytic nutrition are known, even in phototrophic species. Most motile cells possess an eyespot and in one family a lens-shaped structure, an *ocellus*, is present. The nucleus is usually centrally located, and is interesting because no centromere or spindle is present at mitosis (c.f. Euglenophyceae).

Asexual reproduction is by a cell division, or in the case of the non-flagellated genera, or *phytodinads*, by aplanospores or by *Gymnodinium*-type zoospores, a factor that is largely responsible for their inclusion in the Dinophyceae.

Sexual reproduction has been recorded for a few species and involves isogamy with amoeboid or motile gametes. The motile flagellates are divided into two orders, the Gymnodiniales (comprising the unarmoured Gymno-diniaceae and a few other small specialized families) and the Peridiniales (armoured dinoflagellates, comprising a large number of families). The unarmoured dinoflagellates are not well known because they are very difficult to preserve and need to be studied as they are collected (Norris, 1966).

GYMNODINIALES

GYMNODINIACEAE : (*gymnos*, naked; *dinos*, to whirl around)

Although the members lack a cell wall the pellicle is firm enough to give the cells a rigid shape which does vary greatly from one species to another. The cells are generally spherical although ventral or lateral flattening is quite common. From a ventral view the girdle spirals to the left, sometimes going more than once around the cell. The sulcus is usually straight but it too may be twisted around the cell. Cyst formation is quite common. The majority of species lack chloroplasts and are holozoic.

Chloroplasts, when present, are usually radially arranged, and species containing them may possess holozoic nutrition. The two pusules that open to the two flagellar pores are sometimes fused into one. Certain species have been recorded as possessing functional trichocysts or cordocysts. The nucleus is usually centrally located. Cell multiplication is by longitudinal division of either the motile or non-motile encysted form.

In *Gymnodinium* the girdle is usually equatorial with little displacement of

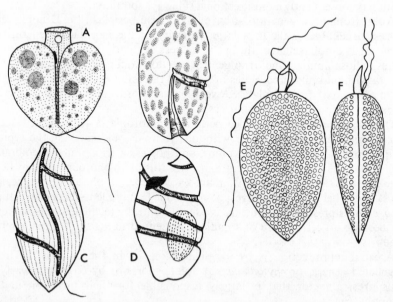

Fig. 7.3 A, *Amphidinium lacustre*; B, *Hemidinium nasutum*; C, *Gyrodinium spirale*; D, *Cochlodinium archimedes*; E, F, *Prorocentrum micans*: E, dorso-ventral view; F, lateral view. (A, B, after Fott; C, D, after Fritsch)

the two ends. In some species the periplast on the non-motile encysted cell may be striated.

In *Amphidinium* (*amphi*, on both sides) (Fig. 7.3) the girdle is anterioly placed so that the epicone is very small. The single chloroplast is arranged peripherally with a large external pyrenoid. The other cell components such as trichocysts, nucleus, pusules, flagella and so on, are typical of dinoflagellates although the trailing flagellum lacks lateral hairs. Many of the species are flattened, either laterally or dorso-ventrally, and are found predominantly in coastal waters. Isogamous sexual reproduction has been recorded for *A. carterae*. After fusion of two individuals, the zygote divides in one plane. The two subsequent divisions are unusual in that they are asynchronous.

In *Hemidinium* (*hemi*, half) (Fig. 7.3), the girdle is incomplete, extending usually about halfway round the cell, and the sulcus is restricted to the hypotheca.

In *Gyrodinium* (*gyros*, spiral) (Fig. 7.3), many species of which show cell elongation, the ends of the girdle are widely separated giving it a spiral appearance, while the sulcus may also be slightly twisted. Most of the species are colourless.

The spiralled appearance of *Cochlodinium* (*cochlea*, screw) (Fig. 7.3) is due to the distinct girdle and furrow.

The three following, widely distributed, colourless dinoflagellates possess an interesting morphology and are accordingly classified in separate families:

Polykrikos (*polus*, many; *krikos*, circle) (Fig. 7.4) is one of the few coenocytic forms, the cell comprising two, four or eight transverse grooves with one, two or four nuclei respectively, because only one nucleus is present for each pair of grooves in the coenocyte. In each zooid the girdle is slightly spiral. The sulcus runs the entire length of the zooid and is continuous with that of each adjacent zooid, since each zooid possesses the same orientation.

Polykrikos is also characterized by the possession of complex functional barbs or nematocysts as well as trichocysts.

Oxyrrhis (*oxus*, sharp; *rhis*, nose) (Fig. 7.4). This genus is considered to be a primitive gymnodinioid member. The transverse furrow and sulcus are both situated at the posterior end of the asymmetrical cell. The latter is rather broad and divided by a lobe-like structure. The species are colourless, holozoic and also phagotrophic. The cells of *Noctiluca miliaris* (*nox*, night; *lucere*, to shine) (Fig. 7.4), the sole species of the genus, are very large, sometimes reaching a diameter of 2 mm. The girdle is non-existent and the transverse flagellum is reduced to a tooth-like projection. The sulcus, however, is deep and prominent with a short flagellum. The most characteristic feature of the cell is a prominent tentacle extending posteriorly from the sulcus. Although the species lacks chloroplasts and is therefore holozoic, the cell has a pinkish hue and it is often a constituent of red tides as well as being one of the luminescent dinoflagellates. Cell division is by fission or multiple fission, producing numerous gymnodinoid swarmers which are thought to be gametes.*

* Isogamous sexual reproduction has recently been shown to occur in *Noctiluca* (Zingmark, 1970).

One family of holozoic, colourless dinoflagellates, the Warnowiaceae, is characterized by an ocellus, a complex lens, with an associated structure of pigmented granules (e.g. *Nematodinium*, *Erythropsidinium*, *Warnowia*), which may be involved in light perception.

In the parasitic colourless Blastodiniaceae the majority are ecto- or endo-parasites of metazoa, though one species has been described as an ectoparasite on the diatom *Chaetoceras*. *Blastodinium* (Fig. 7.4) occurs as an elongate sac-like cyst in the gut of certain copepods. Within the cyst reproductive cells are formed initially from division of the sac cytoplasm. After the first division only one of the resulting two daughter cells undergoes further division, so that eventually there remains one large daughter cell and numerous smaller ones. Upon rupture of the sac envelope the reproductive cells are liberated into the gut and the small ones divide once to give typical gymnodinoid zoospores which come to rest, secrete a new envelope, and thus start a new saccate

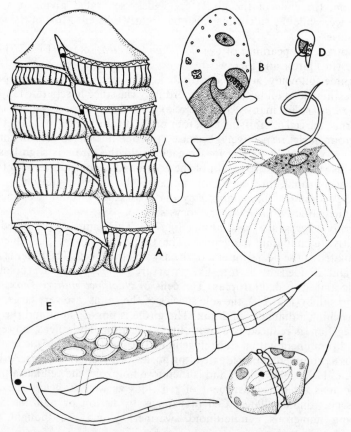

Fig. 7.4 A, *Polykrikos schwartzii*, ventral view; B, *Oxyrrhis marina*; C, D, *Noctiluca miliaris*: C, vegetative cell; D, swarmer; E, *Blastodinium spinulosum* in body of copepod; F, *B. spinulosum*, zoospore. (A–F, after Fott)

individual. The remaining large daughter cell from the first division may secrete a new envelope and in turn undergo a repeat division to give numerous small reproductive bodies. The ultimate fate of the large cell appears to be unknown.

PERIDINIALES

These are the armoured dinoflagellates and get their name from the heavy sculptured plates on the cell wall. The plates form the major distinguishing feature between the Peridiniales and the Gymnodinioideae. The number and arrangement of the plates is of great taxonomic importance. The plates, which are usually arranged in three series on both the epitheca and hypotheca, are also present as a series of small ventral plates on the girdle and sulcus. The sculpturing on the plates is often very intricate, involving numerous tiny depressions (or areolae) with a minute hole, and often with teeth or spines. The girdle, which is usually very distinct, is commonly equatorial or nearly so, and is generally characterized by pronounced ridges on the edges.

In the Peridiniales the flagella emerge through a common slit rather than from two separate ones as in the Gymnodiniales. Also, in certain members the girdle spirals to the right rather than to the left.

The majority of this order possess chloroplasts, although holozoic nutrition is known to occur in some species. Trichocysts have also been recorded.

Reproduction is by oblique cell division of either motile or resting cells always through the point of flagellar insertion. The two parts of the old cell envelope may be retained by the daughter cells (e.g. *Gonyaulax, Ceratium*), or

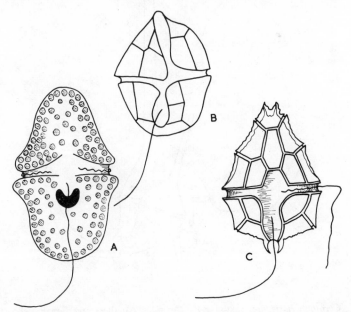

Fig. 7.5 A, *Glenodinium cinctum*; B, *Gonyaulax apiculata*; C, *Peridinium bipes*. (after Fott)

the protoplast may escape completely from the cell envelope, divide and then form a completely new theca, in which the old shell halves are left behind (*Peridinium*). Sexual reproduction (e.g. *Ceratium*) and cyst formation (e.g. *Glenodinium*) have both been recorded for a few species.

Glenodinium (*glene*, eyeball) (Fig. 7.5), a genus of the predominantly freshwater Glenodiniaceae, is probably one of the simpler peridinioids. The cell shape is similar to *Gymnodinium*, but rudimentary armouring of small, delicate, indistinct plates is apparent. Sexual reproduction occurs in which a vegetative cell forms four gymnodinioid isogametes that fuse in pairs to form a resting zygote which eventually germinates to form four new flagellates.

Peridinium (*peri*, around) (Fig. 7.5), family Peridiniaceae, is a common marine and freshwater genus with very highly developed sculpturing. The girdle is typically ridged and very prominent. A ventral concavity is often quite marked giving a kidney shaped appearance from top view.

In *Gonyaulax* (*gonya*, angle; *aulax*, furrow) (Fig. 7.5), family Gonyaulacaceae, the two ends of the girdle are displaced, showing a spiralling tendency and the sulcus is usually quite distinct. As in *Peridinium* there is sometimes formation of antapical horns. In cell division, the separation line follows the sutures of the theca, and in some cases the daughter cells do not separate and

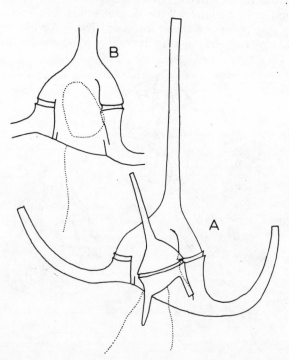

Fig. 7.6 *Ceratium horridum*, sexual cells: A, male (smaller) and female cells; B, male gamete within the female cell. (after von Stosch)

chain-forming colonies may therefore occur. *G. polyedra* is often the predominant species (sometimes 90–95% of cells) in red tides along California.
G. catenella has been known to liberate a toxin in Californian mussels that can
be fatal to humans, while *G. tamarensis* is a toxic species on the Atlantic coast.

One of the most widely distributed and most specialized genera is the
marine and freshwater *Ceratium* (*keras*, horn) (Ceratiaceae) (Fig. 7.6). The
cells are highly asymmetrical and usually flattened dorsiventrally. The most
characteristic feature is the elongated hollow horns. As in most peridinioids
(except *Glenodinium*) there is an apical pore, in this case at the tip of the
apical horn. In the sulcal region is a large ventral area that is very lightly
sculptured. Chains of individuals may be formed as a result of cell division.
Sexual reproduction, involving amoeboid gametes has been reported and has
recently been confirmed by Von Stosch (1964). The female cells are similar
to vegetative cells, but the male micro-swarmers are smaller (Fig. 7.6).

DINOPHYSIALES:

The genera of this order possess a cell wall comprised of two very large
plates and numerous smaller ones. The order is included in the Dinophyceae,
rather than Desmophyceae, because there is a girdle sulcus and theca composed of many definitely arranged plates. The epitheca is usually very small
and lacks the apical pore.

The flagella orientation is similar to that of the Peridiniales. Repro-

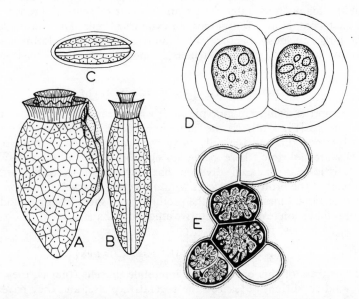

Fig. 7.7 A, B, C, *Dinophysis*, views from various sides; D, *Gloeodinium montanum*; E,
Dinothrix paradoxa. (after Fott)

duction is by longitudinal division, one half of the theca going to each daughter cell. Many species appear to lack chloroplasts, though some species of *Dinophysis* (*phusa*, bladder) have them. In most members of the order the thecal margins around the girdle and suture lines of the valves are flared out to give a wing-like appearance. In nearly all instances the girdle is so far to the front that the wing-like projections give the appearance of a collar. In *Dinophysis*, which is flattened laterally, the upturned margins or wings on the girdle project anteriorly to give the collar a funnel shape (Fig. 7.7).

DINOCAPSALES

This order represents the palmelloid condition in the Dinoflagellates and thus corresponds to the Tetrasporaceae (p. 47), Chrysocapsaceae (p. 167) and Heterogloeales (p. 130). There are but two monotypic genera in the order, *Gloeodinium montanum* and *Rufusiella insignis*. *Gloeodinium* (*gloios*, sticky) (Fig. 7.7), which is usually found in peat bogs, consists of two, four or eight subspherical cells enclosed in several gelatinous sheaths. The individual cells possess a large central nucleus and numerous small chloroplasts scattered throughout the cytoplasm. Reproduction is by cell division or by *Hemidinium*-like zoospores.

DINOTRICHALES

This is another small order with two genera, *Dinothrix* (*thrix*, hair) (Fig. 7.7) and *Dinoclonium* (*clonos*, confused), comprising those members that exhibit the filamentous habit. The plants consist of sparsely branched filaments each containing a few globose or cylindrical cells. The cells possess the typical large nucleus, together with numerous discoid chloroplasts. Reproduction is by gymnodinoid swarmers, formed singly or in pairs in the cells.

DINAMOEBIDIALES (RHIZODINIALES)

This comprises a single monotypic order with one marine species, *Dinamoebidium varians* (Fig. 7.8). As the name implies the vegetative phase is permanently amoeboid and the cells move by means of stumpy pseudopodia. The cells are holozoic, there being no chloroplasts, and there is a large characteristic nucleus. Cell division has never been recorded although encystment of the protoplast occurs within a gelatinous wall. Division of the cyst two or three times results in the production of short-lived gymnodinioid zoospores that soon revert to vegetative amoeboid shape.

PHYTODINIALES (DINOCOCCALES)

The members of this order are all immobile unicells (that is, they are in the coccoid condition) at least in the vegetative phase, and they have small chloroplasts. The majority are freshwater and the cells are either free-floating or attached.

There is no vegetative cell division, reproduction being entirely by two to four motile gymnodinioid zoospores, or two non-motile autospores. *Cystodinium* (*kustis*, bladder) is a free-living form with elongate, extended cells (Fig. 7.8). *Pyrocystis* (*pyro*, fire) is a marine luminescent planktonic genus

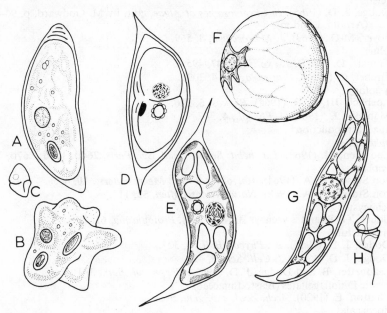

Fig. 7.8 *Dinamoebidium varians*: A, cyst; B, amoeboid form; C, swarmer; D, E, *Cystodinium steninii*, two cells. *Pyrocystis lunula*: F, amoeboid form; G, cyst; H, swarmer. (after Fott)

with large spherical cells, that reproduce by division of the protoplast into new individuals. In *Pyrocystis lunula* (known as *Dissodinium lunula* or *Gymnodinium lunula*) the four to eight daughter cells elongate into crescent shaped cells which are liberated. These cells then produce eight gymnodinioid zoospores within the parent cell (Fig. 7.8). Formation of the large spherical unicell from these zoospores has not been observed.

REFERENCES

GENERAL

Chatton, E. (1952). In *Traité de Zoologie* by P. P. Grassé, Tome **1**, p. 307–390. (Masson et Cie, Paris).

Fritsch, F. E. (1935). *Structure and Reproduction of the Algae*, Vol. **1**, p. 664–720. (Cambridge University Press).

Graham, H. W. (1951). *Manual of Phycology*, edit. by C. M. Smith, p. 105–118. (Chronica Botanica, Waltham).

Lebour, M. V. (1925). *Dinoflagellates of Northern Seas*, p. 1–250. (*Mar. Biol. Ass.*, Plymouth).

Loeblich, Jr., A. R. and Loeblich, III, A. R. (1966). *Stud. Trop. Oceanogr.*, **3**, 1–94.

Pringsheim, E. G. (1962). *Farblose Algen*, (Gustav Fischer Verlag, Stuttgart).
Schiller, J. (1931–37). In *Kryptogamen Flora*, by L. Rabenhorst, Bd. 10.
Flagella
  Leadbetter, B. and Dodge, J. D. (1967). *J. gen. Microbiol.*, **46**, 305.
Nucleus
  Dodge, J. D. (1965). *In Chromosomes of Algae*, edit. by M. Godward, p. 96–115 (Arnold, London).
  Soyer, M-O. (1969). *J. Microscopie*, **8**, 569.
Ocellus
  Francis, D. (1967). *J. exp. Biol.*, **47**, 495.
  Greuet, C. (1968). *Protistologica*, **4**, 209.
Physiology
  Loeblich, III, A. R. (1966). *Phykos*, **5**, 216.
  Mandelli, E. (1968). *J. Phycol.*, **4**, 347.
Sexual Reproduction
Amphidinium
  Cao-Vien, M. (1968). *C.r. hebd. Séanc. Acad. Sci. Paris*, **264**D, 1006; **267**D, 701.
Ceratium
  von Stosch, H. A. (1964). *Helgoländer wiss. Meeresunters.*, **10**, 40.
  von Stosch, H. A. (1968). *Nátúrwissenschaften*, **52**, 112.
Trichocysts
  Bouck, G. B. and Sweeney, B. M. (1966). *Protoplasma*, **61**, 205.
Ultrastructure
  Dodge, J. D. (1967). *Br. Phycol. Bull.*, **3**, 327.
  Dodge, J. D. (1968). *J. Cell. Sci.*, **3**, 42.
  Leadbetter, B. and Dodge, J. D. (1966). *Br. phycol. Bull.*, **3**, 1.
Parasitic Dinoflagellates (Blastodiniaceae)
  Chatton, E. (1920). *Arch. Zool. exp. gén.*, **59**.
Dinococcales
  Baumeister, W. (1957). *Arch. Protistenk.*, **96**, 345.
  Baumeister, W. (1958). *Arch. Protistenk.*, **102**, 1, 21, 241.
  Thompson, R. H. (1949). *Am. J. Bot.*, **36**, 301.
Dinophysidales
  Tai, L-S. and Skogsberg, T. (1934). *Arch. Protistenk.*, **82**, 380.
Gymnodiniales
  Kofoid, C. A. and Swezy, O. (1921). *Mem. Univ. Calif.*, **5**, 1.
  Norris, R. E. (1966). *Endeavour*, **25**, 124.
Peridiniales
  Biecheler, R. (1952). *Bull. Biologie, Suppl.*, **36**, 149.
Amphidinium
  Dodge, J. D. and Crawford, R. M. (1968). *Protistologica*, **4**, 1231.
Ceratium
  Graham, H. W. and Bronikovsky, N. (1944). *Publications of the Carnegie Institution*, **565**, 1–209.
Gymnodinium
  Dodge, J. D. and Crawford, R. M. (1969). *J. Cell. Sci.*, **5**, 479.
Nematodinium
  Mornin, L. and Francis, D. (1967). *J. Microscopie*, **6**, 759.
Noctiluca
  Zingmark, R. (1970). *J. Phycol.*, **6**, 122.
Peridinium
  Messer, G. and Ben Shaul, Y. (1969). *J. Protozool.*, **16**, 272.
Prorocentrum
  Bursa, S. (1959). *Can. J. Bot.*, **37**, 1.

8

CHRYSOPHYTA
BACILLARIOPHYTA

In the Pascherian classification the Xanthophyceae, Bacillariophyceae and Chrysophyceae were grouped into the one division Chrysophyta. Recent biochemical and electron-microscope studies have however prompted a re-evaluation of this concept. While there are some similarities (e.g. storage reserve, heterokont flagella, endogenous cysts, a two-piece cell wall and the occurrence of silica impregnation) there are also some distinct differences. The Xanthophyceae lack both chlorophyll c and the carotenoid fucoxanthin, and show some other pigment differences. The Bacillariophyceae possess a very distinctive cell-wall structure in a silica skeleton (versus scales in the Chryso-phyceae). A second class, the Haptophyceae, has recently been split off from the Chrysophyceae because its members possess a distinctive third organ, the haptonema, and isokont flagella.

It is largely a matter of opinion whether the three or four classes are arranged into one, two or three divisions. The relative importance that is attached to the similarities and differences will be the deciding factor. We have retained three divisions with four classes although many workers still retain only the one division. It seems essential, however, that the Xanthophyceae, should be removed from the Chrysophyta, irrespective of the position of the Bacillariophyceae.

159

The chloroplast pigments of the Chrysophyta and Bacillariophyta are similar (notably chlorophyll *a* and *c*, β-carotene, fucoxanthin and acetylenic carotenoids). The products of photosynthesis are the glycosidic β1:3 type reserves (chrysose, chrysolaminarin or leucosin) and oil. The flagellar arrangement (except in the Haptophyceae) is typically heterokont. Morphologically there is a range from the smallest flagellates to filamentous, and colonial forms within the Chrysophyta. The Bacillariophyta are found only as single or colonial, non-flagellated unicells.

CHRYSOPHYTA

CHRYSOPHYCEAE

Most of the members are unicells but there are colonial representatives and a very few are filamentous. Some representatives, such as the Silicoflagellates, have cell walls impregnated with silica. A characteristic feature of the class is the endogenous cyst (statospore) with silicified walls of two equal or unequal parts and the opening closed by a plug. The motile cells possess one (pleuronematic, flimmergeissel) or two (flimmergeissel and peitschengeissel) flagella. The second flagellum is simple (peitschengeissel) and shorter. In some genera (e.g. *Mallomonas*) it is very greatly reduced, and is regarded as a photoreceptor.

Sexual reproduction has been infrequently recorded, and is isogamous. Cyst formation commonly follows sexual fusion, but cysts can be formed agamically (Fott, 1964).

Most of the numerous species are freshwater planktonic or epiphytic forms, the great majority occurring in temperate or cold waters or else developing in summer when water temperatures rise. Many form a major component of the nannoplankton (p. 404). There are some marine families.

Taxonomic treatment of the group has varied widely. Prior to the erection of the Haptophyceae (p. 170) five orders were generally recognized. These are the Chrysomonadales, Rhizochrysidales, Chrysocapsales, Chrysosphaerales and Chrysotrichales and they illustrate the evolution from flagellate to filamentous form. Bourelly (1957) suggested twelve orders. In 1962 Christensen removed those Chrysophyceae with isokont flagella (with or without an haptonema) to the Haptophyceae and the validity of this class is now generally recognized. The Chrysophyceae (as recognized by Christensen) is now subdivided into 5 (or more) orders. Bourelly (1964, 1968) does not recognize the Haptophyceae, but divides the Chrysophyceae into 4 sub-classes: Acontochrysophycidae, Heterochrysophycidae, Isochrysophycidae, and Craspedophycidae. The Craspedophycidae are the colourless Choanoflagellates and are often assigned to a separate class, Craspedophyceae. They will not be considered here. The Isochrysophycidae, with two orders, correspond to the Haptophyceae. The other two subclasses correspond to the Chrysophyceae. The Acontochrysophycidae are those genera (= Rhizochrysidales) in which zoospores are lacking or in which the zoospores are amoeboid and non-flagellate. The Heterochrysophycidae comprises those genera with one (pleuronematic) or two heterokont flagella. An outline of these two schemes are presented in Table 8.1.

Other workers have adopted different interpretations. Bourelly (1957) recognized twelve orders, one of which has since been removed to the Haptophyceae. Fott (1959) and Round (1965) recognized five orders. The arrangement and number of orders can be expected to change as more work, especially with the electron microscope, is done. We have retained the five order classification (as in Fott and Round), but the student should be aware that other classifications (e.g., Bourelly, 1957, 1968; Parke and Dixon, 1968) also exist.

The five orders, Chrysomonadales, Rhizochrysidales, Chrysocapsales, Chrysosphaerales, Chrysotrichales or Phaeothamniales are analogous to the arrangement in the Chlorophyceae, Xanthophyceae, and Dinophyceae. In Bourelly's classification these five orders are combined into two, the Chromulinales (uniflagellate motile cells) and Ochromonadales (biflagellate motile cells), both in the sub-class Heterochrysophycidae. These two orders in turn are divided into 3 suborders corresponding to the flagellate, coccoid or palmelloid and filamentous construction.

In three orders there is no rigid cell wall of a normal type. The Chrysomonadales, which probably represent some of the more primitive types, are either unicells (*Ochromonas*) or else colonial sheathed forms (*Dinobryon*, *Synura*) or peculiar organisms (Silicoflagellates) with an internal siliceous skeleton. The Rhizochrysidales include amoeboid forms with pseudopodia, that are naked (*Rhizochrysis*), ensheathed (*Lagynion*) or plasmodial (*Myxochrysis*). The third order contains palmelloid (*Chrysocapsa*) or filamentous (*Hydrurus*) analogues of the Chlorophyceae. The remaining two orders comprise species with a rigid cell wall, e.g. the coccoid Chrysosphaerales and the filamentous Phaeothamniales. This last order is of particular interest in relation to the origin of the Phaeophyta.

The pigmented Chrysophyceae are holophytic but there are colourless Chrysomonads, recognized by the typical plugged cysts, and these are heterotrophic, the organic carbon being obtained by diffusion (osmotrophic) or ingestion (holozoic).

CHRYSOMONADALES

This order represents the simplest types, simple flagellates with two, rarely one or four, flagella. The order may be sub-divided into two sub-orders, the Chrysomonadineae and Silicoflagellineae or Silicoflagellates. More recent proposals have suggested three orders: Chromulinales (uniflagellate), Ochromonadales (heterokont biflagellate) and Dictyochales (silicoflagellates). In this latter arrangement the Ochromonadaceae, Synuraceae and Dinobryaceae are referred to the Ochromonadales, while the Chromulinaceae is assigned to the Chromulinales (Table 8.1).

Multiplication commonly takes place by cell division and characteristic cysts are formed during unfavourable conditions. Sexual reproduction has been reported in a few instances.

TABLE 8.1

Christensen (1962)	Parke and Dixon	Bourelly
HAPTOPHYCEAE	HAPTOPHYCEAE	ACONTO-ISOCHRYS-CHRYSO-OPHYCIDAE PHYCIDAE
CHRYSOPHYCEAE	CHRYSOPHYCEAE	CHRYSOPHYCEAE HETEROCHRYSOPH-YCIDAE
Rhizochrysidales		Rhizochrysidales (+3 other orders)
Chrysomonadales	Ochromonadales	Ochromonadales Ochromoiadineae
Chrysomonadineae	Chromulinales	Chromulinales Chromulineae
Silicoflagellineae	Dictyochales	
Chrysosphaerales		Ochromonadales Chrysapionineae
		Chromulinales Chrysosphaerineae
Chrysocapsales	Chrysapiales	Ochromonadales Chrysapionineae
		Chromulinales Chrysosphaerineae
Phaeothamniales	Phaeothamniales	Ochromonadales Phaeothamnineae
	Thallochrysidales	Chromulinales Thallochrysidineae
Prymnesiales Isochrysidales	Prymnesiales Isochrysidales	Prymnesiales Isochrysidales

OCHROMONADACEAE: *Ochromonas* (*ochro*, orange, yellow; *monas*, cell).
Fig. 8.1

The species are mostly planktonic or attached, freshwater organisms, each individual containing one or two chloroplasts with pyrenoids. There is a contractile vacuole and a red eyespot. Flagellar structure is typically hetero-kontic, with fibres passing from the flagellum base to the side of the nucleus. Reproduction takes place by simple fission but siliceous cysts are formed at the end of the vegetative period. Palmelloid stages have been observed in some species. Nutrition of the species varies from phototrophic to phago-trophic to saprotrophic. In some genera cells like those of *Ochromonas* are united into colonies joined by stalks, e.g. *Chrysodendron*, or enclosed in a jelly, e.g. *Uroglena*. *Anthophysa* is a widespread colourless representative. *Didymochrysis* is quadriflagellate, and uninucleate and looks like a monad that has failed to divide (Fig. 8.1).

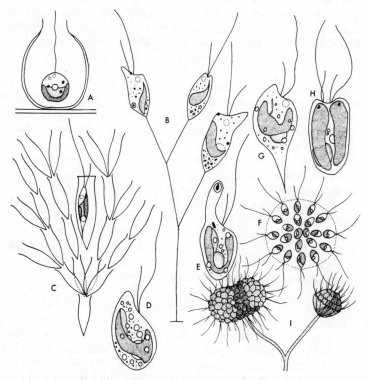

Fig. 8.1 *Chrysophyceae.* A, *Lepochromulina simplex*; B, *Chrysodendron ramosum*; C, *Dinobryon divergens*, colony with one living cell; D, E, G, *Ochromonas* sp; F, *Uroglena americana*; H, *Didymochrysis paradoxa*; I, *Anthophysa vegetans*. (after Fott)

CHROMULINACEAE: *Chromulina* (*chromulina*, yellow).

Whereas *Ochromonas* is a typical representative of the biflagellate group (Ochromonadales), *Chromulina* may be considered typical of the uniflagellate group (Chromulinales). *Chromulina* is a large genus of both marine and freshwater species. The cells typically contain one or two pyrenoid-bearing chloroplasts, an eyespot and contractile vacuole. The cell membrane is non-rigid to allow metabolic movements. The single flagellum is pleuronematic. Reproduction is by cell division. Smooth-walled cysts also occur.

DINOBRYACEAE: *Dinobryon* (*dino*, terrible; *bryon*, moss-like). Fig. 8.1

The members of this family all exist in a cellulosic sheath (lorica) formed from the protoplasm, which is either fine and colourless or thick and impregnated with iron. In some genera the cells are single but in *Dinobryon* bushy colonies are formed by a series of empty sheaths. The cells possess two chloroplasts and contractile vacuoles, an eyespot and heterokont flagella. The cells are attached to the inside of the lorica by a contractile stalk. When longitudinal fission occurs two swarmers emerge at the orifice and two new sheaths are formed. Typical cysts with plugs are formed and as some of these are binucleate they presumably arise as a result of cell fusion. Some species can be abundant in marine and fresh water, but in the latter case they are mainly found in oligosaprobic waters (see p. 39). *Lepochromulina* (Fig. 8.1) is attached to the substrate by the broad base of its sheath.

SYNURACEAE: *Synura* (*syn*, joined; *ura*, tails). Fig. 8.2

This is a floating colonial genus in which the sheaths are united by a pectinaceous material. The individual monads each have two flagella, which are slightly unequal, but heterokont and heterodynamic. The longer flagellum points to the front, the shorter to the rear.

There are two chloroplasts in each cell and a number of contractile vacuoles. The periplast is overlaid by a layer of finely sculptured scales which are both ellipsoidal and siliceous. The monad cell may divide within the colony or without to form a new colony.

In *Mallomonas*, some species of which are colourless, the cells are free and solitary with siliceous scales that often possess a distinct spine.

The second flagellum is greatly reduced, often being barely discernible and is regarded as a photoreceptor. Besides cell division and asexual zoospore formation, isogamous sexual reproduction has been described.

SILICOFLAGELLINEAE: *Dictyocha* (*dicty*, net; *ocha*, vessel). Fig. 8.2

The silicoflagellates constitute a sub-order which is sometimes assigned ordinal status as the Dictyochales. They are phototropic organisms found in marine plankton and possess an internal siliceous skeleton. This skeleton preserves well and hence many genera are known in the fossil condition from the Tertiary onwards. The body is naked and extrudes pseudopodia. There is a single flagellum and the chloroplasts are either in the main body or else in the outer plasmatic layer.

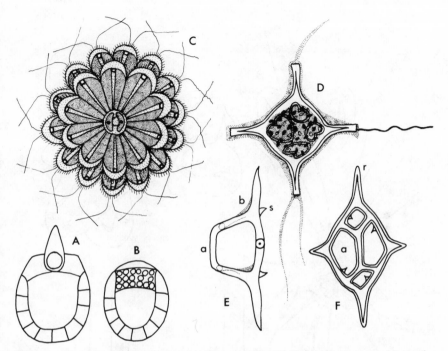

Fig. 8.2 Chrysophyceae. A, *Synura uvella*, cyst with silicified wall; B, cyst without the plug; C, *S. uvella* colony; D, *Dictyocha speculum*, living cell with flagella, pseudopodia and numerous chromatophores; E, skeleton of *D. fibuea* from above; F, skeleton of *D. fibuea* from below. (A, B, D–F, after Fott; C, after Stein)

RHIZOCHRYSIDALES

These amoeboid-like organisms which move by means of broad lobopodia, or thread-like rhizopodia are predominantly fresh water genera. Matvienko (1961) differs from most workers in regarding these organisms as the most primitive of the Chrysophyta and giving rise to the Chrysomonads. As many are colourless the only reason for placing them in the order is that they either form typical cysts or else they synthesise chrysose. Bourelly assigns this order to a family within the Chromulinales.

RHIZOCHRYSIDACEAE: *Rhizochrysis* (*rhizo*, root-like; *chrysis*, golden yellow). Fig. 8.3

These amoeboid organisms with long pseudopodia possess one or two typical chloroplasts but lack flagella or an eyespot. During division a daughter cell may fail to obtain chloroplasts and thus produce a colourless individual. *Chrysamoeba* is a very similar genus but the cells differ in possessing a short flagellum so that the genus is sometimes grouped with the Chrysomonadales. The Lagynionaceae (e.g. *Lagynion*, Fig. 8.3) contain all those sheath-living

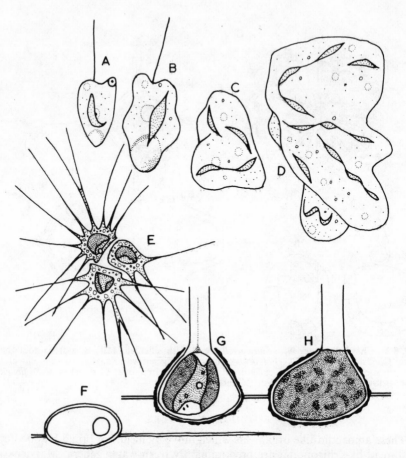

Fig. 8.3 Chrysophyceae. *Myxochrysis paradoxa* A, swarmer; B, swarmer forming pseudo-podium; C, amoeboid form; D, young plasmodium; E, *Rhizochrysis scherffelii*; F, G, H, *Lagynion scherffelii*, three sheathed individuals epiphytic on *Bulbochaete*: F, viewed from above; G, with protoplast, in optical section; H, viewed from the side; sheaths deep brown. (A–D, after Pascher; E–H, after Fott)

members that represent amoeboid analogues of the Dinobryaceae. The peculiar species *Myxochrysis paradoxa* (Fig. 8.3) (Myxochrysidaceae) produces a multinucleate plasmoidal vegetative form with chloroplasts, contractile vacuoles and chrysose. Nutrition is achieved by digesting small unicellular algae. Like *Chrysamoeba* this genus is sometimes referred to the Chrysomonadales.

During reproduction the plasmodium forms a number of cysts which produce chrysomonad-like flagellates. These lose their single flagellum, become amoeboid and fuse with each other to form a new plasmodium.

CHRYSOCAPSALES

In this order, representing the colonial habit, the cells are contained within a gelatinous matrix. In the vegetative phase the cells lack a flagellum but they do possess chloroplasts and chrysose and many also have contractile vacuoles. There are two families: the Chrysocapsaceae, in which the cells are in irregular or spherical gelatinous masses, e.g. *Chrysocapsa* (Fig. 8.4), and the Hydruraceae (*Hydrurus*) (Fig. 8.4) in which the gelatinous material is a simple or branched, erect thallus with an apical growing point. Bourelly does not recognize the order and assigns both families to the Chromulinales.

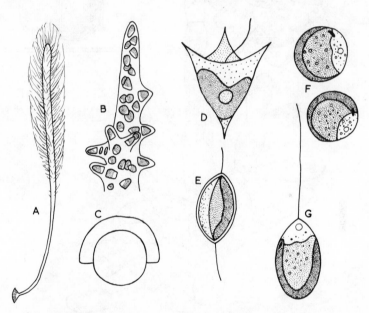

Fig. 8.4 Chrysophyceae. *Hydrurus foetidus*: A, habit; B, apex of a shoot; C, resting cyst from surface above; D, swarmer; E, resting cyst from the side; F, G, *Chrysocapsa* sp. (A–E, after Fott; F, G, after Pascher)

In *Hydrurus foetidus* the vegetative cells, each oriented in the same direction, contain one hemispherical chloroplast with a single pyrenoid, four contractile vacuoles and a large chrysose ball. Vegetative growth of the plant, which occurs in quickly-flowing fresh water, takes place by means of the apical or branch terminal cells. Reproduction is secured by means of tetrahedral zoospores produced singly per cell; each has two flagella, one long, the other short and thick. The long flagellum is locomotory and the short is used to give direction (Fukushima, 1962). Siliceous cysts are also produced.

CHRYSOPHAERALES

The members of this order are non-flagellate coccoid or colonial cells with a distinct cell wall. They represent the morphological analogue of the chlorophycean Chlorococcales (p. 50).

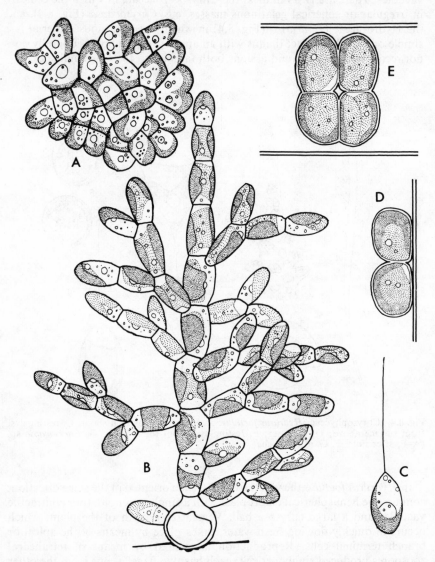

Fig. 8.5 Chrysophyceae. A, *Phaeodermatium rivulare*; B, *Phaeothamnion confervicola*; C, swarmer of *P. rivulare*; D, E, *Chrysosphaera paludosa* on cell of host alga. (A–C after, Pascher; D, E, after Fott)

Chrysosphaera (Fig. 8.5) is one of the commoner members. Besides cell division, reproduction takes place by division of the protoplast within the cell wall into autospores, or by *Chromulina*-type zoospores.

PHAEOTHAMNIALES

This order contains, because of their filamentous organisation, the most highly evolved members. They can be regarded as derived from the previous coccoid order, although Bourelly (1957) considered that the coccoid form was reduced from the filamentous. The plants are simple or branched erect (*Phaeothamnion*) or prostrate (*Phaeodermatium*) threads. The prostrate forms may develop into a pseudo-parenchymatous layer. Reproductive swarmers are of the *Chromulina* (*Phaeodermatium*) or *Ochromonas* (*Phaeothamnion*) type. The species are not so common in nature as other Chrysophyceae but they are both marine and freshwater. There are four families each with but few genera.

In the Phaeothamniaceae, as represented by *Phaeothamnion* (*Phaeo*, brown; *thamnion*, thread) (Fig. 8.5), there is a branched filament arising from a hemispherical basal cell which represents the remains of a typical cyst. When reproducing, four to eight *Ochromonas*-like swarmers are produced per cell. Creeping, branched forms, that later form a pseudo-parenchymatous layer, and which reproduce by *Chromulina*-like zoospores are placed in the Phaeodermatiaceae. In *Phaeodermatium* (*phaeo*, brown; *dermatium*, skin) (Fig. 8.5) the small discs are found on stones in cold, quickly flowing waters. The remaining two families are not well known.

REFERENCES

GENERAL

Bourelly, P. (1957). *Revue algol., Mem. hors-Ser.* 1.
Bourelly, P. (1965). *Revue algol.*, **8**, 56.
Bourelly, P.'(1968). *Les Algues d'eau douce*, Tome **2**, (Boubée et Cie, Paris).
Fott, B. (1959). *Algenkunde* (Jena, Stuttgart).
Fott, B. (1964). *Phykos*, **3**, 15.
Lund, J. W. G. (1962). *Preslia*, **34**, 140.
Parke, M. (1964). *Br. Phycol. Bull.*, **2**, 47.
Parke, M. and Dixon, P. S. (1968). *J. mar. biol. Ass. U.K.*, **48**, 783.
Round, F. E. (1965). *Biology of the Algae*, (Arnold, London).

Chromulina

Belcher, J. H. and Swale, E. M. F. (1967). *Br. phycol. Bull.*, **3**, 257.
Rouiller, C. and Fauré-Fremiet, E. (1958). *Expl. Cell Res.*, **14**, 47.

Chrysocapsa

Belcher, J. H. (1966). *Br. phycol. Bull.*, **3**, 81.

Chrysosphaera

Ettl, H. (1965). *Nova Hedwigia*, **10**, 515.

Dinobryon

Joyon, L. (1963). *Ann. Fac. Sci. Univ. Clermont-Ferrand No.* 22, **1**, 1.

Hydrurus
  Fukushima, H. (1962). *Acta Phytotax, Geobot.*, **20**, 290.
  Hovasse, A. and Joyon, L. (1960). *Revue algol.*, **5**, 1.
  Joyon, L. (1963). *Ann. Fac. Sci. Univ. Clermont-Ferrand No.* 22, **1**, 1.
Mallomonas
  Harris, K. and Bradley, D. E. (1957). *J. R. microsc. Soc.*, **76**, 37.
  Harris, K. and Bradley, D. E. (1960). *J. Gen. Microbiol.*, **22**.
  Warwik, F. (1960). *Arch. Protistenk.*, **104**, 541.
Ochromonas
  Schnepf, E., Deichgraber, G. and Koch, W. (1968). *Arch. Mikrobiol.*, **63**, 15.
Phaeothamnion
  Ettl, H. (1959). *Nova Hedwigia*, **1**, 19.
  Villeret, S. (1951). *Bull. Soc. bot. Fr.*, **98**, 37.
Synura
  Bradley, D. E. (1966). *J. Protozool.*, **13**, 1.
  Schnepf, E. and Deichgraber, G. (1969). *Protoplasma*, **68**, 1.

HAPTOPHYCEAE

This is a relatively new grouping of certain unicellular flagellates (Christensen, 1962) which are predominantly marine. Recent electron microscope studies showed that certain flagellates originally assigned to the Chrysophyceae, often possessed a third whip-like organ in addition to the two flagella. This special organ is called a haptonema, and the terminal portion or the whole organ can contract into a tight spiral and attach the cell to a substrate. The two flagella, in contrast to the Chrysophyceae, are equal, that is, they are isokont. Electron microscopy has also shown that the organisms are covered with minute scales very like those recorded for the Prasinophyceae (see p. 28). These flagellates were originally grouped in the Chrysophyceae to which they are closely allied. Each cell normally contains two brown parietal chloroplasts, with Chrysophycean pigments and oil droplets associated with the typical chrysophycean reserve product, leucosin. Some workers (e.g. Bourelly) retain this group within the Chrysophyceae, as a sub-class Isochrysophycidae.

There are two orders, the Isochrysidales and the Prymnesiales. In the former, the motile phase lacks the haptonema, whereas motile cells of the Prymnesiales possess the haptonema. This latter order includes a group of planktonic organisms known as the Coccolithophorids which possess mineralized scales of two types.

The holococcoliths are composed of submicroscopic crystals, and such scales or crystalloliths are found on *Crystallolithus hyalinus* which is the motile phase of *Coccolithus pelagicus* (Parke and Adams, 1960). The heterococcoliths are larger with more obvious structures built up into plates, ribs or girdles.

Many of the known representatives of the division are not well understood. Until further work has been carried out on certain genera their position in this division or in the Chrysophyceae must remain provisional.

ISOCHRYSIDALES

In this order the haptonema is absent and the motile cells lack the unmineralized scales characteristic of the Prymnesiales.

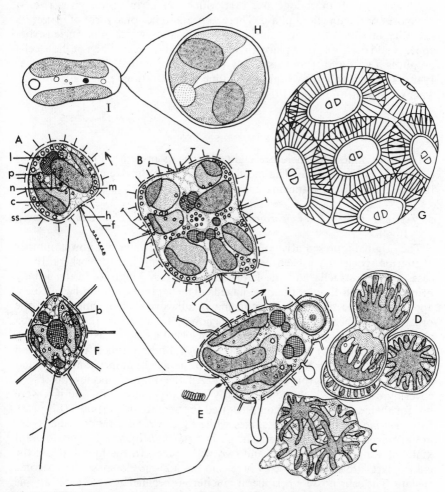

Fig. 8.6 *Haptophyceae.* A, *Chrysochromulina brevifilium* swimming with flagella and haptonema behind body; chromatophores dividing (c = chromatophore; f = flagellum; h = haptonema; l = leucosin vesicle; m = muciferous body; n = nucleus; p = pyrenoid-like body; ss = spined scales); B, same with four pairs of flagella and four extended haptonemata prior to double fission; C, large amoeboid cell of *C. brevifilium* with four deeply lobed chromatophores and four pyrenoid-like bodies; D, *C. ericina*, second fission of a large walled cell almost complete giving four small walled daughter cells; E, same with four chromatophores, swimming, muciferous organelles exuding contents; *Chlorella* cell (i) at non-flagellar pole and an empty wall of a *Chlorella* cell after being ejected from the end of a colourless tube; F, *C. ericina*, anchored cell seen from non-flagellar pole; bacteria (b) in vacuole adjacent to pyrenoid-like body; G, *Coccolithus pelagicus*, external view; H, *C. pelagicus* protoplast; I, *Isochrysis galbana.* (A–F, after Clarke, Manton and Parke; G, H, after Lebour; I, after Parke)

ISOCHRYSIDACEAE: *Isochrysis* (*Iso*, equal; *chrysis*, yellow). Fig. 8.6.

The individual cells which are naked and capable of metabolic movements, are pyriform in shape with two typical chloroplasts placed laterally on each side and with a red stigma at the anterior end. Reproduction takes place by the formation of an endogenous cyst which, on germination gives rise to zoospores or a palmelloid phase. There appears to be some form of primitive sexuality in which two non-motile cells fuse, after which one nucleus disappears and divisions take place giving four daughter cells. Each of these cells produces a cyst and from this a new motile cell eventually emerges (Parke, 1949). The alga will grow rapidly in saline concentrations ranging from 1·5–4 per cent (Kain and Fogg, 1958).

PRYMNESIALES

In the Prymnesiales there are eight families as presently understood, but only three of importance. The motile cells in this order are characterized by the presence of the haptonema and mineralized scales.

PRYMNESIACEAE: *Chrysochromulina* (*Chrysos*, golden; *chromulina*, yellow). Fig. 8.6

The original representative of this genus was freshwater but now a number of marine species have been described by Parke and her co-workers. In all cases the two flagella and the haptonema are very long, the latter usually being longer and thinner than the flagella. The haptonema often has a coiled appearance. The tip of the haptonema is adhesive and by this means the organism can become temporarily anchored. In all cases the haptonema consists of three concentric membranes enclosing a ring of six (*C. strobilus*), seven (most frequent), or rarely eight (*C. kappa*) fibres or tubes. This variation in number of fibres is a distinct contrast to the fixed number that are found in flagella of other groups (see p. 33). All the known species possess a covering of scales, which are apparently of an organic nature, and if it were not for their small size they could be used for taxonomic purposes (Fig. 8.7). In *C. ericina*, the scales have spines while in *C. strobilus* two distinct types are arranged in separate layers. In *C. pringsheimii*, there are four different kinds of scales, two plate-like and two with spines. In the motile phase the chloroplasts are saucer shaped whereas in the non-motile phase they become stellate. The cells possess peripheral muciferous organelles which are capable of expelling their contents (Fig. 8.6). In the motile phase reproduction takes place by means of ordinary fission, the two daughter cells usually being of equal size. In the non-motile phase the cells become amoeboid and may fuse since four plastids can be seen in some larger organisms. Division then occurs and four non-motile daughter cells are produced from each of which a new motile cell escapes. Ingestion of bacteria and other small algal cells by phagotrophy takes place at the non-flagellar end of motile cells (Fig. 8.6). Empty walls of ingested cells are extruded through temporary colourless tubes.

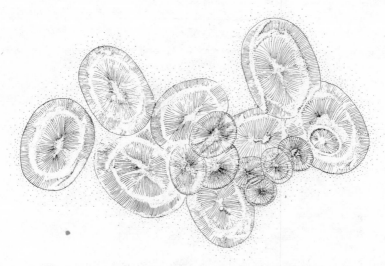

Fig. 8.7 *Chrysochromulina chiton*, field of scales. (after Parke *et al.*)

In *Prymnesium parvum*, the haptonema is much shorter and does not coil. As in *Chrysochromulina* the cell membrane is covered with numerous unmineralized scales. Muciferous bodies are located just inside the membrane. Reproduction is by cell division. Production of cysts has been recorded in natural populations. This flagellate synthesises a variety of ichthytoxins which cause problems in Israeli fish ponds.

PHAEOCYSTACEAE: *Phaeocystis* (*Phaeo*, brown; *cystis*, bladder). Fig. 8.8

The species are planktonic or epiphytic. The individual ovoid or spherical cells are distributed irregularly around the periphery of a gelatinous spherical or irregular colony. Each contains one to four yellow plastids arranged parietally. Reproduction is by means of zoospores with two equal or unequal flagella. In *P. globosa* there is a third short organ (possibly a haptonema). The most important species is *P. poucheti*, the 'baccy juice' of fishermen. This alga, a common food of antarctic crustacea, and thus penguins, synthesises significant amounts of acrylic acid which results in a greatly reduced intestinal flora in the penguins. Herrings are repelled by this organism when it is present in mass, and a vernal maximum off the coast of the Netherlands turns the northward herring migration west towards Britain, and brings about the spring fishery off the coast of East Anglia. The occurrence of an abnormal autumn maximum out of its usual station may completely change the grounds of the autumn fishery during the southward migration. The organism also forms a bloom in the Irish Sea, especially in Liverpool Bay, where it does not appear to have any effect upon other planktonic forms. It has been suggested that the inshore occurrence of such blooms means that some factor associated with terrestrial drainage is mainly responsible (Jones and Haz, 1963).

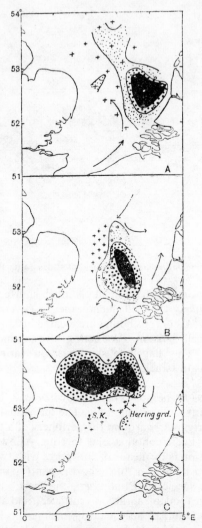

Fig. 8.8 *Phaeocystis* and herrings. A, distribution of *Phaeocystis*, 17–24 April, 1924, normal distribution. +, *Phaeocystis* scarce or absent. ○, stations in *Phaeocystis* zone. Intensity of concentration shown by shading. → assumed herring migrations. B, distribution of *Phaeocystis*, 8–13 April, 1926. Spring fishery interference. C, distribution of *Phaeocystis*, 6–9 November, 1927. Autumn fishery interference. *S.K.* = Smith's Knoll Lightship. (after Savage)

Further work on this organism is clearly necessary because of its economic importance and also because there must be some dormant phase that has not yet been described.

COCCOLITHACEAE: *Coccolithus* (*Coccus*, berry; *lithus*, scale). Fig. 8.6

These small organisms form part of the nannoplankton (see p. 404) and have been known for a long time. *Coccolithus huxleyi* is the commonest coccolithophorid. The characteristic feature is the presence of the holo- and heterococcolith scales. The holococcoliths are composed of a delicate scale comprising rhombohedral or hexagonal calcite crystals in a regular array. Such coccoliths are found on the motile stage (*Crystallolithus hyalinus*) of the non-motile *Coccolithus pelagicus*. The more abundant heterococcoliths are found in *Coccolithus huxleyi* and the non-motile stage of *Coccolithus pelagicus* ('placoliths'). These placoliths, also built of rhombohedral crystals, are composed of two flattened discs joined by a short tube. In *Coccolithus pelagicus* (non-motile) the placoliths appear to be attached on the inner disc to a non-mineralized scale (Manton and Leedale, 1969). As with the scales of *Chrysochromulina* the coccoliths are formed within the golgi apparatus and extruded to the surface. Meanwhile the old external coccoliths are shed. Like other Prymnesiales, the flagellate cells possess two smooth flagella and an haptonema. The chloroplasts, possess the typical Chrysophyceae-Haptophyceae pigments. Chrysolaminarin is the storage reserve. The siliceous cysts typical of the Chrysophyceae are unknown and reproduction is by cell division.

Coccolithus pelagicus demonstrates an unusual life history. A small unicellular motile state (*Crystallolithus hyalinus*), which normally reproduces by division can in older cultures give rise to a large unicellular non-motile stage (*Coccolithus pelagicus*). These non-motile cells then divide twice to give rise again to the motile 'Crystallolithus' stage. In some genera two kinds of coccoliths can be found on the same cell, e.g. *Michaelsarsia*. In the Syracosphaeraceae (e.g. *Cricosphaera carterae*) the coccoliths are of the heterococcolith type (Cricoliths). These are very similar to those of *Coccolithus pelagicus* and as in this organism, unmineralized scales have also been recorded.

In *Cricosphaera*, however, the motile stage divides twice to produce a tetrad, which further divides to give rise to multicellular filamentous stages which lack coccoliths and resemble such Chrysophycean algae as *Chrysosphaera* and *Thallochrysis*. It has been demonstrated that the motile Coccolith stage of *Cricosphaera carterae* is diploid and the benthic filamentous stage is also diploid; that is, an heteromorphic alternation of generations.

REFERENCES

GENERAL

Bourelly, P. (1968). *Les Algues d'eau douce*. Tome **2**, (Boubée et Cie, Paris).
Christensen, T. (1962). *Alger* in *Botanik, Edit. by* T. W. Bocher, M. Lange and T. Sorenson, Bd. 2, (Copenhagen).
Parke, M. (1961). *Br. phycol. Bull.*, **2**, 47.

Haptonema
  Manton, I. (1964). *Arch. Mikrobiol.*, **49**, 315.
  Manton, I. (1967). *J. Cell. Sci.*, **2**, 265.
  Manton, I. (1968). *Protoplasma*, **66**, 35.
Chrysochromulina
  Green, J. C. and Jennings, D. H. (1967). *J. exp. Bot.*, **18**, 359.
  Manton, I. (1966). *J. Cell Sci.*, **1**, 187.
  Manton, I. (1967). *J. Cell Sci.*, **2**, 411.
  Manton, I. and Leedale, G. F. (1962). *J. mar. biol. Ass. U.K.*, **41**, 145 and 519.
  Manton, I. and Parke, M. (1962). *J. mar. biol. Ass. U.K.*, **42**, 565.
  Parke, M., Manton, I. and Clarke, B. (1955–1959). *J. mar. biol. Ass. U.K.*, **34**, 579; **35**, 387; **37**, 209; **38**, 169.
  Parke, M., Lund, J. W. G. and Manton, I. (1962). *Arch. Mikrobiol.*, **42**, 383.
Coccolithophorids
  Paasche, E. (1968). *A. Rev. Microbiol.*, **22**, 71.
Coccolithus and Crystallolithus
  Manton, I. and Leedale, G. F. (1963). *Arch. Mikrobiol.*, **47**, 115.
  Manton, I. and Leedale, G. F. (1969). *J. mar. biol. Ass. U.K.*, **49**, 1.
  Parke, M. and Adams, I. (1960). *J. mar. biol. Ass. U.K.*, **39**, 263.
Cricosphaera
  Manton, I. and Leedale, G. F. (1969). *J. mar. biol. Ass. U.K.*, **49**, 1.
  Rayns, D. G. (1962). *J. mar. biol. Ass. U.K.*, **42**, 481.
  Pienaar, R. N. (1969). *J. Cell Sci.*, **4**, 561.
Isochrysis
  Kain, J. M. and Fogg, G. E. (1958). *J. mar. biol. Ass. U.K.*, **37**, 781.
  Parke, M. (1949). *J. mar. biol. Ass. U.K.*, **28**, 255.
Phaeocystis
  Jones, P. G. W. and Haz, S. M. (1963). *J. Cons. perm. int. Explor. Mer.* **28**, 8.
Prymnesium
  Manton, I. (1964). *J. R. Microsc. Soc.*, **83**, 317.
  Manton, I. (1966). *J. Cell Sci.*, **1**, 375.
  Manton, I. and Leedale, G. F. (1963). *Arch. Mikrobiol.*, **45**, 285.
  Vlitzur, S. and Shilo, M. (1970). *Biochem. Biophys. Acta*, **201**, 350.

BACILLARIOPHYTA

BACILLARIOPHYCEAE

These unicellular algae, the diatoms, are abundant as isolated or colonial forms in marine or freshwater plankton, and also as epiphytes on other algae and higher plants. They form a large proportion of the bottom flora of lakes and ponds and occur widely on salt marshes, although certain diatoms are said to be very sensitive to the degree of salinity in the medium. Diatoms are also found in the soil where they can form quite a substantial proportion of the soil flora. In the rain forests of the tropics they can be associated with Cyanophyceae on the leaves of trees. Most species exist as single cells but there are some colonial species in which the cells are attached to each other by mucilage or else are enclosed in a common mucilage envelope.

The Bacillariophyceae were originally classified with the Chrysophyceae in the Chrysophyta, as there are many important similarities between the two groups. The chloroplasts are olive green to brown in colour and possess chlorophylls *a* and *c* with fucoxanthin and acetylenic carotenoids. The products of photosynthesis are oil and chrysolaminarin. The gametes possess a pleuronematic flagellum and silicification is present on the cell wall. The most characteristic feature of the class is the highly ornamented and sculptured silica wall, overlaying an inner pectin layer and organic matrix.

The diatoms are generally classified into two orders, the Pennales and Centrales. The Pennales are characterized by a bilateral symmetry, their gliding motion and sexual reproduction (when present) by an amoeboid isogamy. The Centrales are radially symmetrical, lack the gliding motion and demonstrate anisogamy or oogamy with motile male gametes. Floating and predominantly marine forms are found in the Centrales as against mainly mud, soil and epiphytic forms in the Pennales. The further classification is based almost solely on the structure of the siliceous walls, which are highly characteristic and magnificently sculptured. Special terms are used to describe the cell structures and these have been defined by various workers (Desikachary, 1957; Reimann, 1960; Helmcke and Krieger, 1961) and the more important ones are described below.

Each shell or frustule is composed of two halves of varying shape, the older and larger epitheca, fitting closely over the younger and smaller hypotheca. Each half consists of a valve with a connecting band. The connecting band of the epitheca overlaps that of the hypotheca and the overlap region is referred to as the girdle (Fig. 8.9). Consequently a dorsal or ventral view is referred to as a valve view; a side-on view is called a girdle view. The whole arrangement can be pictured as resembling a pill box or petri-plate. The shells in nearly all cases contain many tiny pores; these perforations (punctae) which are of a very complex structure form from 10–30% of the valve area (Fig. 8.10). These pores can be more or less occluded by outgrowths or by thin perforated siliceous membranes (sieve membranes). Often the punctae are very closely spaced in linear array (striae) giving the cell a striated appearance. The striae are either radial from a central point (Centrales) or in longitudinal rows (Pennales). In the Pennales the two rows of striae on each valve may be separated by a cleft or groove (raphe) or a clear area (pseudoraphe). In transverse section the raphe appears as a V-shaped fissure, with the two arms of the fissure (or V) opening to the inside and outside of the wall.

In the Pennales the raphe can be one of two kinds. In the one it consists of two medium longitudinal slits following a straight or sinuous course, while in the other (canal raphe) it is a canal lodged in a crest or keel. At each pole of the cell, adjacent to the end of the raphe there is a nodule or thickening of the valve wall (polar nodule). A similar thickening occurs in the centre of the cell (central nodule) and this divides the raphe in half.

The modifications of this general pattern of the silica-wall structure are multitudinous. Within the Centrales various morphological forms are represented. Cells in valve view may be circular (*Coscinodiscus*, *Cyclotella*, *Melosira*), oval (*Biddulphia*, *Chaetoceras*), or triangular (*Triceratium*). In girdle view the cells may be appressed (*Coscinodiscus*, *Cyclotella*) or cylindrical (*Melosira*, *Biddulphia*). In some cases (e.g. *Chaetoceros*) long horn-like

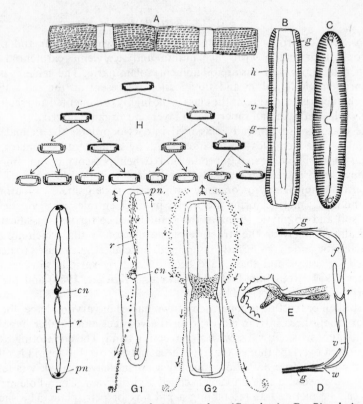

Fig. 8.9 Bacillariophyceae. A, *Melosira granulata* (Centricae); B, *Pinnularia viridis* (Pennatae), girdle view; C, same, valve view; D, *P. viridis*, union of valve and parts of adjacent girdle bands; E, *P. viridis*, termination of the two parts of the raphe in the polar nodule; F, *P. viridis*, diagrammatic view showing the two raphes; G, movement of *P. viridis* as shown by sepia particles: 1, in valve view; 2, in girdle view; H, diagram to illustrate successive diminution in size of plant; the half-walls of the different generations are shaded appropriately. (cn = central nodule, f = foramen, g = girdle, h = hyptheca, pn = polar nodule, r = raphe, v = valve, w = wall of valve) (A, H, after Smith; B–G, after Fritsch)

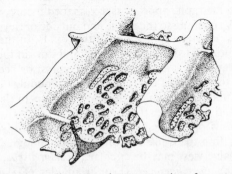

Fig. 8.10 *Surirella gemma*, diagrammatic reconstruction of a pore cavity. (after Fott)

projections extend from the four corners in girdle view. Centric diatoms may be solitary (*Coscinodiscus*) or form chains (*Melosira, Chaetoceras*) and in occasional instances form colonies. *Coenobiodiscus* produces a uniseriate colony of *Thallasiosira*-like cells all interconnected by a filamentous matrix.

The Pennales are often subdivided according to the presence or absence of raphe and pseudoraphe:

The Araphidineae (e.g. *Rhabdonema, Thalassiothrix*) have a pseudoraphe, while the Raphidioidineae (e.g. *Eunotia*) have a rudimentary raphe. If there is a raphe on one valve and a pseudoraphe on the other the genera are assigned to the Monoraphidineae (e.g. *Cocconeis*). The Biraphidineae (e.g. *Cylindrotheca, Phaeodactylum, Gomphonema*) houses those genera with a raphe on both valves. Cell form may range from oval (*Cocconeis*), boat-shaped (*Phaeodactylum*) and long and narrowly fusiform (*Nitzschia*). *Phaeodactylum* is noteworthy in being very weakly silicified.

Colonial forms as filaments (e.g. *Navicula*) or regular aggregates (*Asterionella*) are known.

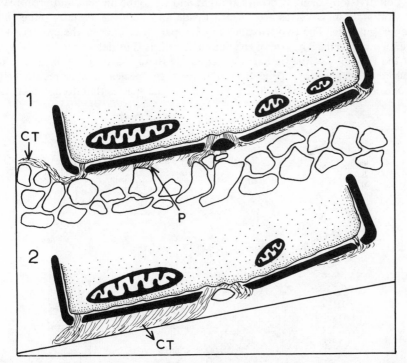

Fig. 8.11 Schematic diagram of diatom raphe systems during locomotion. 1, L.S. through two raphe systems of the ventral valve of a naviculoid diatom moving over particulate substratum; locomotor secretion material is being deposited by the posterior raphe system on all particles over which the diatom moves; if particles are close together a continuous trail (CT) results; if not, a small pad of material is left on each isolated contact-adhesion site. 2, L.S. as in (1), but the diatom is pictured as when moving over a smooth surface: locomotor secretion is deposited in a continuous trail as the diatom moves forward. (CT = continuous trail, P = pad) (after Drum and Hopkins)

The pennate diatoms that have a raphe are capable of movement but a great deal more work is necessary before the mechanism is fully understood. There may indeed be more than one type of mechanism, especially as some diatoms can move when either the girdle or the valve is in contact with the substrate. A recent important study by Drum and Hopkins (1966) has shown that linear streaming always in the opposite direction to that of locomotion, is necessary in one or more raphe systems and that movement cannot occur without adhesion at a point of contact between the raphe system and the substratum (Harper & Harper, 1967). A secretion is produced during locomotion and this forms a mucous trail on the substrate (Fig. 8.11). The nature of the mucoidal material is not known. The source of the secreted material are crystalloid bodies found only in the cytoplasm of motile diatoms. These motile diatoms have a fibrillar contractile system that is capable of continuous, rhythmic contraction and which produces the necessary pressure for forced secretion and streaming. This fibrillar system is activated by a photophobic, phototactic, mechanical or nutritional stimulus.

Cell division normally occurs at night and when the nucleus and protoplast have divided new valves are formed inside and then the parent connecting bands separate. The two frustules of each parent cell act as the epitheca for the two new cells. The size of the daughter cells is thus determined by the size of the frustules of the parent. Consequently there is a dimunition in size of successive generations (Fig. 8.9). The average species size is restored by auxospore formation (Fig. 8.12). There are some species that do not decrease

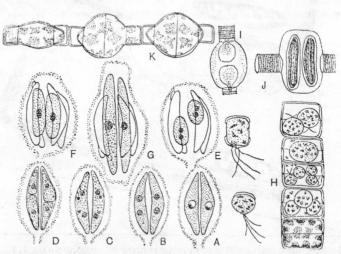

Fig. 8.12 Bacillariophyceae. A–G, auxospore formation by two cells in the pennate diatom, *Cymbella lanceolata*: A, synaptic contraction; B, after first division; C, second division of meiosis with functional and degenerating pairs of nuclei; D, division of each protoplast into two uninucleate gametes; E, young zygotes; G, zygotes elongated to form auxospores; H, microspore formation in *Melosira varians*; I, J, auxospore formation in *Rhabdonema arcuatum*; K, asexual auxospores in *M. varians*. (A–H, K, after Smith; I, J, after Fritsch)

in size and it is thought that in such cases the plasticity of the cell wall and the construction of the girdle in the form of an open hoop allows cells to resume their original size on division.

The vegetative cells of all diatoms are diploid and meiosis occurs at gamete formation. Apart from *Rhabdonema adriaticum*, all species studied are monoecious.

The centric diatoms are oogamous, the young oogonia being recognized by increase in size of nucleus and extension of the cell. Normally only one ovum matures while the other three nuclei formed during meiosis disintegrate. In *Isthmia* (Steele, 1966) the nucleus divides in two, and one nucleus aborts. The remaining nucleus again divides, with one of the daughter nuclei degenerating.

In *Biddulphia mobilensis* two ova are produced per oogonium. Spermatogonia can be formed in a variety of ways: (1) a vegetative cell may become a spermatogonium directly (*Cyclotella tenuistriata*); (2) a vegetative cell may divide to give several spermatogonia (*Melosira varians*); (3) spermatogonia may remain enclosed within the mother cell or they may complete their development after liberation. Each spermatogonium, after meiosis, produces four spermatozoids, each with a single flimmergeissel flagellum in which the two central strands may be missing (Manton and Stosch, 1965). Fertilization occurs in the female cell (auxospore mother cell).

In those centric diatoms where no sex cells have yet been seen it is assumed that sexual reproduction is by means of intracellular autogamy. In this case (e.g. *Cyclotella*), the nucleus of one cell divides to give four haploid nuclei, two of these degenerate, while the remaining two fuse within a mass of protoplasm to give the 'zygote'.

The pennate diatoms (e.g. *Cocconeis, Surirella*) are isogamous and the gametes amoeboid. Usually two or three of the four meiotic nuclei degenerate. The cells come together in pairs and connect through a conjugation tube. The female gamete usually remains in the parent cell. The frustules of the male cell separate and the male gamete migrates to the female cell where fertilization occurs. In *Gomphonema* autogamy has been reported. In this case two nuclei from meiosis degenerate. The protoplasm divides to give two 'gametes' each with one undegenerated and one degenerated nucleus. These fuse within the cell.

In both diatom groups the zygote starts to grow at once into an auxospore, which eventually becomes about three times the width or length of the mother cell. Two mitotic divisions then occur, at each of which one daughter nucleus aborts. A new shell is formed after each division (Fig. 8.12).

In *Rhabdonema adriaticum* (see above), the sole oogamous pennate diatom, only one ovum is produced per oogonium. The male mother cells produce spermatagonia that are dispersed passively in the water and become attached by mucilage pads to the oogonia. They then give rise to naked, non-flagellated microgametes from which only the nucleus enters the ovum. It is thought that this species may represent the kind of transitional type that formerly linked the centric and pennate diatoms (Lewin and Guillard, 1963).

Phylogenetically the centric diatoms are regarded as more primitive and they also have an older geological history, occurring first in the Jurassic.

Among the pennate diatoms the Araphidineae (those with only one pseudo-raphe) provide some transitional stages. They presumably evolved into the Monoraphidineae and then the Biraphidineae.

It is quite clear from the single pleuronematic flagellar structures, and from wall silicification and pigment composition, that there is a close relationship between this division and the Chrysophyta. There is rather less affinity with the Xanthophyta.

REFERENCES

GENERAL

Hendey, N. I. (1959). *J. Queckett. Micr. Club.* Ser 4. **5**, 147.

Lewin, J. C. and Guillard, R. R. L. (1963). *A. Rev. Microbiol*, **17**, 373.

Motility

Drum, R. W. and Hopkins, J. T. (1966). *Protoplasma*, **62**, 1.

Harper, M. A. and Harper, J. F. (1967). *Br. phycol. Bull.*, **3**, 195.

Nultsch, W. (1956). *Arch. Protistenk.*, **101**, 1.

Sexuality

Geitler, L. (1957). *Biol. Rev.*, **32**, 261. (General).

Drebes, G. (1966). *Helgoländer wiss. Meeresunters.* **13**, 101. (*Stephanopyxis*).

Manton, I. and von Stosch, H. A. (1966). *Jl. R. Microsc. Soc.*, **85**, 119. (*Lithodesmium*).

Migita, S. (1967). *Bull. Jap. Soc. Scient. Fish.*, **33**, 392. (*Skeletonema*).

Migita, S. (1967). *Bull. Fac. Fish. Nagasaki Univ.*, **23**, 123. (*Melosira*).

Rozumek, K. E. (1968). *Beitr. Biol. Pfl.*, **44**, 365. (*Rhabdonema*).

Schultz, M. E. and Trainor, F. R. (1968). *J. Phycol.*, **4**, 85. (*Cyclotella*).

Steele, R. L. (1966). Quoted in *Marine Botany*, by E. Y. Dawson, p. 31–33, (Holt, Rinehart, Wilson, New York).

Wall Structure

Desikachary, T. V. (1957). *Proc. Indian Acad. Sci.*, **46**, 54.

Drum, R. W. and Pankratz, H. S. (1964). *Am. J. Bot.*, **51**, 405.

Helmcke, J. G. and Krieger, W. (1961). *Diatomeenschalen im elektronenmikroscopischen Bild.* Pt III. (Cramer, Weinheim).

Reimann, B. E. (1960). *Nova Hedwigia*, **2**, 349.

Centrales

Hasle, G. R. (1962). *Nova Hedwigia*, **4**, 299. (*Cyclotella*).

Holmes, R. W. and Reimann, B. E. (1966). *Phycologia*, **5**, 233. (*Coscinodiscus*).

Loeblich, III, A. R., Wight, W. W. and Darley, W. M. (1968). *J. Phycol.*, **4**, 23. (*Coenobiodiscus*).

Reimann, B. E., Lewin, J. C. and Guillard, R. R. L. (1963). *Phycologia*, **3**, 75. (*Cyclotella*).

Pennales

Drum, R. W. (1963). *J. Cell Biol.*, **18**, 429. (*Nitzschia*).

Drum, R. W. and Pankratz, H. S. *J. Ultrastruct. Res.*, **10**, 217. (*Gomphonema*).

Hasle, G. R. (1964). *Skr. Norske Videnske. Acad. Oslo, Mat. — Naturv. Kl. N. S.* **16**, 1–48. (*Nitzschia*).

Hasle, G. R. and De Mendiola, B. R. E. (1967). *Phycologia*, **6**, 107. (*Thalassiosira*).

9

PHAEOPHYTA

PHAEOPHYCEAE

The algae composing this class range from minute discs to thalli 100 metres or more in length and are characterized by the presence of the carotenoid, fucoxanthin, which masks the chlorophyll. The class can be divided into a number of orders and families which can be treated independently (Fritsch, 1945; Scagel, 1966), or the families may be placed into three groups as proposed by Kylin (1933). These groups are based upon the type of alternation of generations, though the classification involves difficulties so far as the families Cutleriaceae and Tilopteridaceae are concerned and it is therefore best not to use a formal classification. The three groups are:

(*a*) **Isogeneratae:**
   Plants with two morphologically similar but cytologically different generations in the life cycle (e.g. Ectocarpaceae, Sphacelariaceae, Dictyotaceae).
(*b*) **Heterogeneratae:**
   Plants with two morphologically and cytologically dissimilar generations in the life cycle: (i) Haplostichineae: Plants with branched threads, which are often interwoven, and usually with trichothallic growth (e.g. Chordariaceae, Mesogloiaceae, Elachistaceae, Spermatochnaceae, Sporochnaceae, Desmarestiaceae). (ii) Polystichineae: Plants built up by intercalary growth into a

parenchymatous thallus (e.g. Punctariaceae, Dictyosiphonaceae, Laminariales).

(c) **Cyclosporeae:**

Plants possessing a diploid generation only (e.g. Durvilleales, Ascoseirales, Fucales).

In any phylogenetic consideration there is really very little difference between the two classifications. Most of the recent significant proposals have involved revision at the class and division levels, and there have been few significant changes in the lower taxa.

The Phaeophyceae are extremely widespread and are confined almost entirely to salt water, being most luxuriant in the colder waters, though the genera *Heribaudiella, Pleurocladia* and *Bodanella*, and *Sphacelaria fluviatilis* occur in fresh water. All of these except the *Sphacelaria* belong to the Ectocarpales; many of the records come only from China.

Some species exhibit morphological variations and it has been shown that these may depend on both the season of the year and the nature of the locality.

Although the unicellular and siphonous habits are unknown, extremely varied morphological forms do occur, and those range through simple filament (*Acinetospora*), heterotrichy (*Myrionema, Scytosiphon*), interwoven filaments or cable type (*Myriogloia*), hollow parenchymatous sphere (*Leathesia*), cortication (*Sphacelaria*), and simple (*Punctaria*) or complex parenchymy (Laminariales, Fucales). The large size of many Phaeophyceae and their location in the surf zone require additional mechanical support. This can be seen in (a) increased wall thickness (*Stypocaulon*), (b) twisting and rolling together of threads, (c) occurrence of anchorage devices ('root' branches) or *haptera* (Laminariales, Fucales).

Growth in the primitive brown algae is by an intercalary meristem at the base of a hair or filament (trichothallic) growth. In some of the advanced orders trichothallic growth has been replaced by an apical cell or group of apical or marginal cells. Hyaline hairs occur in many species. Cells vary greatly in size but they always have distinct walls, which are usually composed of cellulose on the inside and pectin outside, and although they are usually uninucleate, they occasionally become multinucleate.

When the fucoxanthin-protein complex of the plastids has been denatured the thallus becomes green. Usually each cell contains several chloroplasts, which are commonly parietal and discoid though there is some diversity of form. Although pyrenoids have been described for some species they are not wholly comparable with those in other algae as they appear as peripheral bulges on the chloroplast. Chlorophyll *b* is replaced by chlorophyll *c*, but unlike the other Chromophyta these algae have no acetylenic carotenoids (see p. 3). The products of assimilation are sugar alcohols, fats and the complex polysaccharides, laminarin and fucoidin. A characteristic feature of the Phaeophyceae is the presence of colourless, highly refractive vesicles known as fucosan vesicles or physodes. These contain a phenolic material known as fucosan, and they are particularly abundant in tissues where active metabolism or division is taking place.

Vegetative reproduction may take place by splitting of the thallus or else by the development of special propagules (*Sphacelaria*). In some orders (Dictyotales, Fucales) perennation is achieved by means of a perennial, prostrate attachment organ (see pp. 196, 239).

Asexual reproduction is commonly secured by means of biflagellate, typically heterokont zoospores which are normally produced in specialized sporangia. In the Dictyotales tetraspores replace zoospores. In *Zonaria*, and at times in *Pocockiella*, eight spores are produced instead of the usual four. In the Tilopteridales asexual reproduction is by means of uni- to quadrinucleate monospores. Monospores are best regarded as forerunners of tetraspores, especially when considered in relation to the vegetative characters. Sexual reproduction ranges from isogamy (with both pear-shaped gametes characteristically bearing two flagella inserted laterally), to anisogamy and finally to oogamy.

The ova are not normally retained on the parent plant and fertilization therefore takes place in the water, though the ovum may remain attached by a long mucilaginous stalk (cf. p. 251). The change from isogamy to anisogamy is accompanied by a corresponding differentiation of the gametangia. Electron-microscope studies of the antherozoids of several species have revealed the fact that the flagellum possesses the typical $9 + 2$ arrangement (cf. p. 33). The longer flagellum is of the flimmergeissel type (pleuronematic or pantonematic). The shorter one trails and is a peitschengeissel type, although in some Fucales, the reverse situation occurs. In some Fucalean antherozoids (Fucaceae) there is a peculiar transparent beak-like proboscis which is part of the flagellar apparatus. Another feature in the antherozoids of *Himanthalia* and *Xiphophora* is a single spine on the front flagellum near the distal end. There is morphological evidence that the uniflagellate *Dictyota* antherozoids have been derived from biflagellate ancestors (Manton, 1959).

Plurilocular organs can be borne on both diploid and haploid plants, whereas unilocular sporangia are normally confined to diploid plants. Plurilocular sporangia found on diploid plants represent an accessory means of reproduction, while those found on the haploid generation are gametangia. In isogamous forms gametes are produced from plurilocular structures, so that in anisogamous forms the antheridia and female gametangia must be regarded as modified plurilocular structures.

In general the presence of a unilocular or tetrasporic sporangium indicates a diploid thallus, and as meiosis occurs in the formation of the contents the swarmers are invariably haploid. The chief evolutionary features to be observed within the Phaeophyceae, at least as far as reproduction is concerned, are:

(1) The elimination of accessory reproduction (the plurilocular sporangia) in the sporophyte.
(2) A tendency to soral aggregation of sporangia on the sporophyte, e.g. Sphacelariales, Dictyotales, Laminariales.
(3) A tendency towards anisogamy and finally oogamy.
(4) A tendency towards an increase in the basic haploid number of eight

chromosomes, e.g. Ectocarpales, 8–12; Dictyosiphonales, 8–10 (sometimes 18–26); Sphacelariales, and Dictyotales, 16; Cutleriales, 24;
Laminariales, 30; Fucales, 32 (Cole, 1967).

Most species show an alternation of generations, but this is by no means
regular, and the phenomenon in some orders at least is therefore better·
termed a life history.

In some orders (e.g. Laminariales) a very marked heteromorphic alternation of generations exists. Within the Heterogeneratae, where the gametophyte is much reduced, the regular alternation may be masked by such
phenomena as parthenogenesis, e.g. *Laminaria, Alaria*. Those members of the
Heterogeneratae (excluding the Laminariales) which exhibit this type of
alternation have a fully developed diploid or delophycée form which is
common in summer, and a much reduced haploid or diploid adelophycée stage
which usually appears during the winter months.

ECTOCARPALES

The members of this order, which are the least specialized of the Phaeophyceae, are essentially branched heterotrichous filaments with prostrate and
erect portions of the thallus; one or other portion may subsequently become
lost as in *Phaeostroma*. A number represent forms that have become reduced
as a result of either an epiphytic or parasitic habit. Despite the views of
Russell (1964) it is regarded here as more desirable to retain this order in a
restricted sense and to exclude plants that can be placed in the Chordariales
and Dictyosiphonales.

ECTOCARPACEAE: *Ectocarpus* (*ecto*, external; *carpus*, fruit). Fig. 9.1
The plants are composed of sparsely or profusely branched uniseriate
filaments. The aerial portion arises from a rhizoidal base, which in some
epiphytic species occasionally penetrates the host. *E. fasciculatus* grows on the
fins of certain fish in Sweden, but the nature of this relationship is not clear.
The branches of certain genera (not *Ectocarpus*) terminate in a colourless
hair; in young plants of *E. siliculosus* pseudo-hairs may occur, but later, with
increasing age, they disappear through truncation (Russell, 1966). The erect
filaments of some species have an intercalary growing region, while in other
species growth is diffuse; the rhizoids increase in length by means of apical
growth. Each cell, which contains one nucleus together with brown, disc or
band-shaped chloroplasts, possesses a wall that is composed of three pectic-
cellulose layers. Generally two kinds of reproductive structures are present,
the plurilocular gametangia and sporangia and unilocular sporangia. The
unilocular structures always occur on diploid plants and they give rise, after
meiosis, to numerous haploid zooids which may either function as gametes or
else develop without undergoing a fusion. The sporangia are sessile or stalked
and vary in shape from globose to ellipsoid, the mature ones dehiscing
through the swelling in the centre layer in the wall. The plurilocular structures, which are either sessile or stalked, range from ovate to siliquose in

shape and are present on haploid or diploid thalli. In *E. siliculosus* they
represent modified lateral branches. The plurilocular organs are divided into
a number of small cells, each one of which gives rise to a swarmer. When the
swarmer is ripe, dehiscence takes place by means of a pore and the contents
either germinate directly or else behave as isogametes, depending on the
genetic status of the plant. Recently Müller (1968) has shown that what are
probably female gametes exude a volatile odiferous substance to attract
gametes of the opposite sign.

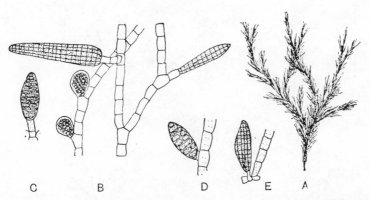

Fig. 9.1 *Ectocarpus.* A, *E. confervoides*, plant; B, *E. tomentosus*, unilocular and plurilocular
sporangia; C, meiosporangium, *E. virescens*; D, megasporangium, *E. virescens*; E, micro-
sporangium, *E. virescens*. (C–E, after Kniep)

In *Giffordia secunda* (*E. secundus*) there is well-marked anisogamy, the
smaller gametangia being antheridia and the larger female gametangia.
The contents of the latter are sometimes capable of parthenogenetic development. In *E. padinae* the unilocular sporangia are absent and there are three
kinds of plurilocular organs. One type, which has very small loculi, represents
the antheridia, while there are also medium-sized or 'meiosporangia', and
larger organs. One of these must represent the female reproductive organs.

The life histories of the species are full of interest, especially in view of what
has been discovered for *E. siliculosus*. Knight (1929) found that the plants in
the Isle of Man occurred in early spring and late autumn and were all diploid.
They bore unilocular and plurilocular structures, the former producing
gametes while the latter gave rise to zoospores (Fig. 9.2). In the Bay of Naples,
on the other hand, the plants were all haploid and only bore plurilocular
organs. The zooids from these behaved as gametes. Berthold recorded a
microscopic form which has since been regarded as diploid because unilocular
sporangia were found on it.

Subsequently Schussnig and Kothbauer (1934) and Müller (1966) revealed
the existence of unilocular structures. It would seem, therefore, that the
somewhat complex schema of these later workers is the more correct, at any
rate so far as the Neapolitan form is concerned.

It has been suggested that the differences between plants from the two localities are due to differences in the tides, light conditions or temperature, with perhaps more emphasis on the last. Müller (1962) has since shown that in one form of *E. siliculosus* unilocular structures are formed at 13° C and plurilocular at 19° C and that there is a lunar periodicity in the liberation of

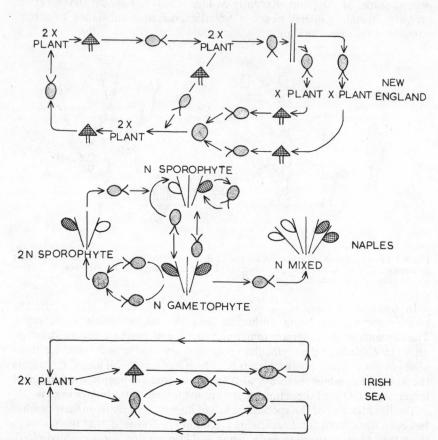

Fig. 9.2 Diagram to show different life histories for *Ectocarpus siliculosus* from different localities.

the swarmers. The later results indicate that the earlier work of Knight in the Irish Sea needs to be repeated. Yet another study of this species has been carried out in America by Papenfuss (1935). In America the diploid plants were found growing epiphytically on *Chorda* or *Spartina* and these either bore pluri- or unilocular structures independently, or else both could be found on the same thallus. The unilocular structures occurred only in summer while the plurilocular were present throughout the whole year. Although the zooids from both types of structure acted as zoospores and germinated directly, nevertheless meiosis always took place in the unilocular structures, the zooids of which developed into the sexual plants that were found growing as obligate epiphytes on *Chordaria*, in some cases the nearest asexual plants being twenty miles distant. The plants growing on *Chordaria* were dioecious and only bore plurilocular gametangia (Fig. 9.2).

This extremely large genus is now subdivided, and a number of new genera have been established, leaving two main species complexes, *E. siliculosi* and *E. fasciculati* within the genus (Russell, 1966). The view of Russell (*loc. cit.*) that the *E. siliculosus* complex represents a clone is in agreement with the variations found in the life histories.

The allied genus *Pylaiella* is very similar to *Ectocarpus* morphologically. Two species, *P. littoralis* and *P. (Bachelotia) fulvescens* are widespread, the former occurring often as an epiphyte while the latter is generally saxicolous. The species are readily distinguished from *Ectocarpus* by the position of the sporangia because these bodies are nearly always intercalary, very rarely terminal, and when this latter is the case it is frequently due to the loss of the terminal vegetative portion. A study of British populations has shown that there are two entities present that may differ genetically and physiologically (Russell, 1964).

TILOPTERIDALES

The members of this order comprise Atlantic deep-water algae with incompletely known life histories. They would appear to represent a transition from the Ectocarpalean type to the Dictyotalean type of reproduction, though retaining the morphology of the Ectocarpales. Asexual reproduction takes place by means of characteristic, large, motionless quadrinucleate monospores. Sexual fusion has not so far been reported within the order though what are probably plurilocular male gametangia and oogonia are known. Recent work has shown that the genus *Acinetospora*, formerly placed in this order, is properly a member of the Ectocarpales. A proposal by Christensen (cf. Scagel, 1966) to treat this order as a family within the Dictyosiphonales has not so far been accepted.

TILOPTERIDACEAE: *Haplospora* (*haplo*, simple; *spora*, seed). Fig. 9.3
The plants, which arise from a basal disc, are filamentous with irregular pinnate branching from the main axes. The lower portion becomes multiseriate by septation and then resembles *Sphacelaria* although the growth is always intercalary. The sexual plants develop intercalary, tubular, plurilocular organs (gametangia) which are produced by the transformation of one or more cells of the main filament. The swarmers from these organs are thought to be male gametes. Besides these organs there are larger, spherical, sessile, uninucleate oogonia partly immersed in the branches. These plants were formerly known as *Scaphospora speciosa*. The asexual plant reproduces by means of quadrinucleate spores formed singly in stalked or sessile monosporangia which are terminal or intercalary. Meiosis has been reported as occurring in these sporangia and this would be expected if they were primitive tetrasporangia. It seems therefore that the plants known as *Haplospora*

globosa and *Scaphospora speciosa* are simply alternate phases of one and the same species, the latter being the sexual generation (Sundene, 1966).

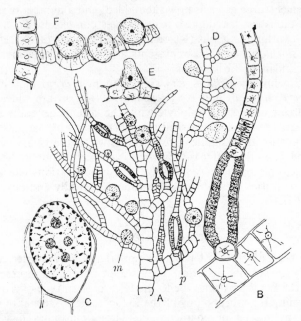

Fig. 9.3 *Haplospora globosa*. A, portion of plant with uninucleate sporangia, *m* (oogonia?), and plurilocular microgametangia, *n*; B, plurilocular microgametangium; C, monosporangium with quadrinucleate monospore; D, monosporangia; E, F, unilocular sporangia (oogonia?). (A–C, after Oltmanns, D–F, after Tilden)

SPHACELARIALES

The next three examples belong to the Sphacelariales, an order frequently known as the 'Brenntalgen' because its members possess a very characteristic large apical cell with dense brown, tanniferous contents and the detailed classification of the group is based primarily upon the behaviour of this apical cell at branch formation. Structurally they can be regarded as strengthneed multiseriate filaments, *Sphacella* being one of the more primitive members of the group with a non-corticate monosiphonous axis. Electronmicroscope studies have shown that in *Sphacelaria* the microfibrils are randomly arranged.

In most members of the order two types of lateral branches occur, those of limited and others of unlimited growth. The plants exhibit heterotrichy with a well-developed prostrate system, which is perennial and from which new erect systems can arise each season. In those cases where it has been studied there is an alternation of isomorphic generations. The order is most widely represented in the southern hemisphere.

SPHACELARIACEAE: *Sphacelaria* (gangrene). Fig. 9.4

The plants grow attached to stones or other algae by means of basal discs or rhizoids that have spread down from the lower cells of the axis. The basal disc may give rise to horizontal stolons and so enable the plant to spread.

The erect system is filamentous and the type of branching is termed hemiblastic in which the laterals, whether of definite or indefinite growth, arise from the whole height of a cell representing the upper segment derived from the division of a primary segment (Fig. 9.4, E). If the branching is pinnate practically every superior segment produces branch initials, though in some cases the initials may remain dormant and the branching is then less luxuriant. Branches of limited growth bear colourless hairs which are cut off as small cells to one side of the apical cell (Fig. 9.4, B–D). As the apical cell elongates the hair comes to occupy a lateral position. These hairs may disappear with age. This type of hair formation is regarded as sympodial, the hair initial being the true apical cell while the so-called branch axis is regarded as a subsidiary branch. Branching, which occurs in this way is said to be holoblastic.

Growth in length takes place by transverse divisions of the apical cell, and when the basal cell so cut off (s in Fig. 9.4, E) has reached a certain length it

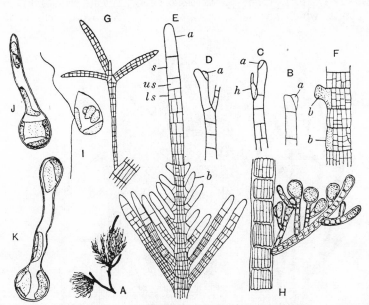

Fig. 9.4 *Sphacelaria.* A, plant of *S. cirrhosa*; B, apical cell (a) of *S. cirrhosa*; C, *S. cirrhosa*, origin of hair (*h*), (*a* = apical cell); E, apex of thallus of *S. plumigera* showing branches, (*b*); single segment (*s*), which later divides into upper (*us*) and lower (*ls*) segments, (*a* = apical cell); F, origin of branch (*b*); G, bulbil of *S. cirrhosa*; H, unilocular sporangia, *S. racemosa*; I, zoospore of *S. bipinnata*; J, K, germinating spore of *S. bipinnata*. (A × ½, B–G × 52·5, H–K × 1200) (B–G, after Oltmanns; H, after Taylor; I–K, after Papenfuss)

divides into a superior and inferior segment. These segments exhibit no further enlargement, though they divide transversely and longitudinally to form the corticated thallus characteristic of the genus. This type of vegetative system is said to be leptocaulous, and the filament diameters are essentially uniform throughout.

Vegetative reproduction takes place in this genus by means of modified branches or propagules which vary in shape from wedge-like to di- or triradiate with long or short pedicels. Unilocular and plurilocular structures are borne terminally or laterally on simple or branched laterals and are either sessile or shortly pedicellate. The life cycle of *Sphacelaria bipinnata* has been worked out in some detail and is probably typical of the genus.

The asexual generation bears both uni- and plurilocular sporangia. Zooids from the former are said to fuse in clumps, but those from the latter germinate directly and represent a means of accessory reproduction. The cytology of the clumps has not been studied though meiosis occurs in the unilocular sporangia. These clumps must give rise to the isomorphic sexual generation which reproduces by means of isogametes liberated from plurilocular gametangia. *S. harveyana* and *S. hystrix* are anisogamous. Male and female gametangia are born on separate plants in *S. harveyana* while in *S. hystrix* they can occur on the same plant.

CLADOSTEPHACEAE : *Cladostephus* (*clado*, shoot; *stephus*, crown). Fig. 9.5
The plants, which are bushy in appearance, arise from well-developed

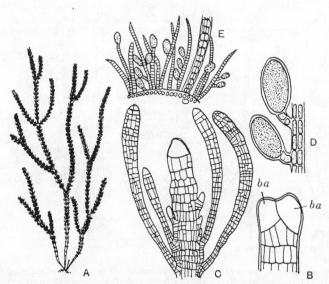

Fig. 9.5 *Cladostephus verticillatus.* A, plant; B, apex to show origin of branch (*ba* = branch apex); C, thallus showing cortication; D, unilocular sporangia; E, part of thallus with unilocular sporangia. (A × ½, C × 225, D × 45) (A, D, E, after Newton: B, C, after Oltmanns)

holdfasts and are characterized by ecorticate branches arranged in whorls with tufts of hairs just below their apices. Cells just below the apex divide to give a number of branch segments, this type of branching being known as polyblastic. The first segment of these laterals of limited growth cuts off a cell or cells that form a cushion of tissue belonging to the main axis. Later the spaces between laterals may be filled by tissue so that the bases are completely enveloped. As a result of this septation the older branches, although hemiblastic in origin, appear to arise only from the tops of the original superior segments. The lower portions of the segments may later give rise to secondary whorls which are said to arise meroblastically. The main axis is corticate because both upper and lower primary segments undergo abundant division. In addition the cells so formed enlarge in width and length so that the axis increases in size from the apex downwards. Such a form of growth is termed auxocaulous. *Cladostephus* also forms long branches which arise in a hemiblastic manner.

Both unilocular sporangia and plurilocular gametangia are formed on separate plants on special branchlets which arise from the rhizoidal cortex in the internodes between the whorls of vegetative branches.

STYPOCAULACEAE: *Halopteris* (*halo*, sea; *pteris*, fern). Fig. 9.6

The pinnate frond arises from a distinct basal pad, the plants in summer having the appearance of shaggy tufts, while in winter the branching appears more regular as surplus branchlets are shed. The inner cortex of the central axis is composed of a number of cubical cells and there is also an outer cortex of rhizoidal cells (pericysts), the whole forming a pseudo-parenchyma. Because there is no enlargement of the original cells, *Halopteris* is leptocaulous. Pericysts are only found in the main axis and in branches of unlimited growth. Branching in *Halopteris* is holoblastic (cf. hair production in *Sphacelaria*). At an early stage the branch initial cuts off a small upper axillary cell. This cell is regarded as the true apical cell so that the branch is really a lateral of a diminutive branch axis. In some species the axillary cell may later give rise to a tuft of hairs.

Vegetative reproduction can take place by fragmentation, the fragments developing rhizoids and a new axis.

Sexual reproduction is by means of antheridia and oogonia and asexual by unilocular sporangia and zoospores. The reproductive organs are borne in the axils of laterals, when they must be regarded as terminal on axillary shoots, or on branchlets arising from pads derived from the axial initial. In *H. scoparia* the sori are united into spikelets, the lower laterals being sterile and only the upper being fertile. A number of species are known, especially from the South Pacific, but their identification is not always easy.

CUTLERIALES

This order is characterized by trichothallic growth, regular alternation of generations, and a well-marked anisogamy. It is generally placed in the Isogeneratae, even though this leads to a difficulty because the two generations are not isomorphic in *Cutleria*, although they are isomorphic in *Zanardinia*. In *Microzonia* only the diploid generation is known.

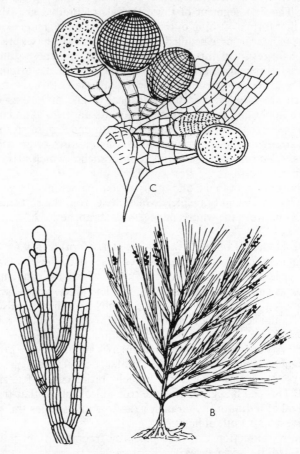

Fig. 9.6 *Halopteris.* A, apical portion of *H. funicularis*; B, fertile plant of *H. hordacea*; C, oogonia and antheridia of *H. hordacea*. (A × 180, B × 1, C × 126) (A, B, after Lindauer, C, after Moore)

CUTLERIACEAE: *Cutleria* (after Miss Cutler). Fig. 9.7

The gametophyte and sporophyte generations differ in their seasonal occurrence, the former being a summer annual while the latter is a perennial reaching its maximum vegetative phase in the Northern hemisphere in October and November with a peak fruiting period in March and April. The gametophyte is an erect, flattened thallus with irregular branching brought about by periodic failure of the marginal filaments to fuse together. In cross-section the plant consists of larger cells in the centre and smaller ones in the epidermis. The thallus and apices are clothed with tufts of hairs, each with a basal growing region, while the female gametangia and antheridia, which are borne on separate plants, occur in sori on both sides of the thallus. The antheridia are borne normally on much-branched threads. The mature plurilocular antheridium contains about 200 antherozoids that are much smaller than the mature female gametes.

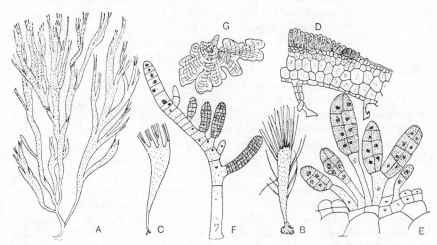

Fig. 9.7 *Cutleria multifida*. A, plant; B, young plantling; C, plantling slightly older to show branching; D, t.s. diploid thallus with unilocular sporangia; E, female gametangia; F, male gametangia; G, 'Aglaozonia' stage. (A × ⅓) (B-D, G, after Oltmanns; E, F, after Yaman-ouchi)

The female gametangia, with which hairs are sometimes associated, are also formed from superficial cells that divide into a stalk cell and a game-tangium initial. The ripe gametangium contains fewer female gametes which, after liberation, remain motile for a period of from five minutes to two hours, while the antherozoids can remain active for about 20 hours. No fertilization occurs while the female gametes are motile. After settling they round off and attract large numbers of antherozoids chemotactically though this power is lost within 3–4 hours.

Upon germination a small columnar structure is first formed and then a flat basal expansion grows out from its base to form the adult sporophyte, which is a prostrate expanded thallus attached by means of rhizoids. It is morphologically different to the gametophyte and was initially thought to be a separate genus, *Aglaozonia*. The female gametes may germinate partheno-genetically to give sterile haploid *Aglaozonia* plants. The sporophytic thallus is composed of large cells in the centre with superficial layers of small cells. The sessile unilocular sporangia borne on stalk cells and sometimes accom-panied by deciduous hairs, are clustered in palisade-like sori or are scattered irregularly on the upper surface of the thallus. Eight to thirty-two haploid zoospores are formed in each sporangium. The zoospores on germination give rise to *Cutleria* plants.

DICTYOTALES

This order is characterized by a well-marked regular alternation of iso-morphic generations. Morphologically the plants are parenchymatous and consist of one or more medullary layers of large cells and a cortical layer of small cells. The species are also characterized by apical growth, e.g. *Dictyota*, or by a marginal meristem, e.g. *Padina, Taonia*. In the tropical genus *Padina* there may be considerable deposition of lime on the thallus. Some of the genera, e.g. *Dictyopteris*, have a distinct mid-rib. Asexual reproduction is by tetraspores produced in superficial tetrasporangia. The antheridia and oogonia are borne in sori on separate plants. All the reproductive cells develop from surface cells which enlarge and protrude above the surface of the thallus. In some genera a stalk cell may be cut off. At meiosis in the tetrasporangia, segregation of the sexes takes place so that two of each group of four tetraspores will give rise to male plants and two to female. In certain genera, e.g. *Taonia, Padina*, the tetrasporangia and sex organs are borne in zonate bands across the thallus, and there is some evidence of a correlation between their development and diurnal tidal cycles (e.g. *Taonia*) or with spring tidal cycles, (e.g. *Padina*). A period is evidently required between the initiation of each new crop so that the plant can accumulate the necessary food material. In *Taonia* and *Dictyota* asexual plants are more abundant than the sexual, at least in some localities. This is partly accounted for by the persistence of a sporophytic rhizoidal portion that regenerates new plants, but more commonly the contents of the tetrasporangium fail to divide before they are liberated. These undivided contents give rise to new plants, which are more resistant and vigorous than plants produced from normal tetra-spores. The order is primarily a tropical one but some species, e.g. *Dictyota dichotoma*, are widely distributed in temperate waters.

DICTYOTACEAE: *Dictyota* (like a mat). Fig. 9.8

In *D. dichotoma* the flattened thallus exhibits what is practically a perfect dichotomy because there is always a median septation of the apical cell. In transverse section the thallus is seen to be composed of three layers, a central one of large cells and an upper and lower epidermis of small assimilatory cells from which groups of mucilage hairs arise.

The sex structures are borne in sori on separate plants, the male sorus being composed of as many as 300 plurilocular antheridia surrounded by an outer zone of sterile cells. At the formation of an antheridium a superficial cell divides into a stalk cell and an antheridium initial, which then partitions into individual antheridial cells. The mature antherozoid is pear-shaped with only one flagellum, though two basal bodies can be seen, one of them being vesti-gial. Fewer ova are produced and it has been estimated that there are about 6,000 antherozoids available for each ovum. In the oogonial sorus the large, fertile oogonia, 25–50 in number, are situated in the centre and surrounded by sterile cells on the outside. The oogonia likewise arise from superficial cells that divide into a stalk cell and oogonium initial, and each oogonium produces one ovum. Fertilization occurs in the water, and the eggs may be chemotactic. Unfertilized ova may develop parthenogenetically; such plants, however, always die in culture, though it is possible that in nature they may persist. The sex organs are produced in regular crops, the new sori appearing between the scars of the old, and when the whole of the surface has been used up the plant dies.

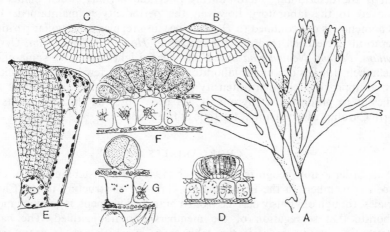

Fig. 9.8 *Dictyota dichotoma.* A, portion of plant showing regular dichotomy; B, apical cell; C, apical cell divided; D, group of antheridia surrounded by sterile cells; E, single antheridial cell and a sterile cell; F, sorus of oogonia; G, tetrasporangium. (A–D, F, G, after Oltmanns; E, after Williams)

The zygote develops into a morphologically similar plant which reproduces by means of spores that are normally formed in tetrads in superficial sporangia. At sporangium formation an epidermal cell swells up in all directions, and after a stalk cell has been cut off the sporangium initial divides meiotically to give the four tetraspores. In some cases the contents of a tetrasporangium fail to divide into four spores but germinate as a whole. In other cases, such as in *Zonaria* and sometimes in *Pocockiella*, eight spores may be produced instead of the usual four. While the sex organs are produced in rhythmic crops there is no such periodicity in the case of tetraspores.

Three kinds of rhythmic periodicity for the sex organs of *Dictyota* have been described from different localities:

(a) In Wales the sori require 10 to 13 days to develop while in Naples 15 or 16 days are necessary, the gametes being liberated about once a fortnight in both areas.

(b) In North Carolina liberations occur once a month, at the alternate spring tidal cycles, although only eight days are required for the development of the sex organs.

(c) In Jamaica the successive crops take a very long time to mature, e.g. very little change can be seen even after 22 days. This results in almost continuous fruiting with two successive crops overlapping. There is obviously a considerable difference in the behaviour of species of *Dictyota*.

Wherever the plants occur, the bulk of the gametes (60–70 per cent) are usually liberated in a single hour around daybreak. It has been suggested that light is the determining factor during intertidal periods; when plants are removed to the laboratory however, the periodicity is maintained. The phenomenon is not confined to *Dictyota* because regular or irregular periodic liberations have been recorded for species of *Halicystis, Sargassum, Cystophyllum, Padina, Nemoderma* and *Ectocarpus*. Light does have some effect (Vielhaben, 1963) since under laboratory conditions a few hours of artificial light at approximately lunar intensity (0·3 lx) at any period of the night are sufficient to produce the rhythm. It has been observed that the plants in North Carolina always fruit at the time of full moon.

CHORDARIALES

This order is not recognized by Fritsch (1945) or Christensen (1962), who place the members in the Ectocarpales. The type of structure in the Chordariales, though obviously derived from a branched filament, is so much more elaborate that segregation of the members appears justified. The basic construction in this order is the cable type, in which one or more erect parallel strands arise from a prostrate basal portion. These give rise to interweaving lateral branches and the whole is often enclosed in a mucous matrix. In fully developed species there are three principal zones that can be recognized in the thallus: (*a*) a medulla composed of one or more lon-threads accompanied by off-shoots of the first order; (*b*) a subcortex composed of off-shoots from the medulla; (*c*) a cortex of peripheral assimilatory filaments and colourless hairs.

Growth is commonly trichothallic but apical growth occurs in two families.

There is also a certain amount of secondary tissue which in some parts may be rhizoidal in character. Modified versions of this cable type of construction are to be found in some genera (*Leathesia, Elachista*), while in the Myrionemataceae there appears to be considerable reduction, though the structural pattern remains clear.

In most advanced members (*Chordaria, Scytothamnus australis*) a pseudo-parenchymatous construction may tend to obscure the basic underlying morphology of the order (Price, 1969).

In nearly all the species investigated the basic life cycle consists of a macroscopic diploid sporophyte alternating with a microscopic gametophyte. In some cases the diploid plant arises as a lateral outgrowth from a microscopic plant. The number of families in the order varies according to the taxonomic treatment. The Ralfsiaceae, for example, are sometimes placed in the Ectocarpales even when recognition is accorded the Chordariales.

CHORDARIACEAE: *Mesogloia* (*meso*, middle; *gloia*, slime). Fig. 9.9

In this genus there is a single central strand terminating in a hair and having a distinct intercalary meristem just below the apex. The cortex is formed of short lateral filaments with somewhat globose, terminal cells that are packed in a gelatinous material. The hairs, which are frequently worn away in the

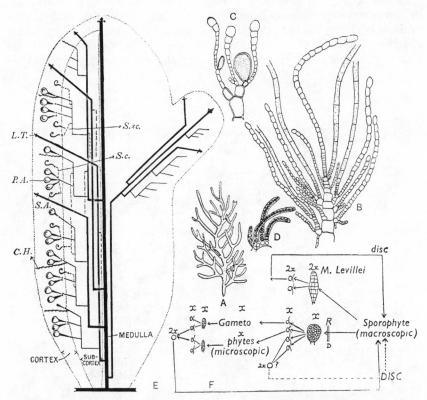

Fig. 9.9 *Mesogloia vermiculata*. A, plant (sporophyte); B, apex of filament with branches and beginning of cortication; C, unilocular sporangia; D, plurilocular gametangia on gametophyte; E, diagram to illustrate construction of thallus (central thread type), (C.H. = colourless hair, L.T. = leading thread with intercalary growth zone, P.A. = primary assimilator, S.A. = secondary assimilator, S.c. = secondary cortex, S.sc. = secondary sub-cortex); F, diagram to illustrate life cycle. (B × 135) (A, C, D, after Tilden; B, E, F, after Parke)

older parts of the thallus, occupy a lateral position, but owing to inequalities of growth they may appear to be terminal. The unilocular sporangia are ovoid and are borne at the base of the cortical filaments, but the elongate plurilocular sporangia, which are only known for *M. levillei*, replace the terminal portion of the assimilatory hairs and are therefore always stalked. Meiosis takes place in the unilocular sporangia during zoospore formation, and the zooids from the unilocular sporangia germinate into a minute winter gametophyte that bears plurilocular gametangia. The zooids from these gametangia fuse and the zygote develops into the characteristic basal disc from which the central erect filament of the macroscopic plant arises. There is thus an alteration of morphologically distinct generations in this species.

In *Eudesme* and *Myriogloia* the branched mucilaginous plants possess a dense growth of hair and differ from *Mesogloia* in having more than one central strand in the medulla; there are also developmental differences.

In the genus *Chordaria* development has proceeded a stage farther and the branched cartilaginous fronds possess a firm, pseudo-parenchymatous medulla of closely packed cells that have become elongated in a longitudinal direction. The cortex is composed of crowded, radiating assimilatory filaments, which are either simple or branched, the whole being embedded in a

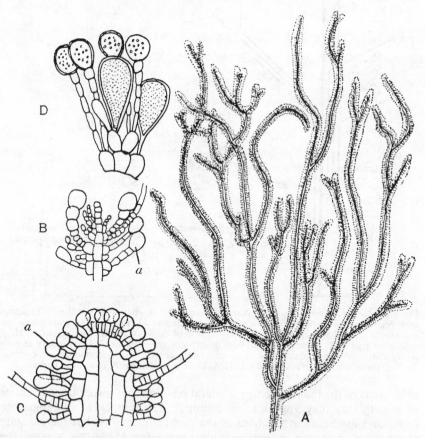

Fig. 9.10 *Chordaria divaricata.* A, plant; B, apex of young plants showing commencement of cortication; C, apex of older plant of *C. flagelliformis* showing structure of thallus (*a* = assimilator); D, unilocular sporangia. (A × ⅔, C × 300) (A, D, after Newton; B, C, after Oltmanns)

thick layer of jelly which makes the plant slimy to touch. This type of structure, even though the growth is still confined to the apex, marks the highest development of the cable type of construction. Only unilocular sporangia are known.

In recent years culture work has led to the discovery of a dwarf generation, (Caram, 1955; Kornmann, 1962). In *C. flagelliformis* the dwarf generation produced plurilocular reproductive structures, which are probably asexual accessory bodies since no fusion of products was observed. Kornmann (1962) does not regard the dwarf generation as gametophytic. It seems that the zoospores of the unilocular sporangia can either germinate into this dwarf thallus which diploid and filamentous or into a new diploid macro-scopic plant. Which occurs appears to be dependent on sea temperature; at

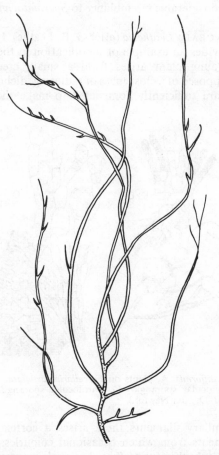

Fig. 9.11 *Acrothrix novae-angliae* (× 0·7). (after Taylor)

high temperatures the macroscopic plant develops and at low temperatures the dwarf filament.

ACROTRICHACEAE: *Acrothrix* (*acro*, top; *thrix*, hair). Fig. 9.11

Species of this genus are found in the northern hemisphere. The plants arise from a disc-like holdfast and are basically uniaxial in construction, the main axial filament being terminated by a hair. Growth in length is normally trichothallic though the hair may later be lost leaving a pseudo-apical meristem. Cells cut off from the meristem enlarge and produce whorls of primary laterals of limited growth. The basal cells of these laterals increase in size and give rise to short perimedullary threads that form the cortex which in older parts of the thallus may separate from the axial filament. As this latter undergoes great alteration or disintegrates the thallus can become hollow. Only unilocular sporangia are known and they are borne at the base of the assimilators. Except for the intercalary growth and terminal hair *Acrothrix* shows considerable resemblance to *Spermatochnus* (p. 206).

CORYNOPHLAEACEAE: *Leathesia* (after G. R. Leathe). Fig. 9.12

This genus provides an example of modification in the cable type of construction. The young plant arises from a small, creeping, rhizomatous portion and is composed of a close mass of radiating, dichotomously branched filaments which are sufficiently compacted to make the plant mass solid.

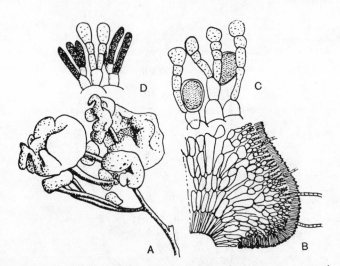

Fig. 9.12 *Leathesia difformis.* A, plants on *Furcellaria fastigiata*; B, t.s. to show thallus construction; C, unilocular sporangia; D, plurilocular sporangia. (B × 24, C, D × 336) (A, after Oltmanns; B–D, after Newton)

From these medullary filaments there arises a cortex of densely packed assimilatory filaments from which occasional colourless hairs emerge. The young plants are subspherical at first, but with increasing age the central medullary filaments commence to disintegrate and as a result the mature thallus becomes hollow and irregularly lobed. Plurilocular and unilocular sporangia are known, the zoospores from the ovoid unilocular sporangia germinating to disc-like plantlets on which plurilocular structures ultimately appear. These represent accessory reproductive bodies since the swarmers give rise to other similar plantlets. A new adult plant arises as a vegetative bud so that the dwarf plants must be diploid. No dwarf gametophytes have been reported. The zoospores from the plurilocular sporangia of the macroscopic plant produce dwarf plants that can reproduce by means of plurilocular sporangia or can give rise to buds which develope into new adult plants.

ELACHISTACEAE: *Elachista* (very small). Fig. 9.13

Juvenile plants possess a horizontal portion from which a number of erect filaments arise, so that in the early stage *Elachista* is comparable morphologically to *Eudesme*. In the older plant the erect filaments develop to form a cushion composed of densely branched filaments matted together and only

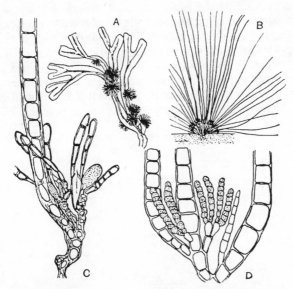

Fig. 9.13 *Elachista fucicola*. A, plants on *Fucus vesiculosus*; B, single plant in section showing penetrating base, crowded sporangia, short paraphyses and long assimilators; C, unilocular sporangia; D, plurilocular sporangia at base of assimilation thread. (A ×0·36, C × 120, D × 220) (A–C, after Taylor; D, after Kylin)

becoming free at the surface. The mature thallus comprises three distinct regions. In the dense central portion there is a mat of profusely branched threads which gives way to a looser peripheral zone of assimilatory filaments (sometimes termed paraphyses). Outside this is an outermost zone of free, long assimilatory filaments with chloroplasts. Morphologically it is therefore a reduced and modified *Eudesme* structure.

The species are epiphytic on other algae, the genus being represented in both hemispheres. In most species unilocular sporangia arise from the base of the assimilators, the distal portions of the latter being modified to form plurilocular sporangia. The zooids from the unilocular sporangia germinate in late autumn to give a branched, thread-like, microscopic gametophyte which persists throughout the winter. In late winter and spring plurilocular gametangia develop on the minute gametophytes, and when the gametes have been liberated fusion occurs and the zygote germinates on the host into a new macroscopic *Elachista* plant.

Exceptions to the normal type of life cycle are found in *E. fucicola* and *E. stellaris*. Only unilocular sporangia are known and the zoospores develop into plantlets that give rise to new sporophytes as lateral outgrowths. Plurilocular sporangia have recently been described in the creeping basal layer of *E. fucicola*, but they are probably accessory organs.

MYRIONEMATACEAE: *Myrionema* (*myrio*, numerous; *nema*, thread). Fig. 9.14

This represents one of the highly reduced genera, but the fundamental basic cable construction of prostrate thallus and erect threads can be found, though the erect threads do not branch. The genus is of wide distribution, the commonest species being *M. strangulans* epiphytic on *Ulva* and *Enteromorpha*. The various species form thin expansions or minute flattened cushions or discs that are very variable in shape and from which numerous, closely packed, erect filaments and hairs arise. The basal monostromatic portion of the thallus, which has a marginal growing region, is composed of crowded,

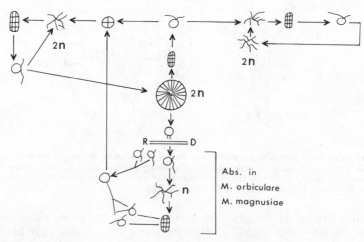

Fig. 9.14 Life history diagram for *Myrionema strangulans* and *M. feldmanni*. (after Loiseaux)

radiating filaments that may, on rare occasions, penetrate the host plant. The unilocular sporangia, which are not borne on the same plants with plurilocular sporangia, give rise to haploid zoospores, and these develop into a thread-like gametophytic plant bearing long filaments. The gametophyte phase is absent in *M. orbiculare* and *M. magnusii*. Although *M. strangulans* and *M. feldmanni* possess a life cycle typical of the Chordariales (Loiseaux, 1964), it has been suggested that some of the other species placed in the genus or in allied genera may have isomorphic alternation, in which case they may need to be removed to the Ectocarpales.

RALFSIACEAE: *Ralfsia* (after G. Ralfs). Fig. 9.15
Some authorities place this family in the Ectocarpales, but structurally it seems very closely related to *Myrionema*.
In this genus reduction has proceeded to the point where the erect filaments from the prostrate thallus have fused together to produce a parenchymatous leathery crust (Fig. 9.15). The various species are found on rocks between high and low-tide mark. Reproduction is by means of unilocular sporangia

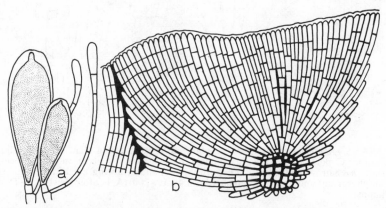

Fig. 9.15 *Ralfsia verrucosa*. A, younger and older sporangia with paraphyses; B, vertical section of a crust at right angles to the direction of growth, showing filamentous structure.

that are borne laterally in sori on newly developed free filaments. Plurilocular sporangia are terminal in position. Chen *et al.* (1968) have shown that *R. clavata* zoospores produce a *Petalonia* thallus, whose swarmers in turn produce a *Ralfsia* thallus. One species of *Ralfsia* is thus the alternate generation of a member of the Dictyosiphonales. Wynne (1969) has similarly suggested that *Ralfsia californica* might be the alternate generation of *Petalonia fascia* (see p. 214).

SPERMATOCHNACEAE: *Spermatochnus* (*sperma*, seed; *chnus*, fine down). Fig. 9.16
This is essentially a corticated type. The filamentous, cylindrical, branched thallus is derived from a single central axis with a definite apical cell. Each

individual cell of this axis filament segments at one end and so definite nodes
are formed. The corticating filaments arise from the nodes, and growth of
the cortex is secured by tangential division of the primary corticating cells,
though later more filaments may grow on top of them. The outermost layer
of the cortex bears the assimilatory filaments and hairs. As the plants become
older mucilage develops internally and forces the cortex away from the
primary central filament, although a connection is maintained by the threads
from each node. Unilocular sporangia, together with clavate paraphyses,

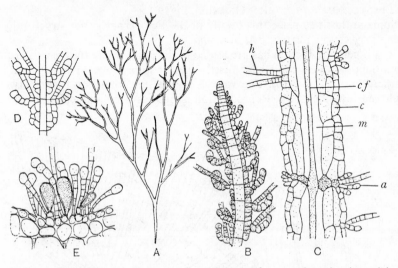

Fig. 9.16 *Spermatochnus paradoxus.* A, plant; B, apex of young plant showing origin of
cortication; C, portion of old thallus filament (*a* = assimilator, *c* = cortical cells, *cf* = central
filament, *h* = hair, *m* = mucilage); D, portion of thallus showing cortication and pairs;
E, paraphyses and unilocular sporangia. (A × 0·44, E × 200) (A, E, after Newton; B–D, after
Oltmanns)

develop in sori, the sporangia arising from the base of the sterile threads. The
life cycle has not yet been worked out, but if it is at all comparable with the
other closely related genera then the zoospores should give rise to a microscopic plant phase or sexual generation.

STILOPHORACEAE: *Stilophora* (*stilo*, point; *phora*, bear). Fig. 9.17
    The plants are bushy and attached by a holdfast disc. Structurally they are
multiaxial with growth taking place by divisions of the apical cells of the
axial filaments. This condition has probably arisen through the loss of
trichothallic growth. A parenchymatous cortical envelope is formed as in
Spermatochnus, the peripheral cells of this cortex dividing to form a small-
celled superficial layer that bears colourless phaeophycean hairs and simple
or branched assimilators arranged in localized areas. The lower cells of the
assimilators give rise to unilocular or plurilocular (accessory) sporangia that

appear macroscopically as wart-like sori. Zoospores from the unilocular
sporangia germinate to a dwarf ectocarpoid gametophyte that reproduces by
means of plurilocular gametangia. Zoospores from the plurilocular sporangia
give rise to cushion-like protonemal thalli that bear unilocular and plurilocular sporangia (Dangeard, 1968).

SPLACHNIDIACEAE: *Splachnidium* (*Splanchnos*, spleen-like). Fig. 9.17
This is a monotypic genus of considerable phylogenetic interest. The single
species, *S. rugosum*, in found in South America, South Australia, South
Africa and New Zealand. The plants occur scattered intertidally on the rocks
in the lower half of the littoral zone. They are tubular, sparsely branched and
contain a large amount of mucilage. Structurally the plant is multiaxial but
the axes become widely separated and only at the periphery do the lateral

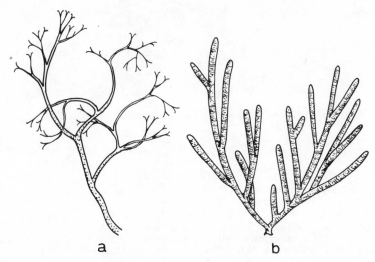

a   b

Fig. 9.17 a, *Stilophora rhizodes*; b, *Splachnidium rugosum*. (a, after Taylor, b, after
Hopkins)

branches proliferate to form a pseudo-parenchymatous cortex. At the apex of
the axis or a branch there is sometimes a cell of the green endophyte *Codiolum
kuckuckii*, and this was formerly regarded as the apical cell. The apical parts
are clothed with short, clavate assimilators that later become shed so that the
surface is then formed by a compact group of small cells. Reproduction takes
place by means of unilocular sporangia that develop in sori with tufts of hairs
in shallow conceptacles. The zoospores after liberation swim about for a
short time and then settle on a substrate where they germinate into a dwarf,
branched, filamentous, ectocarpoid generation that reproduces by means of
plurilocular organs; this is probably the gametophyte (Price and Ducker, 1967).

NOTHEIACEAE: *Notheia* (a spurious thing). Fig. 9.18

The filiform sporophyte of the sole species, *N. anomala*, grows out parasitically from the base of conceptacles of the Fucalean *Hormosira banksii*
(see p. 252) and from its own conceptacles. There is an apical growing zone
of three cells that give rise to filamentous rows of cells. Later divisions in
different planes lead to a pseudo-parenchymatous thallus in which epidermis,
cortex and medullary tissue can be distinguished. The apical filamentous
development is characteristically Chordalian. True branching is rare, such
branches being recognized by the lack of any basal constriction. Most of the

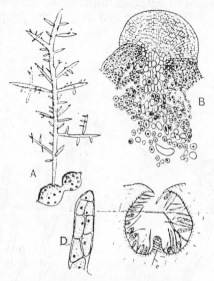

Fig. 9.18 *Notheia anomala.* A, plant growing out from *Hormosira*; B, point of entrance of
parasite into host; C, conceptacle with megasporangia and branch shoot (*s*); D, mature
megasporangia with eight ova. (B, C × 40, D × 180) (A, after Oltmanns; B–D, after Williams)

apparent branches, which are constricted at the base, represent new plants
that have developed from zygotes produced on a filamentous gametophyte
that grows inside the conceptacle. In those portions of *Hormosira* that are
attacked by the parasite the hollow of the vesicle-like internode becomes
filled up by new tissue formed as a result of the stimulation, but the parasite
is apparently unable to attack *Hormosira* unless the host is growing in areas
where it is continuously submerged.

The fertile conceptacles contain either mega- or microsporangia, the
former being much more frequent. Each megasporangium gives rise to eight
pyriform zooids (originally believed to be non-motile ova) that germinate
apparently in the conceptacle to a branched filamentous gametophyte with
erect colourless hairs. Cells of erect branches can each produce a swarmer
which is presumably a female gamete. It seems likely from the construction

of the thallus that some of these germinate inside the conceptacle to give a new macroscopic plant. Others must escape and infect *Hormosira* plants. The microsporangia each give rise to 64 zooids, the fate of which is not known. They could presumably either develop to a dwarf-male gametophyte or else function as a male gamete. These features are of special importance in considering the origin of the Fucales (see p. 338).

SPOROCHNALES

This is a small order characterized by the aggregation of the unilocular sporangia into swollen, oval or elongate receptacles that bear a cluster of hairs at their apex. Structurally they are pseudoparenchymatous with an intercalary, dome-shaped meristem lying beneath the group of hairs that terminates each growing branch. The gametophytes are microscopic plants which are believed to reproduce oogamously.

The chief centre of distribution is in the southern hemisphere where several species of *Sporochnus* occur together with the genera *Bellotia*, *Perithalia*, *Encyothalia* and *Pseudosporochnus*, which are confined to this region. All the species tend to grow in deep water and are only obtained in the castweed or else by diving or dredging. The northern European representative is *Nereia*.

SPOROCHNACEAE: *Sporochnus* (*sporo*, offspring; *chnus*, fine down). Fig. 9.19

The plants are moderately large and grow by means of a horizontal meristem that lies beneath the terminal tuft of assimilatory hairs. Young plants consist of a simple erect thread anchored by rhizoids. The lower part represents a pedicel which plays no further part in development. The upper part forms the primary hair and the cell between the upper and lower parts is the archi-meristem from which the adult thallus is produced by longitudinal divisions in various planes. Threads which grow downwards are produced and coalesce to give the pseudoparenchymatous body. The sporangia are borne laterally on special, branched, fertile threads in the soral region. The gametophytes are ectocarpoid with small antheridia and swollen terminal cells that are believed to be oogonia.

DESMARESTIALES

The plants belonging to this small order are often of large size with leaf-like lateral branches, which terminate in branched uniseriate filaments, though these may also occur laterally in tufts along the branches. The filaments in *Desmarestia* and *Arthrocladia* are deciduous in the late autumn so that the plants have a definite summer and winter aspect. There is regular alternation of generations between the macroscopic diploid sporophytes and microscopic gametophytes. Cells of *Desmarestia* readily liberate small quantities of free sulphuric acid on death, and if collected they should always be kept separate from other material. The various species are restricted to colder waters.

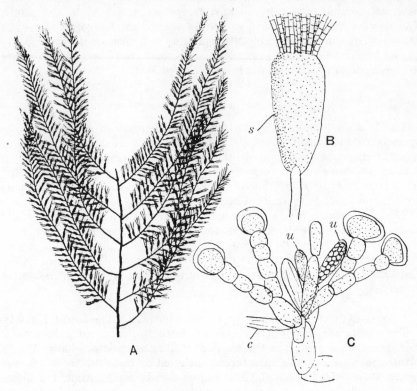

Fig. 9.19 *Sporochnus pedunculatus.* A, plant; B, fertile branch with receptacle (*s* = sorus); C, unilocular sporangia (*u*) (*e* = empty sporangium) (after Oltmanns)

DESMARESTIACEAE: *Desmarestia* (after A. G. Desmarest). Fig. 9.20

The plants are bushy and usually of some size (0·5–1 m), especially the Pacific species. They sometimes bear swellings (galls) which are caused by a copepod. The erect, cylindrical or compressed thallus arises from a disc-like holdfast and exhibits regular pinnate branching, often with a distinct mid-rib. The thallus is composed of a single prominent central row of large cells with an intercalary meristem at the base of the terminal hair. The large central cells, the transverse walls of which are pitted, are surrounded by cortical cells, the primary cortex being derived from the basal cells of the primary laterals. These divide periclinally to give a many-layered cortex, the cells of which are smaller toward the periphery. In the older parts, small hyphal cells are interspersed in the cortex; these originate as a result of secondary activity from the cortical cells.

The sporophytic plants have only unilocular sporangia and these are slightly modified cortical cells. The zoospores escape in a mass and germinate to produce dioecious filamentous gametophytes which are heteromorphic. The smaller male plants produce terminal antheridia from each of which is liberated a single antherozoid; the larger female plants produce swollen oogonia. Each oogonium contains a single ovum which escapes, but as fertilization and germination take place just outside the pore of the oogonium the young sporophyte develops as far as the monosiphonous stage while still

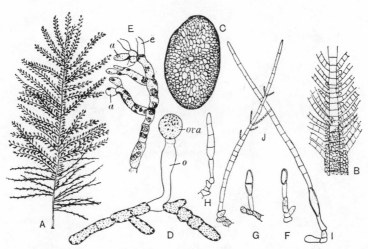

Fig. 9.20 *Desmarestia*. A, plant with summer and winter appearance; B, apex showing cortication; C, t.s. stipe; D, female gametophyte (*o* =oogonium); E, male gametophyte (*a* =antheridium, *e* =empty antheridium); F–J, stages in sporeling germination. (A × ⅓) (A, after Newton; B, C, after Oltmanns; D–J, after Schreiber)

possessing a primitive attachment in the shape of the empty oogonium. Under cultural conditions fertilization of the ovum occurs in the oogonium, the new sporophyte commencing life attached to the gametophyte. Cortication, which is best observed near the apex of old plants, commences in the young plants after a few weeks.

DICTYOSIPHONALES

The members of this order are fundamentally parenchymatous in construction, the thallus either being of a simple or branched radial habit or else a flattened leafy structure with intercalary growth. There is usually a medulla of large cells with a narrow cortical zone of smaller cells on which are borne the assimilatory hairs and sporangia. In some cases (*Asperococcus, Colpomenia*) there is subsequent degeneration of the large internal cells giving a tubular or saccate thallus. The members show heteromorphic alternation, the macroscopic plant being diploid, usually arising as a lateral outgrowth from a dwarf thallus or from swarmers produced in a dwarf thallus.

MYRIOTRICHIACEAE: *Myriotrichia (Myrio,* many; *trichia,* hairs). Fig. 9.21
The minute epiphytic species of the genus *Myriotrichia* are the simplest

H

plants in the Dictyosiphonales. The main axis bears the sporangia, hairs and assimilators, though the last-named are apparently not always present. Uni- and plurilocular structures generally occur on different individuals and this suggests alternation of isomorphic generations, but there is not sufficient cytological evidence to support this view.

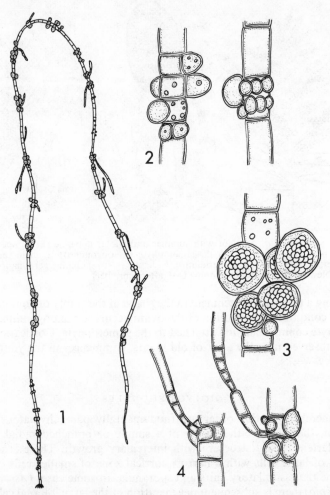

Fig. 9.21 *Myriotrichia adriatica.* 1, habit; 2, young unilocular sporangia; 3, mature unilocular sporangia; 4, hair formation. (1 × 120, 2 × 300, 3 × 400, 4 × 300) (after Lindauer *et al.*)

STRIARIACEAE: *Striaria* (*Stria*, a furrow). Fig. 9.22

Striaria attenuata is a widespread sublittoral epiphyte with a branched cylindrical thallus. The branches are usually opposite and taper to their apex and base but terminate in a hair. The thallus can be regarded as morphologically more advanced than that of *Myriotrichia*. In the old thallus a central cavity arises and this is surrounded by two layers of cells. The plants bear unilocular sporangia associated with unicellular paraphyses and phaeophycean hairs in bands to give the characteristic striations. According to Caram (1964) two types of life history occur in *Striaria attenuata*.

In one case there is an alternation between the normal thallus (with unilocular sporangia) and a filamentous gametophyte with plurilocular gametangia. The life history is therefore of the dimorphic diplohaplontic

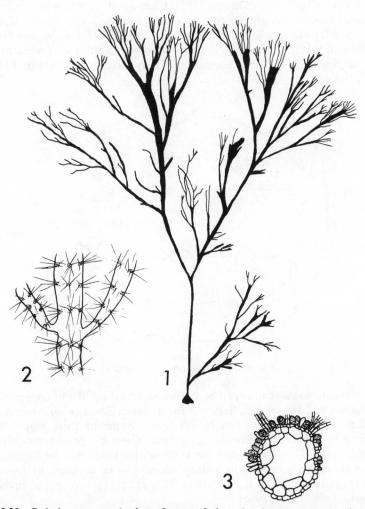

Fig. 9.22   *Striaria attenuata*. 1, plant; 2, part of plant showing arrangement of hairs and sporangia; 3, t.s. of frond with sporangia and hairs. (1 × 1, 2 × 250, 3 × 100) (after Lindauer *et al.*)

type (see p. 345). In the other type of alternation the swarmers from the unilocular sporangia may fuse and develop again into the sporophyte, as does a
monomorphic diplont (Caram, 1965).

PUNCTARIACEAE: *Petalonia* (*Phyllitis, Ilea*). Fig. 9.23

The fronds are unbranched, expanded, membranous, leaf-like structures
with an internal medulla composed of large, colourless cells interspersed with
hyphae, and an outer layer of small, superficial, assimilatory cells. Unilocular
sporangia are not known, but plurilocular sporangia arise from surface cells
and produce swarmers that germinate to give a creeping basal thallus from
which a new plant arises (Caram, 1965). Chen *et al.* (1968) have extended
these observations and shown that *Petalonia* swarmers give rise to *Ralfsia
clavata* (p. 205) whose zoospores in turn give rise to *Petalonia*. Wynne (1969)
has indicated that *Ralfsia californica* may also be the alternate generation of
Petalonia fascia. Light and temperature may determine the life history
(Hsiao, 1970).

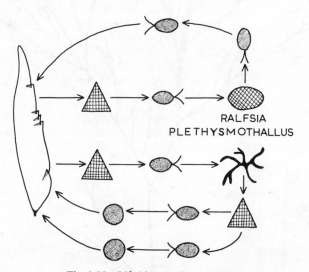

RALFSIA
PLETHYSMOTHALLUS

Fig. 9.23   Life history of *Petalonia.*

The creeping basal thallus can be branched or *ralfsia*-like in appearance.*
In this genus and the related, hollow, tubular genus *Scytosiphon* it seems that
two kinds of dwarf plants can be produced. Some of these dwarf thalli
reproduce by means of unilocular sporangia. These are presumably diploid
plethysmothalli. This means that the macroscopic plant may be haploid. In
Scytosiphon lomentaria the life history appears to be different in Japanese
and European waters and it is possible there are really two species involved
(Tatewaki, 1966; Lund, 1966).

* Iodide concentrations may determine whether blade, protonema, plethysmothallus or
Ralfsia-thallus develops (Hsiao, 1969).

The only other group of brown algae with a similar morphological life history is the Cutleriales. Until fusion of swarmers has been seen it could only be assumption that these '*ralfsia*' plants are sporophytes. It may be that *Petalonia* and *Scytosiphon* need to be removed to a separate order (Nakamura, 1965). In one instance, at least (Wynne, 1969), the '*ralfsia*' stage appears to give rise directly to the *Scytosiphon* form.

Other dwarf thalli reproduce by means of plurilocular organs which could be gametangia, though fusion of swarmers has not been seen. Further work on the life history of these algae is required in order to determine whether reduction division does occur.

DICTYOSIPHONACEAE : *Dictyosiphon* (*dictyo*, net; *siphon*, tube). Fig. 9.24

The plants arise from small lobed discs and have either a few or many branches, the younger ones commonly being clothed with delicate hairs. In young plants there is the usual apical hair with a basal meristem. Later the

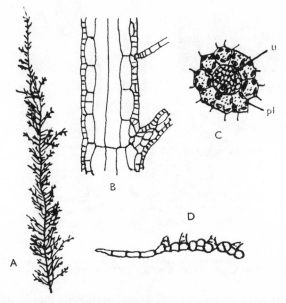

Fig. 9.24 *Dictyosiphon*. A, plant; B, longitudinal section of thallus; C, sporangium and thallus cells in surface view; D, gametophyte. (A, after Oltmanns; B, C, after Murbeck; D, after Sauvageau)

hair is shed leaving an apical growing cell. Subsequently a central medulla with four rows of large elongated cells develops and this may be penetrated by hyphae from the smaller cortical cells; in old plants it is often ruptured and the axis becomes partially hollow. Only unilocular sporangia occur, and the zoospores germinate to form microscopic plantlets; these represent the gametophytic generation and reproduce by means of plurilocular gametangia.

The gametes either develop parthenogenetically into a new plantlet or else two gametes from different gametangia, form a zygote which develops into a small ectocarpoid plant. This may either reproduce itself by means of plurilocular sporangia or else gives rise directly to the adult sporophyte. In *D. chordaria* meiosis appears to have been suppressed as the unilocular zooids give rise to a new sporophyte.

ASPEROCOCCACEAE: *Asperococcus* (*aspero*, rough; *coccus*, berry). Fig. 9.25
The structure of the adult plant is essentially the same as that of the two preceding genera except that the central filaments degenerate and the centre becomes filled with a gas. The fronds are simple or branched and bear small superficial cells with sporangia and mucilage hairs scattered over the surface in sori. Plurilocular and unilocular sporangia occur on the same or on different

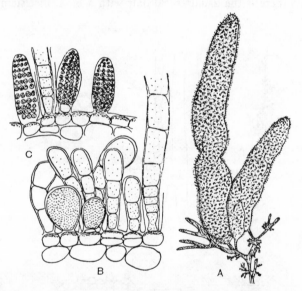

Fig. 9.25 *Asperococcus bullosus.* A, plant; B, unilocular sporangia; C, plurilocular sporangia. (B, C, × 225) (A, after Oltmanns; B, C, after Newton)

plants, the sori with unilocular sporangia containing sterile paraphyses. In *A. compressus* the life cycle is simple, the zoospores from the unilocular sporangia germinating directly into a dwarf phase; this can reproduce itself by swarmers from uni- and plurilocular sporangia until the advent of favourable conditions enables the development of the macroscopic phase to take place once more. Only the diploid generation is therefore known. In *A. fistulosus* meiosis occurs in the unilocular sporangia. If the swarmers fuse, the zygote develops into a 'streblonema' phase, so named after the brown alga it resembles, from which a new adult plant can arise. In this case there is no evidence for the existence of a gametophytic generation, nor has any evidence been obtained to show that such streblonemoid plants can reproduce them-selves. If no fusion of the swarmers takes place in *A. fistulosus* and *A. echinatus*, the '*streblonema*' phase is produced parthenogenetically, but under these circumstances plurilocular sporangia are formed which give rise to a new '*streblonema*' generation. No investigator, under such conditions, has however succeeded in obtaining macroscopic plants again and so it has been suggested that sex has been inhibited in these plants. (Dangeard, 1968).

In *A. bullosus* zooids from plurilocular sporangia on the macroscopic thallus do not fuse but germinate directly to give rise to a series of plantules bearing plurilocular sporangia: these tide over the winter season, and then in spring young *Asperococcus* plants develop in place of the sporangia on the ectocarpoid plantules. In the unilocular sporangia meiosis takes place and zooids develop into minute gametophytic plants that produce plurilocular gametangia. The gametes produced can fuse to produce a zygote that develops into a plantule from which a new macroscopic plant arises, but if there is no fusion then they merely develop into a new gametophytic genera-tion. Sauvageau also reported that the zooids from the unilocular sporangia may give rise to creeping filaments which later produce young plantules as *Asperococcus*. This direct reproduction of the macroscopic plants can only be explained by work on the cytology of the different generations. It will be evident that direct alternation of generations is obscured through the number of possible independent circuits and short-cuts. In *Colpomenia peregrina* (*C. sinuosa* of Japan), a member of a closely allied genus, there are two morphologically similar generations. The dioecious gametophytes appear in spring and reproduce by means of anisogametes that are formed in dissimilar plurilocular gametangia. The zygote gives rise to a new adult sporophyte that reproduces in autumn by means of plurilocular sporangia. This life cycle is somewhat different from the others that have been described and a re-investigation would seem to be desirable, especially in view of the absence of unilocular sporangia, which is the usual place of meiosis in brown algae.

REFERENCES

GENERAL

Cole, K. (1967). *Can. J. Genet. Cytol.*, **9**, 519.

Fritsch, F. E. (1945). *Structure and Reproduction of the Algae*, Vol. **2**, p. 19–191. (Cambridge University Press).

Kylin, H. (1933). *Lunds. Univ. Arsskr. N.F. Avd.*, **2**, 29.

Papenfuss, G. F. (1951). *Manual of Phycology*, edit. by G. M. Smith, p. 119–158. (Chronica Botanica).

Scagel, R. F. (1966). *Oceanogr. Mar. Biol. Ann. Rev.*, **4**, 123.

Wynne, M. (1969). *Univ. Calif. Publs. Bot.*, **50**, 1.

Ultrastructure

Evans, L. V. (1966). *J. Cell Sci.*, **1**, 449.

Manton, I. (1957). *J. exp. Bot.*, **8**, 294.

Manton, I. (1959). *J. exp. Bot.*, **10**, 448.

Chordariales

Christensen, T. (1962). *Br. phycol. Bull.*, **2**, 269.

Kylin, H. (1940). *Lunds. Univ. Arsskr., N.F. Avd.*, **2**, 36.
Cutleriales
Hartmann, M. (1950). *Pubbl. Staz. Zool. Napoli*, **22**, 120.
Desmarestiales
Kornmann, P. (1962). *Helgoländer wiss. Meeresunters.*, **8**, 287.
Dictyosiphonales
Papenfuss, G. F. (1947). *Bull. Torrey bot. Club*, **74**, 398.
Ectocarpales
Russell, G. (1964). *Br. phycol. Bull.*, **2**, 322.
Tilopteridales
Sundene, O. (1966). *Nature, Lond.*, **209**, 937.
Myrionemataceae
Loiseaux, S. (1967). *Revue gén. Bot.*, **74**, 329.
Scytosiphonaceae
Dangeard, P. (1963). *Le Botaniste*, **46**, 1.
Striariaceae
Caram, B. (1966). *C.r. hebd. Séanc. Acad. Sci., Paris*, **262**, 2333.
Asperococcus
Dangeard, P. (1968). *Le Botaniste*, **51**, 59.
Chordaria
Caram, B. (1955). *Bot. Tidsk.*, **52**, 18.
Kornmann, P. (1962). *Helgoländer wiss. Meeresunters.*, **8**, 277.
Sundene, O. (1963). *Nytt Mag. Bot.*, **10**, 159.
Colpomenia
Kunieda, H. and Suto, S. (1938). *Bot. Mag., Tokyo*, **52**, 539.
Dictyota
Vielhaben, V. (1963). *Z. Bot.*, **51**, 156.
Elachista
Blackler, H. and Karpitia, A. (1963). *Trans. Proc. bot. Soc. Edinb.*, **39**, 392.
Ectocarpus
Knight, M. (1929). *Trans. Roy. Soc. Edinb.*, **56**, 307.
Müller, D. G. (1966). *Planta*, **68**, 57.
Müller, D. G. (1968). *Planta*, **81**, 160.
Papenfuss, G. F. (1935). *Bot. Gaz.*, **96**, 421.
Russell, G. (1966). *J. mar. biol. Ass. U.K.*, **46**, 267.
Russell, G. (1967). *J. mar. biol. Ass. U.K.*, **47**, 233.
Schussnig, B. and Kothbauer, E. (1934). *Ost. bot. Z.*, **83**, 81.
Halopteris
Moore, L. B. (1951). *Ann. Bot., N.S.*, **15**, 265.
Leathesia
Dangeard, P. (1965). *Le Botaniste*, **48**, 5.
Myrionema
Loiseaux, S. (1964). *C.r. hebd. Séanc. Acad. Sci., Paris*, **258**, 2383.
Loiseaux, S. (1967). *Rev. gén. Bot.*, **74**, 529.
Myriotrichia
Dangeard, P. (1965). *Le Botaniste*, **49**, 79.
Notheia
Nizamuddin, M. and Womersley, H. B. S. (1960). *Nature, Lond.*, **187**, 673.
Petalonia
Chen, L., Delstein, T. and McLachlan, J. (1968). *Int. Seaweed Symp., Barcelona.*
Dangeard, P. (1963). *Le Botaniste*, **46**, 1.

Dangeard, P. (1964). *Phycologia*, **4**, 15.
Hsiao, S. I. C. (1969). *Can. J. Bot.*, **47**, 1611.
Hsiao, S. I. C. (1970). *Can. J. Bot.*, **48**, 1359.
Nakamura, Y. (1965). *Bot. Mag., Tokyo*, **78**, 109.
Pylaiella
Russell, G. (1963). *J. mar. biol. Ass. U.K.*, **43**, 469.
West, J. A. (1967). *J. Phycol.*, **3**, 150.
Ralfsia
Edelstein, T., Chen, L. and McLachlan, J. (1968). *J. Phycol.*, **4**, 157.
Splachnidium
Price, I. R. and Ducker, S. C. (1967). *Phycologia*, **5**, 261.
Stilophora
Dangeard, P. (1968). *Le Botaniste*, **51**, 95.
Striaria
Caram, B. (1964). *C.r. hebd. Séanc. Acad. Sci., Paris*, **259**, 2495.
Caram, B. (1965). *Vie et Milieu*, **16**, 21.
Scytosiphon
Lund, S. (1966). *Phycologia*, **6**, 67.
Tatewaki, M. (1966). *Phycologia*, **6**, 62.
Scytothamnus
Price, I. R. (1969). *Phycologia*, **8**, 37.
Zonaria
Biddle, L. (1968). *J. Phycol.*, **4**, 298.
Neushul, M. and Biddle, I. (1968). *Am. J. Bot.*, **55**, 1088.

IO

PHAEOPHYTA

PHAEOPHYCEAE

LAMINARIALES

The Laminariales are principally temperate, the bulk of the species being confined to colder waters, and there are, in particular, a number of monotypic genera confined to the North Pacific. The thallus, representing the large, conspicuous sporophytic generation, is nearly always bilaterally symmetrical with an intercalary growing zone, while the gametophytes are microscopic ectocarpoid filaments. The sporophytes reproduce by means of zoospores formed in unilocular sporangia, commonly arranged in sori with paraphyses, while the gametophytes reproduce by means of ova and antherozoids that are borne on separate microscopic plants. The gametophytes only mature under low water temperatures. The number of zoospores produced per sporangium ranges from 16 in *Chorda* to 128 in *Saccorhiza*.

There is considerable variation in habit between the sporophytes of the different genera. They range from the long whip-like thallus of *Chorda* (Fig. 10.1) to the giant *Macrocystis* (Fig. 10.10). Between these extremes are found the sea palm, *Postelsia* and the sea fern, *Thalassiophyllum* (Fig. 10.6). In the northern hemisphere *Laminaria* is a typical representative of the genus and in the southern, *Ecklonia*, while *Macrocystis* is a useful teaching representative in the Pacific and in South Africa.

The genera *Saccorhiza* and *Alaria* are of interest because the fronds possess shallow conceptacles or cryptostomata and resemble those found in the Fucales, those in *Saccorhiza* even possessing the typical tuft of hairs.

There are five families in the order:

(*a*) **Chordaceae**: the plants are long, whip-like and with deciduous hairs. It is a monotypic family.

(*b*) **Laminariaceae**: the species either possess a simple blade or one that is segmented longitudinally, but the splits do not extend into the transition zone so that there is only one intercalary meristem.

(*c*) **Himantothallaceae**: branching is distichous and lateral from the primary stipe. There is a unique internal assimilatory tissue associated with conducting vessels. Reproduction is more like that of *Desmarestia* than *Laminaria*.

(*d*) **Lessoniaceae**: the splits of the primary blade extend into the transition zone so that each segment has a stalk and an intercalary meristem.

(*e*) **Alariaceae**: plants have special lateral sporophylls.

CHORDACEAE: *Chorda* (a string). Fig. 10.1

The long whip-like thallus, which is clothed in summer with mucilage hairs, arises from a small basal disc on the growing region situated just above the holdfast. The hollow fronds are simple with diaphragms at intervals, the thallus being like that of a multiseptate cable whose construction is derived

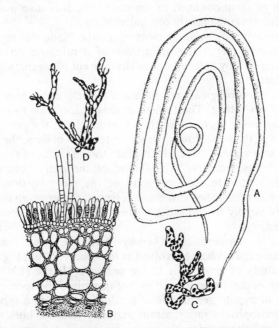

Fig. 10.1 *Chorda filum.* A, plant; B, t.s., highpower, with sporangia; C, female gametophyte; D, male gametophyte. (A × ½, C × 145, D × 175) (B, after Oltmanns; C, D, after Kylin)

from the multiaxial chordarian type by further segmentation of descending hyphae to form a pseudo-parenchyma. The epidermal layer is ultimately clothed with sporangia, paraphyses and deciduous mucilage hairs, while the central cells become elongated and support the filaments that go to form the diaphragm. The zoospores on germination give rise to small filamentous gametophytes, the male plants being composed of small cells, each with two to four chloroplasts, and the female of larger cells with more numerous chloroplasts. The gametangia are borne laterally or terminally on short branches, but the plants do not become fertile for at least three months or more often six months after their formation. After fertilization the zygote remains attached to the wall of the oogonium. The macroscopic plant is an annual.

LAMINARIACEAE: *Laminaria* (a thin plate). Figs. 10.2–10.4

This genus has a very wide distribution in the waters of the north temperate and Arctic zones. The expanded lamina has no mid-rib and is borne on a stipe that arises from a basal holdfast which can vary greatly in form. In *L. angustata* populations may have either ovate or lanceolate laminae though sporelings of the former type frequently disappear at an early stage of their existence. The simplest transition area from stipe to lamina is quite plain, but folds, ribs or callosities are also found in that position, which is the region of intercalary growth. In *Laminaria sinclairii* three types of growth can be recognized, all of them confined to the stipe, and it is also possible to find all three processes taking place in one individual:

(1) The ordinary growth and extension of the blade during the growing season. This hardly merits the description of continuous physiological regeneration given to it by Setchell unless the concept of regeneration is to have a wider significance.

(2) Periodic physiological regeneration which represents the annual formation of a new blade. The transition area bulges, due to new growth in the medulla and inner cortex, and then ruptures from the pressure, to leave the frayed ends of the non-growing outer cortex with collars, the upper one of which rapidly wears away. After the rupture the new cells of the medulla and inner cortex elongate rapidly. The failure of the outer cortex to grow is probably associated with the proximity of the inner cortical cells to the medullary hyphae where they can monopolize all the growing materials, thus cutting off any supply to the outer cortex. There may, of course, be other factors involved.

(3) Restorative regeneration whereby branches arise from wounded surfaces, the same tissues being involved as in process (2). (cf. Fig. 10.2.)

A detailed study of growth in *L. saccharina* by Parke (1948) has shown that in this species the longevity and rate of growth of the sporophyte depend on the season of zygote germination, depth in the sea and type of habitat. Thus winter sporophytes rarely attain maturity and the bulk of the populations in Great Britain originate in the spring, except on sheltered coasts where summer plants thrive equally well. The total life span of this species and also of *L. japonica* rarely exceeds three years but that of some others may be longer. Growth is seasonal, more rapid growth occurring between January and June and slower growth afterwards. In the Barents Sea the blade of *L. digitata* only grows from spring to July whereas the stipe grows all summer. In Norway growth of the new lamina commences in December as the waters are rather warmer than those of the Barents Sea.

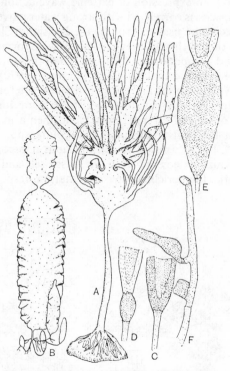

Fig. 10.2 *Laminaria.* A, *L. hyperborea*; B, *L. rodriguezii*; C–E, normal regeneration; C, rupture just commencing; D, E, the new tissues are more heavily shaded; F, wound regeneration. (A–F × ½) (A, B, after Oltmanns; C–F, after Setchell)

The attachment 'roots' or haptera, however, attain their maximum growth in the autumn. In *L. saccharina*, *L. hyperborea* and *L. digitata* the seasonal changes in growth rate are indicated in the stipe by alternate zones or 'rings' of lighter and darker tissue, the darker being formed during the periods of slow growth. In *L. saccharina* the maximum height and greatest growth is achieved during the second rapid growing season. In *L. digitata* four annual growth rings appears to be the maximum but most plants usually have less. In *L. hyperborea* the number of growth rings is usually seven but Scottish plants apparently live longer and for that reason form taller underwater 'forests' than the same species farther south (Kain, 1963).

Making use of transplants and measuring growth by means of holes punched in the lamina, Sundene (1964) has shown that the shape of the frond in *L.*

digitata is greatly influenced by growth conditions, and in particular by temperature. At the period of maximum growth (February-April in South Norway,
March-May in North Norway) the mean growth increment is equivalent to
about 14 cm per 30 days. Growth takes place all over the thallus but the greatest growth occurs in the basal meristematic region (first 10 cm of lamina). The
width of the thallus seems to depend on the degree of protection so that *forma
stenophylla* is simply an expression of extreme wave action. The splitting of
the fronds in this species is correlated with strong water movement.

Many of the species are used as food by the Russians, Chinese and Japanese. Furthermore, these kelps, as they are called, were formerly valuable as a
source of iodine and as a potassium fertilizer. Their present main commercial
use is as a source of the important alginates (Chapman, 1970).

Morphologically both lamina and stipe are divided into four regions. On the
outside there is the actively dividing meristoderm or limiting layer, and
inside there is an outer and an inner cortex and then a mass of interwoven
threads that form the central medulla. The young sporeling is at first a
uniseriate filament, but as a result of cell divisions it becomes a flat monostromatic plate. Rhizoids emerge from the basal cell and the thallus then
becomes distromatic, the periclinal divisions that are responsible starting at

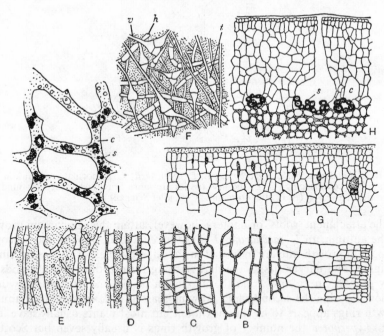

Fig. 10.3 Laminariaceae. A–F, portions of the stipe of *Macrocystis* passing successively
from the epidermis, A, through the outer cortex, B–F, to the medulla. *h* = hypha, *v* = connecting thread, *t* = 'trumpet' hyphae; G, stages in development of mucilage canals, *L. hyperborea*;
H, mucilage canal of *L. hyperborea* in t.s. (*c* = canal, *s* = secretory cells); I, mucilage system
in *L. hyperborea* (*c* = canal, *s* = secretory cells). (after Oltmanns)

the basal end. When the holdfast, stipe and blade have been differentiated, further growth is restricted to the intercalary meristem at the base of the blade. When a distromatic lamina has been produced a central group of cells, which form the primary tubes of the medulla, are cut off and separate the two outside layers. Although the origin of the medulla in the stipe is clear, it is not so for the lamina. The intercalary growing zone forms a ring around the base of the lamina, the central tissues not being meristematic.

Later the primary limiting layer in the stipe is replaced by a secondary meristem which is four to eight cells deep in the cortex. This is responsible for the subsequent growth in thickness by production of secondary annual cortical tissue.

The medulla consists of primary medullary tubes, and two types of lateral connecting branches which are the connecting threads and the hyphae. The former arise first in the course of development as papilloid outgrowths from individual cells. They meet and fuse at their tips and so give rise to a system of cross connections. With further growth the cells divide and elongate, but even when mature they are composed of relatively few cells. The hyphae, which arise later as branches of small cells cut off from the original vertical cells, remain free but may branch. Ultimately they contain numerous cells which subsequently elongate very considerably. One of the most characteristic

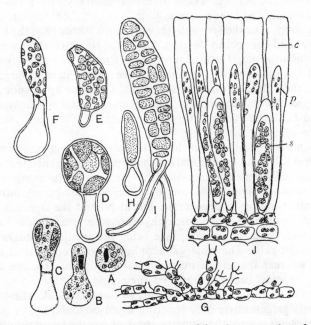

Fig. 10.4 *Laminaria*. A–F, stages in development of female gametophyte from a spore; G, male gametophyte; H, I, first two stages in development of young sporophyte; J, sporangia (*s*), paraphyses (*p*) and mucilage caps (*c*). (A–D × 1333, E–F × 600, G × 533) (A–I, after Kylin; J, after Oltmanns)

features of the genus, and indeed of the order, is the presence in the medulla of 'trumpet-hyphae', which arise as modifications of some of the cells, usually in the connecting threads, more rarely in the hyphae. At a transverse cell wall the ends of both cells swell out to form bulbs, the upper bulb always being larger. The transverse wall is perforated to form a sieve plate, and a callus develops on each side, both callus and sieve plate being traversed by protoplasmic strands. Apart from the sieve plates the trumpet hyphae also possess spiral thickenings which appear as striations. The function of these trumpet hyphae is still unknown. In some species many of the other cells also contain pits with a thin membrane across the opening and these presumably facilitate the transport of storage reserves (Fig. 10.4).

Most of the genera possess systems of anastomosing mucilage ducts lined with secretory cells. In *L. hyperborea* these are confined to the stipe and in *L. digitata* to the frond, but in *L. saccharina* they occur in both. They arise lysigenously through an internal splitting of the thallus caused by cell disintegration; this is followed by a differential growth so that the canals become more and more submerged in the thallus. The haptera have an apical growth and differ from the rest of the thallus in that there are no connecting hyphae nor is there any medulla.

The sporangia and paraphyses are borne in irregular or more or less regular sori on both sides of the lamina. They arise from superficial cells that grow out from the surface. These divide tangentially into a basal cell and paraphysis initial, and later a sporangium is cut off from the basal cell (Fig. 10.4). In *L. saccharina* the sporophytes only reach maturity at the end of their second year, and for sori to develop the distal tissue of the frond must be at least six months old. The actual number of reproducing plants in a population and the duration of the reproductive period depend largely on the numbers of the different season and age groups making up the population. In general *L. saccharina* in Great Britain fruits in summer, *L. hyperborea* in winter, and *L. digitata* from April to November with maxima in spring and autumn.

The zoospores liberated from the unilocular sporangia germinate to form minute gametophytes. On germination they first put out a tube that terminates in a bulbous enlargement into which the contents of the zoospore migrate. There the nucleus divides and one daughter nucleus passes into the tube while the other is said to degenerate, but at present the significance of this phenomenon is obscure. The gametophytes show a great deal of variation in shape and size, the male gametophyte being the smaller throughout as it is built of smaller cells. The male gametophyte always consists of more than three cells whereas the female may consist of only one cell, the oogonium; both however may become branched filaments. The antheridia arise terminally or intercalarily and produce only one antherozoid, whereas any cell of the female gametophyte may function as an oogonium. The male gametophyte degenerates after the gametes are shed but the fertilized oogonium is normally retained on the female gametophyte which persists during the early growth stage of the sporophyte.

Reproductive organs are only formed at low temperatures, such as,

2° C–6° C, and above 12° C–16° C they are rarely produced; this fact accounts for their temperate and arctic distribution. It is also known that the eggs may develop parthenogenetically to give a haploid sporophyte of irregular shape.

Kain (1964) has shown that in the first few days saturating light only just enables photosynthesis to balance respiration. Saturation light varies with temperature, being 35 μg cals/cm²/sec. at 10° C and 100–200 μg cals/cm²/sec. at 17° C. At 10° C in saturating light gametophytes of *L. saccharina* mature in 8–9 days, those of *L. hyperborea* in 10 days and those of *L. digitata* in 13–14 days (Kain 1969).

It has been possible to produce hybrids between *L. japonica* and *L. religiosa* or *L. ochotensis* and also between *L. religiosa* and *L. ochotensis*. Crosses between other species produced abnormal sporophytes of types not found in nature. It has also been possible to produce hybrids between *Laminaria* and *Alaria* (Yabu, 1965).

The young sporophyte first produces numerous rhizoids of limited growth, but these are later covered by a disc-shaped expansion. In older plants this disc is replaced by the haptera which arise as papillate outgrowths from the stipe.

Saccorhiza polyschides differs from *Laminaria* in several important respects:

The persistent lamina arises from a flat compressed stipe with wavy edges which, as a result of unequal growth, is twisted through an angle of 180° near

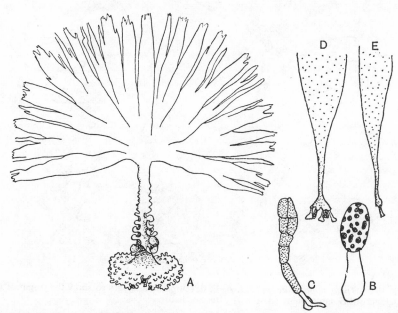

Fig. 10.5 *Saccorhiza polyschides*. A, plant; B, female gametophyte; C, young sporophyte; D, E, young plants of *S. dermatodea* to show origin of bulb (A × ⅓) (A, after Tilden; B, C, after Kniep; D, E, after Oltmanns)

...he base. It differs from *Laminaria* in lacking cortical mucilage ducts and sieve-tube like elements (Norton and Burrows, 1969). The young sporophyte is attached at first by a small cushion-like disc, but later a warty expansion, the rhizogen, develops above it and forms a bulbous outgrowth which bends over and attaches itself to the substrate by means of haptera. As a result of the development of this adult holdfast the juvenile disc may be lifted completely off the substratum (Fig. 10.5).

The development of the sporophyte to maturity in both known species requires only one year so that the plants are mostly true annuals. Some plants occasionally overwinter, and the bulb certainly does (Norton and Burrows, 1969). *Saccorhiza polyschides* is found on the Atlantic coasts of North and West Europe whereas *S. dermatodea* is circumpolar.

In the Pacific genus, *Thalassiophyllum* (Fig. 10.6), the perennial sporophyte is apparently composed of a spirally twisted, fan-shaped lamina unrolling

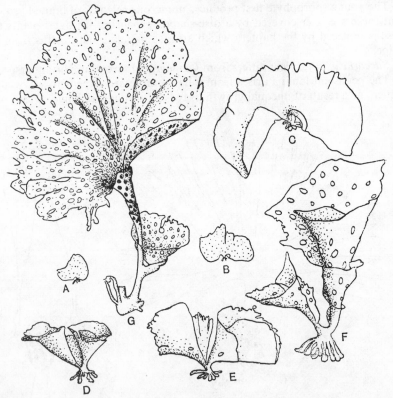

Fig. 10.6 *Thalassiophyllum clathrus*. A–F, developmental stages to show the origin of the single scroll; G, adult plant. (A–F × ⅔) (after Setchell)

from a one-sided scroll without any mid-rib. A study of the embryonal stages, however, shows that the young plant is flat and bilaterally symmetrical. The two edges then curl up and the plant tears down the centre giving rise to two lateral scrolls each unrolling from a thickened outer margin; one of the scrolls soon ceases to develop and the mature plant only possesses one scroll borne on a solid bifid stipe with the vestigial scroll on one of the branches. Slitting is represented by rows of small holes which start to develop when the first tear has taken place.

HIMANTOTHALLACEAE: *Phyllogigas* (*phyllos*, leaf; *gigas*, giant). Fig. 10.7

This is a monotypic family with but a single genus and species, *P. grandifolius* (Skottsberg and Neushul, 1960), though other workers have segregated the different growth forms into separate species. It is a sublittoral alga of the Antarctic continent. In young plants the branching is from the primary stipe and is distichous. In older plants only stumps of branches may be left. As the plants age an abundance of hapteres are produced from the basal part of both stipe branches so that a complex rhizome-like body is formed. When only two

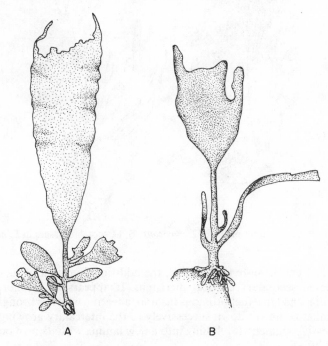

A   B

Fig. 10.7 *Phyllogigas grandifolius* (Gepp.). A, young plant; B, somewhat older plant. (A, B, × ½) (after Skottsberg)

branches of almost equal size are left, branching then appears to be dichotomous. In many plants the stipe is spirally twisted. Growth is by means of an intercalary growth zone and there is also a superficial meristem. The genus differs from other members of the Laminariales in that there are large,

elongated, internal conducting vessels each surrounded by a sheath of assimilatory cells filled with plastids. The reproductive bodies are also distinct. They
comprise unilocular sporangia with a corona-like structure at the apex,
mixed with sterile, two-celled paraphyses, the whole being part of a modified
epidermis. In this respect the plant shows a resemblance to *Desmarestia*
(p. 210). Ultimately, as Neushul (1963) suggests, it may be necessary to place
this genus in a separate order.

LESSONIACEAE: *Lessonia* (after R. P. Lesson) Fig. 10.8
   This family contains some of the largest known algae. Species of *Lessonia*
form 'forests' in relatively deep waters off the shores bounding the southern
Pacific. The stipe of *L. flavicans* is extremely stout and rigid, 2–3 m. long and

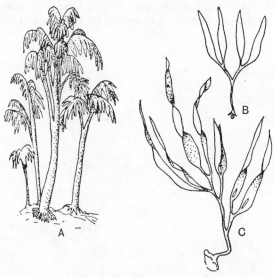

Fig. 10.8 *Lessonia*. A, adult plants of *L. flavicans*; B, C, sporeling stages in *L. nigrescens*.
(after Oltmanns)

sometimes 15 cm or more in diameter, the additional growth being due to a
deeper seated secondary cortical meristem. It appears to be more-or-less
regularly branched in a dichotomous fashion, a feature which is brought about
by the lamina being slit down successively to the intercalary growing region,
each successive segment developing into a new lamina with its own portion of
stipe (Fig. 10.8).
   Two interesting genera are *Nereocystis* (Fig. 10.9) and *Pelagophycus*. Both
arise from small holdfasts but attain stipe lengths from 15 m (*Pelagophycus*)
to 50 m (*Nereocystis*). The upper 2 to 3 m of the stipe is expanded and
hollow, forming a long thick-walled (1–2 cm) pneumatocyst that terminates
in a bulbous sphere. In *Nereocystis* the maximum diameter is about 15 cm
but in large *Pelagophycus porra* diameters of 30 cm may be reached. The total

volume may approach 4 dm³ and the gas mixture inside is less than atmospheric. The gas contains oxygen, nitrogen, carbon dioxide and up to 12%
carbon monoxide (usually 1–5%). The top of the pneumatocyst subtends one
(*Pelagophycus*) or four (*Nereocystis*) extensions of the stipe. These dichoto-
mise in a pattern characteristic for each genus, one branch of each dichotomy
bearing one blade. The blades in *Pelagophycus* are very large, up to 10 m long
and 1·5–3 m wide. The blades of *Nereocystis* are more numerous (up to 64
in number) and smaller. The characteristic branching pattern accounts for the
names Elk-horn Kelp (*Pelagophycus*) and ribbon Kelp or sea-otters cabbage
(*Nereocystis*).

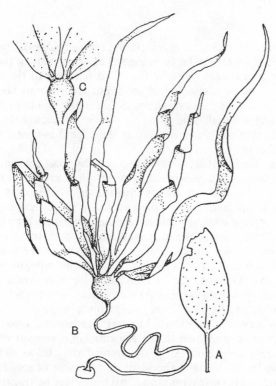

Fig. 10.9 *Nereocystis luetkeana*. A, young plant; B, mature plant; C, branching from
bladder. (after Oltmanns)

The three known species of *Pelagophycus* represent a morphological series
(Table 10.1) and Parker & Bleck (1966) have argued that *P. gigantea* is an
autopolyploid of *P. porra*, while *P. intermedia* is a hybrid.

A third fossil genus *Julescraneia* has recently been described (Parker &
Dawson, 1965) which is morphologically intermediate between *Nereocystis*
and *Pelagophycus*.

TABLE 10.1

MORPHOLOGICAL CHARACTERISTICS OF *Nereocystis*

AND *Pelagophycus* SPP.

	Nereocystis luetkeana	*Pelagophycus porra*	*Pelagophycus intermedia*	*Pelagophycus gigantea*
Stipe length	36 m	27 m	10 m	3 m
No. of Blades	32–64	12–21	5–8	6–10
Size of Blades (Average)	0·17 × 4 m	0·4 × 6 m	0·4 × 8 m	0·4 × 10 m

LESSONIACEAE: *Macrocystis* (*macro*, large; *cystis*, bladder). Fig. 10.10

The perennial fronds may be 60 m long. The upper part of the plant floats on the water surface, kept there by the basal bladders of the leaves. In the juvenile plant the stipe is simple and solid, but later it branches one to three times in a dichotomous fashion, although ultimately the branching becomes unilateral and sympodial. The growing region on each branch is basally situated in the terminal blade, and along the edge it is continuous with the stipe, and it is in the former area that splitting takes place to form the individual laminae. The splitting is brought about by local gelatinization of the inner and middle cortex together with a cessation of growth in the epidermal area; this forces the adjacent tissues into the gelatinized areas until finally the epidermis is ruptured. One of the two segments formed in this splitting process stops growing while the other, the outer, continues growth.

The internal structure of the stipe and fronds, with the exception of the 'sieve' tubes (see below), is typical of the order as described under *Laminaria* (see p. 226). The rate of growth of this plant, during the growing season, is quite spectacular. On the Pacific coast of North America the terminal portion grows at an average rate of 7·1 cm per day, which works out at about 1–1¼ blades daily. At a depth of 20 m whole fronds can grow as much as 45 cm per day, the most rapid plant growth known (Clendenning, 1964). The amount of organic matter accumulated in the growing tip is equivalent to that produced and translocated from 35 blades immediately below. There are very many more blades than this in the plant during most of its life, so that the remaining surplus of photosynthesized material must be translocated to the stipe and haptera and stored as reserve. Tagging experiments have shown that the average age of an individual frond with its blades is about six months, so that the rate of turnover is considerable.

Macrocystis and *Nereocystis* possess 'sieve' tubes in the cortex as well as trumpet hyphae in the medulla. The 'sieve' tube callus has been found to resemble callose from the sieve plates of vascular land plants. The plates are penetrated by strands of material and there must be some kind of mass flow as Parker (1965) has shown that the rate of movement of photosynthate is

65–78 cm per hour. Neighbouring tubes are connected to one another by lateral filaments, the cross walls of which are also perforated.

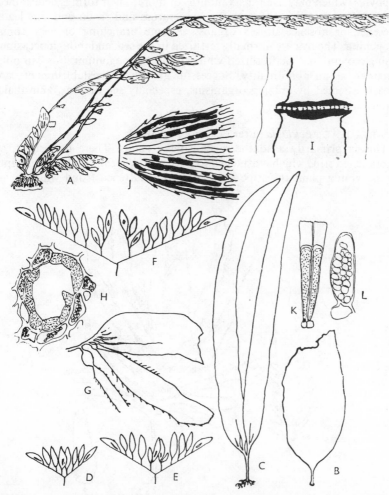

Fig. 10.10 *Macrocystis pyrifera*. A, entire plant; B, young plant; C, slightly older plant with primary slit and two secondaries; D–F, schematic drawings showing branching; G, shoot apex; H, mucilage duct; I, 'sieve tube' and plate with callus; J, part of fertile blade (sori black); K, two paraphyses; L, unilocular sporangium. (A × $\frac{1}{150}$, B × $\frac{1}{2}$, C × $\frac{1}{8}$, D–G × $\frac{1}{3}$, H × 200, I, L × 290, J × $\frac{1}{4}$, K × 140) (A, B, D–L, after Skottsberg; C, after Oltmanns)

Reproduction is by means of zoospores produced on sporophylls which are generally dichotomously branched and with or without a bulb, near the base of the plant. Although two kinds of zoospore have been recorded, recent work has failed to substantiate this finding. Enormous numbers of zoospores are produced, the actual number being related to the size of the frond. Each sporangium produces 32 zoospores and on a large frond as many as 700,000 zoospores may be liberated per square millimetre of frond (both surfaces). On germination the zoospores give rise to the characteristic laminarian game-tophytes, which may live for six months or more. Their form is determined by nutrient supply, bacterial growth, temperature and so on and this leads to presence or absence of sex organs, extreme branching or very compact branching. The ova are normally fertilized in the sea, and though occasionally a single ovum may be fertilized while still in the oogonium, this can only be regarded as an abnormality. Successful growth and establishment can be greatly affected by grazing organisms, especially sea urchins, and abalones (*Haliotis*).

ALARIACEAE: *Ecklonia* (after Ecklon). Fig. 10.11

This is primarily a southern-hemisphere genus. There is a short, solid unbranched stipe which is attached to the rocks by means of the usual haptera. In the young plant the stipe expands into a flattened frond which in most

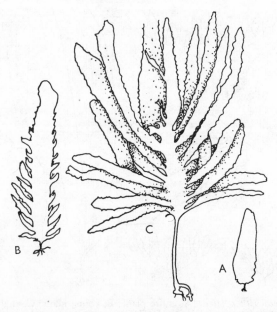

Fig. 10.11 *Ecklonia radiata*, young, juvenile and adult plants (× ¼). (after Bergquist)

species is initially broadly dentate. Later the terminal portion disappears and the basal portion produces numerous lateral outgrowths or sporophylls. In its internal morphology the structure is comparable to that described for *Laminaria*. There is evidence that the shape of the frond in *E. radiata* is determined by the degree of water movement so that var. *richardiana*, which only partially leaves the juvenile condition, is an ecological form associated with strong water movement. The genus tolerates rather higher sea temperatures than do other genera of the Laminariales and is therefore found at the north of New Zealand; *E. kurome* also penetrates south into the warmer waters of Japan. Hybrids have been successfully produced between this species and species of *Laminaria*.

In the northern hemisphere the Alariaceae is represented by the widely distributed genus *Alaria* (Fig. 10.12). There is a short, solid, unbranched stipe, naked below, with an intercalary growing zone that allows for con-

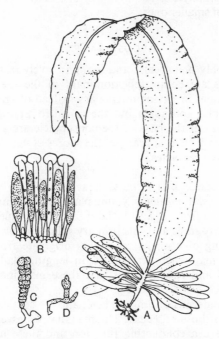

Fig. 10.12 *Alaria esculenta*. A, plant of *A. oblonga* with sporophylls; B, sporangia and paraphyses; C, germinating sporophyte; D, female gametophyte. (B × 200, C × 100, D × 80) (A, after Oltmanns; B–D, after Newton)

tinual renewal, while above the growing region the stipe expands into a flattened rachis which each year bears a fresh crop of marginal rows of sporophylls. The frond terminates in an expanded, sterile lamina which has a well-marked mid-rib and is produced annually. In addition to intercalary growth there is also marginal growth that imparts a wavy appearance to the terminal frond. The sterile frond bears so-called cryptostomata, although these are barely more than tufts of hairs arising in slight depressions. Growth of *A. esculenta* is not good above a temperature of 16° C and the sporophyte dies above 20° C. The distribution of this species is therefore restricted to waters with a mean August sea-temperature of less than 16° C. Below this temperature, growth of the blade is also reduced if the salinity is lowered. The plant does not therefore normally occur around estuaries.

The gametophytes are protonemal in form, simple or sparingly branched.

The ovum is fertilized on emergence from the pore of the oogonium and the young sporophyte develops *in situ* without the characteristic early appearance of a holdfast.

An extension of the *Alaria*-type can be seen in the North Pacific American genus, *Egregia*. Each branch is strap-shaped and bears three types of outgrowths: ligulate sterile leaflets, small fertile leaflets, and conspicuous stipitate bladders.

The female gametophyte is composed of one or two large cells, while the male is composed of smaller ones.

ASCOSEIRALES

This order has only one species and is found solely in the Antarctic. The macroscopic plants are completely dominant in the life history but they differ from the Fucales in having intercalary instead of apical growth. There are no true unilocular sporangia but the so-called 'sporangial' cells in the surface of the thallus at the base of the conceptacle are at present regarded as homologous to unilocular sporangia, the spores of which develop *in situ* to produce the fertile conceptacles.

ASCOSEIRACEAE: *Ascoseira* (*Ascus*, sack; *seira*, chain). Fig. 10.13

In the single species, *A. mirabilis*, young plants have a single blade and there is a short stipe arising from a pad-like hold-fast. The single blade splits down the middle at an early stage and this primary split is followed by more or less symmetrical splitting of both halves. Splits originate from the tips of the blades. Old plants may reach nearly 2 m. in length. Blades of large plants show constrictions and it is thought that these may indicate seasonal regeneration.

Structually the epidermis or limiting layer and the cortical cells beneath are very comparable to those of *Fucus*. The medulla possesses elongated tubes, which probably have a conducting function and these are surrounded by thin-walled elongated cells and hyphae which can be united by connecting branches, very much as in the medulla of *Laminaria*. The conducting cells are surrounded by a ring of small assimilatory cells containing brown plastids. In this respect *Ascoseira* is like *Phyllogigas* (see p. 229).

The laminae have the same structure as the stipe, but in older, well-developed plants there are, in addition, conceptacles characteristic of the Fucales and comparable to those found in *Notheia* (p. 208) and *Splachnidium* (p. 207). Each conceptacle has a narrow canal which opens to the surface and in the early stages is filled with hairs that have developed from the floor. Eventually chains of cells arise from special cells, 'sporangial cells', in the boundary walls, and when ripe each cell of the chain divides to give eight smaller cells, so that chains of eight-celled bodies are formed. The 'sporangial cells' that give rise to the fertile chains could be interpreted as unilocular sporangia that germinate *in situ* to produce the fertile chains. The exact nature of the fertile bodies remains unknown. In very old plants white patches can be seen on the thallus and these contain what appear to be unilocular sporangia that give rise to discrete bodies. This may be some form of accessory reproduction.

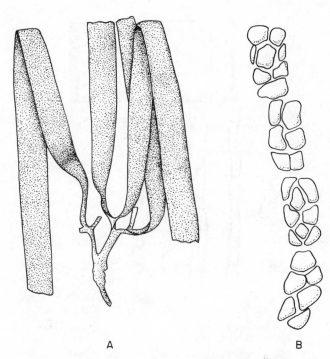

A   B

Fig. 10.13 *Ascoseira mirabilis.* A, plant; B, row of reproductive organs divided into fertile bodies. (A × ⅓, B × 1200) (after Skottsberg)

DURVILLEALES

This order was recently established by Petrov to include the genus *Durvillea* which differs from Fucales in lacking an apical growing cell, in having endogenous conceptacles and in having oogonia borne on branched filaments. The embryo produces only a single rhizoid and lacks apical hairs and an apical groove (Nizamuddin, 1968).

DURVILLEACEAE: *Durvillea* (after J. D. D'Urville). Fig. 10.14

The diploid plant is dark olive-brown or black in colour and looks very much like a *Laminaria*. This applies to all four species (five if *Sarcophycus potatorum* of Tasmania is included). The large, solid stipe arises from a scutate holdfast and very soon passes into a flat, expanded, fan-shaped lamina, which later splits into segments although no definite appendages are produced from the frond. The New Zealand *D. willana* is characterized, however, by lateral fronds proliferating from the stout stipe. The attachment disc bears a resemblance to the primary disc found in some members of the Laminariales, and it is formed in much the same way by tangential divisions of the outer layer or meristoderm. The growing region is at the margin. In-

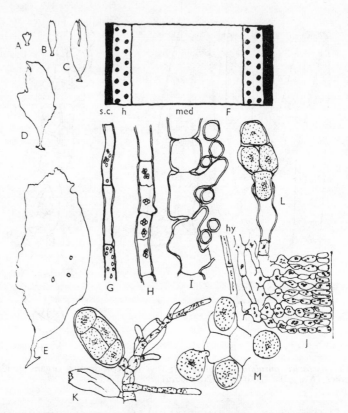

Fig. 10.14 *Durvillea antarctica.* A–E, young plants; F, diagram of tissues: black = secondary cortex, dots = unswollen hyphae, plain = medulla; G–I, stages in swelling of longitudinal hyphae; J, R.L.S. of meristoderm and cortex (hy = hypha); K, oogonium; L, M, oogonium liberating ova. (A–E × 1⅖, G, H × 260, I, × 160, J × 170, K–M × 180) (after Naylor)

ternally the thallus possesses small hyphae which anastomose and fuse with one another just as in the medulla of *Laminaria* (see p. 226). These hyphae extend right up to the apex, thus differing from genera such as *Fucus*. There is also a mechanism for splitting the thallus as there are thallus-splitting mechanisms in the Laminariales. Before splitting, new cortical tissue is formed.

In the most widespread species, *D. antarctica*, the adult lamina is characterized by large, internal air spaces separated from one another by septa. These air spaces are thought to result from an internal splitting process which may or may not be associated with degeneration of the tissues. In this genus, growth is localized at the tips of the laminae, the surface layer forming the primary meristematic layer.

The ends of the older laminae become frayed and broken off by wave action, while the holdfast may attain a diameter of 60 cm through the annual addition of new tissue. If this secondary growth did not occur the plant would soon be torn from its moorings because the holdfast is continuously riddled with holes made by boring molluscs. The oogonia and antheridia, which are borne in conceptacles on different plants occur over the whole of the lamina. The oogonia (macrosporangia) are usually borne on branched hairs, whereas in all other genera they are borne directly on the wall. This condition in *Durvillea* is comparable to the antheridial structure generally in the Fucales and must be regarded as the primitive condition. The ripe oogonium contains four ova. The plant is known as the 'bull kelp' and forms submarine forests in deep waters off New Zealand, the Antarctic islands and the southern part of South America.

FUCALES

The diploid plants are wholly dominant in the life cycle but although diploid there is no apparent asexual reproduction as the plants always reproduce by means of ova and antherozoids. In most genera it is likely that the plants exist over more than one season: this has been clearly established for species of *Fucus*, *Ascophyllum*, and *Marginariella urvilliana*. There is considerable tissue differentiation, and in their external features the plants exhibit much more variation than is to be found in the Laminariales. The thallus is normally differentiated into a discoid holdfast, stipe and frond, and growth in length is due to the activity of one or more apical cells. The genus *Bifurcaria* is anomalous in that the discoid holdfast is replaced by a basal rhizome.

Morphologically, an external meristoderm, cortex and inner medulla can be recognized. In the genera commonly occurring in the intertidal regions the walls of the central or medullary cells become heavily gelatinized and there is a greater profusion of longitudinally running hyphae; *Hormosira* is an apparent exception to this in that the swollen segments become hollow. In the genus *Halidrys* sieve plates occur on the walls of the medullary cells and there is an indication of similar plates in the lateral medullary connections of *Bifurcaria*. Scattered over the surface are minute cavities, the *cryptostomata*, which bear sterile hairs or paraphyses.

Reproduction is by means of oogonia and antheridia borne in conceptacles, which are derived from enlarged cryptostomata and are localized in regions known as receptacles. Some workers consider that the structures called oogonia and antheridia are really macro- and microsporangia producing mega- and microspores which germinate before they are liberated from the sporangium, so that while the reproductive bodies originate as spores, nevertheless the liberated products are gametes. This view is significant on a phylogenetic basis. In the primitive condition eight ova are produced in each oogonium and 64 antherozoids in each antheridium. Meiosis takes place during the first two divisions in the formation of microspores, and as there is often a pause after the second division the first four nuclei have been regarded as the functional microspores, each of which subsequently undergoes four mitotic divisions so that they can be said to germinate to a 16-celled gametophyte where each cell functions as an antherozoid. In the macrosporangium the first four nuclei formed have been regarded as the functional megaspores, and each of these is considered to germinate subsequently to a two-celled female gametophyte where each cell functions as an ovum. In those species where less than eight mature ova are produced it must be assumed that some of the megaspores abort or else do not develop.

If this is the correct interpretation, and it would seem to be more satisfactory than any other theory in comparison with other members of the Phaeophyceae, then not only is there a cytological alternation of generations but there is also a morphological alternation, although the sexual generation is even further reduced from that found in the Laminariales. This would really form a basis for placing the Fucales in the Heterogeneratae. The alternative interpretation is that the sexual generation has been completely suppressed and is solely represented by the gametes, so that while there is a cytological alternation of generations there is only one morphological generation (cf. Chapter 14).

The flask-shaped depressions of the thallus (conceptacles), are lined with sterile hairs or paraphyses, and open to the surface by means of an ostiole. The plants of the different species may be dioecious or monoecious. In a number of the fucoids the ova, when they are extruded through the ostiole, remain attached by means of a thin gelatinous stalk. These stalk-forming fucoids, e.g. *Scytothalia, Marginariella, Sargassum, Bifurcaria*, have been studied in some detail and they have certain features in common:

(1) The stalk is formed from the mesochiton or middle wall of the oogonium (see p. 245); (2) they are inhabitants of deep water or pools near low water of spring tides (*Bifurcaria brassicaeformis* of South Africa is an exception); (3) the conceptacles are unisexual; (4) they have only one ovum per oogonium (*Bifurcariopsis capensis* is an exception); (5) maturation takes place at a late stage; (6) the fertilized ovum begins development within the jelly of the attachment stalk. In addition, those belonging to the family Fucacae (*Marginariella*) have the following features: (7) Conceptacles are borne on specialized laterals; (8) the gelatinous stalks are hollow; (9) they are endemic genera or species, of restricted distribution.

The number of primary rhizoids in the embryo depends first on the species and also on the size of the rhizoidal cell, which in turn bears a relation firstly to the size of the egg, and secondly to the complexity of the thallus. On this basis a series of increasing embryonal complexity may be traced, e.g. *Fucus — Ascophyllum — Pelvetia — Cystoseira — Sargassum*. The eight-egg condition has generally been regarded as the more primitive. On the other hand a study of the antherozoids, using the electron microscope (Manton, 1964) suggests that *Cystoseira* is more primitive than Fucus. Views on evolution within the order will obviously need re-examining. The Fucales are classified into four groups, the classification being based primarily upon the structure of the apical growing cell or cells:

(1) **Fuco-Ascophyllae**. Growth is determined in the adult stage by one four-sided apical cell. Recently it has been found that in *Marginariella urvilliana* the apical growing cell is three-sided in the adult condition. The apical growing cell of the salt marsh fucoids (see p. 373) has also been shown to persist in the three-sided condition. Despite these exceptions, however, the basis of the classification remains very useful.

(2) **Loriformes** (Himanthaliaceae): Growth is due to *one three-sided* apical cell which gives rise to a long and repeatedly forked, whip-like thallus.

(3) **Cystoseiro-Sargassaceae**: The apical cell is again *three-sided* but there is copious branching which results in bilateral, radial and bilaterally-radial thalli.

(4) **Anomalae**: Composed of one genus, *Hormosira*, which is confined to the Antipodes. Growth is brought about by a group of cells, not a single cell.

In most of the Fucales the apical cell is sunk in an apical pit or groove. In the Fucaceae proper, this groove is usually parallel to the plane of flattening of the thallus, but in *Halidrys* and *Himanthalia* it is at right angles to it. In the adult the hairs in the young apical pit are deciduous, but in *Himanthalia*, *Bifurcaria* and *Xiphophora* hairs can be found growing out from the apical groove of mature plants.

FUCO-ASCOPHYLLEAE

FUCACEAE: *Fucus* (a seaweed). Figs. 10.15–10.17

This genus contains a number of species that are widely scattered in the northern hemisphere, many of them exhibiting a wide range of form with numerous so-called varieties. When two or more species occur in the same area they are generally present in different zones on the shore, probably dependent on the degree of desiccation that they can tolerate (cf. p. 426). The plants are attached by means of a basal disc and there is usually a short stalk, which continues to form the mid-rib of the frond in those regions where the expanded wings or alae are developed, these latter being of varying width with either entire or serrate margins. Branching is generally dichotomous and the degree of branching increases as the rate of growth decreases. In many species the branches bear swollen vesicles or *pneumatocysts*. In at least one species, *F. vesiculosus*, these are known to be seasonal in formation, but the nature of the habitat is also involved as more are produced in quiet than in rough waters. Their production appears to be linked with high photosynthetic activity in the spring and summer as the proportions of contained oxygen and nitrogen vary accordingly; carbon dioxide seems to be absent. In at least one species, *F. serratus*, the overall growth rate not only varies with season but also with the degree of shelter.

F. ceranoides will not grow in pure seawater and this species is therefore restricted to the brackish water of estuaries.

With increasing age the lower portions of the algae may be frayed off by wave action, leaving only the mid-rib, which then has the appearance of a stipe. The whole of the expanded thallus is covered with sterile pits or cryptostomata similar to those of *Saccorhiza*, but in fruiting plants it is only the ends of the branches (receptacles) that become swollen and studded with the fertile conceptacles. In *F. spiralis* these conceptacles are hermaphrodite, containing both oogonia and antheridia; in *F. vesiculosus* and *F. serratus* the plants are dioecious, the two types of sex organs occurring on separate plants, while in *F. ceranoides* either state may be found. It has recently been shown that the auxin content of male receptacles of *Fucus* is higher than in female

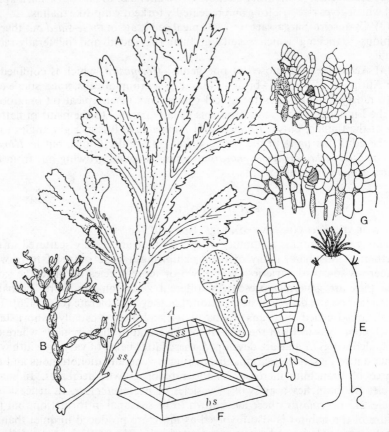

Fig. 10.15 *Fucus*. A, plant of *F. serratus*; B, a marsh form of *F. vesiculosus*; C-E, seedling stages of *F. vesiculosus* showing origin of rhizoids and apical tuft of hairs; F, diagram to show method of segmentation of apical cell, *A* (*bs*=basal segment, *ss*=side segments); G, apical cell of young thallus; H, apical cell of old thallus. (A, B, ×0·30) (A, B, after Taylor; C–H, after Oltmanns)

receptacles. In *F. vesiculosus* the reproduction peak occurs in spring and summer, but in *F. serratus* in autumn and winter. This may possibly be related, as in *Laminaria*, to differences in germination time and survival of plants (see p. 227), though no such evidence has been reported. The mortality rate of sporelings is extremely high in the early stages of growth (this also may possibly be seasonal), largely due to limpets. A number of very peculiar forms have been found which commonly occur on salt marshes: these rarely fruit, reproduction being secured by means of vegetative proliferations (cf. p. 373). Other forms suggestive of hybridization occur on rocky coasts, as well as probable hybrids between *F. vesiculosus* and the brackish-water species, *F. ceranoides*. The age of *Fucus* plants has not been studied in much detail but the following figures (Table 10.2) may be cited from the results of tagging plants.

TABLE 10.2

LIFE-SPAN OF FUCOID THALLUSES

Species	F. spiralis	F. serratus	F. vesiculosus	Ascophyllum nodosum
Max. age (yrs.)	4	4	2–1/2	19 (12–13 in Norway
Av. age (yrs.)	1–1/2	2	1	3–4

Recent work from Russia suggests that the maximum age of *Fucus* plants may be rather greater than that given above. In the Barents Sea, *F. vesiculosus* plants reach their maximum size at the greatest depth for the species (3 m.)

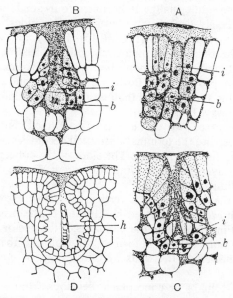

Fig. 10.16 *Fucus.* A–C, origin of conceptacles in *F. serratus* (*b* =basal cell, *i* =initial); D, juvenile conceptacle of *Cystoseira* (*h* =hair). (after Oltmanns)

Morphologically the primary thallus is built up by the activity of the apical growing cell (see below) and by the surface layer of cells, the limiting layer or meristoderm. Below the limiting layer is a cortex composed of several layers of parenchymatous cells which become more and more elongate and mucilaginous towards the centre, and these probably form the storage system. In the very centre the cells are extended into hyphae which are interwoven into a loose tangled web, but they do not extend right up to the apex. This central tissue (medulla) probably acts as a conducting system, because the transverse walls of the hyphae are frequently perforated with the same type of pit that is to be found in some of the Laminariaceae. The primary medullary hyphae are relatively thin-walled, but when secondary growth of the thallus takes place the new hyphae which result are very thick-walled and are probably mechanical in function. Secondary growth is due to the activity of the limiting layer and the inner cells of the cortex, which also forms the secondary hyphae (cf. Fig. 10.16), and these penetrate between the primary medullary hyphae and finally outnumber them. There is a greater development of secondary thickening in the stipe and mid-rib than there is in the frond, while in very old parts of the thallus the limiting layer may die off and then the underlying cortical cells take over its function. Development of secondary hyphae is most pronounced near the basal disc, which ultimately consists of a mass of hyphae.

Growth in length takes place by means of an apical cell which lies at the bottom of a slit-like depression that has resulted from the more rapid growth of the surrounding limiting layer. The apical cell is three-sided in young plants while in the adult thallus it becomes four-sided, the new segments being cut off successively from the base and four sides, after which they develop into the various tissues (Fig. 10.15). Injury, and also the stimulus provided when the thallus lies on marsh soil, induces new growth in the neighbouring cells, and in this way proliferations are formed which may also serve for vegetative propagation. Both cryptostomata and conceptacles arise as depressions in the surface of the thallus, and it is now known that they can arise in one of two ways (Fig. 10.15):

(1) A linear series of two or more cells is formed, but their horizontal activity then ceases, and the terminal initial cell becomes sunk in a depression as the surrounding tissues grow up. The sides of the conceptacle are derived from the limiting layer and underlying cortex, e.g. *Fucus*, though in *Himanthalia* only the limiting layer is involved. *Pelvetia canaliculata* forms an exception in that the first division is vertical and not horizontal. The floor of the conceptacle, however, is derived by division of the original basal cell of the linear series. Finally, around the remnants of the one or more initial cells, a central mucilaginous column is formed, stretching to the neck of the conceptacle and connected to the walls by thin strings of mucilage, which are later ruptured.

(2) The cryptostomata or conceptacle develops from a single initial cell that divides transversely into two unequal cells; the upper or tongue cell degenerates while the lower one gives rise to the walls of the conceptacle by lateral divisions, except at the very top where they came from adjacent tissue. This method of formation occurs in *Sargassum*, *Bifurcaria*. The tongue cell and apical cell of the linear series are regarded as vestigial hairs so

that the conceptacles really originate from the basal cell of a hair. The cryptostomata or hair pits may be regarded as a permanent juvenile stage of the fertile conceptacle.

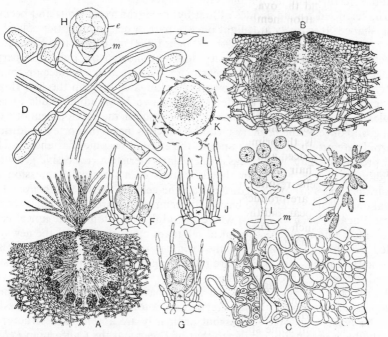

Fig. 10.17 *Fucus.* A, t.s. 'female' conceptacle of *F. spiralis* var. *platycarpus*; B, t.s. 'male' conceptacle of *F. vesiculosus*; C, portion of thallus of *F. spiralis* to show structure; D, origin of hyphae 1 cm below apex, *F. spiralis*; E, antheridia (microsporangia); F, young, and G, old oogonium (megasporangium); H, liberated ova (*e* = endochiton, *m* = mesochiton); I, ova being liberated (*e* = endochiton, *m* = mesochiton); J, empty sporangium showing torn exochiton; K, ovum being fertilized; L, antherozoid. (C × 125, D × 235) (C, D, after Pennington; A, B, E–L, after Oltmanns)

In the mature fruiting conceptacles there are branched hairs or paraphyses with the antheridia borne terminally on the branches near the base; or else the paraphyses are unbranched and associated with the oogonia, which are either sessile or borne on a single stalk cell. Each oogonium characteristically contains eight ova when mature. In those species where the conceptacles are hermaphrodite all these structures occur together. The expulsion of the gametes normally takes place at low tide, because the conceptacle is then full of mucilage and the loss of water makes the thallus shrink, so that the ripe ova and antherozoids in their envelopes are forced through the ostiole to the surface. When the tide returns the inner wall bursts and so liberates the antherozoids. The antheridium wall is two-layered but the oogonium wall has three layers, the endochiton, mesochiton and exochiton. When the ova are ripe the exochiton ruptures and liberates the eggs through the ostiole enclosed in the meso- and endochiton. Next, the mesochiton ruptures and inverts itself, to expose the endochiton which is finally dissolved. The eggs are then set free into the water where fertilization takes place, the antherozoids clustering around the ova, until eventually one penetrates and fertilizes the ovum. A fertilization membrane is then produced.

The fertilized zygote divides at first by three transverse walls and becomes club-shaped. It is likely that light plays a part in determining the polarity of the zygote, one to two hours of weak light (2000 lx) producing a detectable effect. Increasing the light intensity does not shorten the induction period. Other possible determinants of polarity are contact with a surface or point of entry of the antherozoid. The lowest cell elongates to form the first rhizoid while the uppermost cell divides twice by two longitudinal walls at right angles to give a quadrant. Further divisions result in the development of a central group of cells, the primary medulla, which is inside the cortex. Additional rhizoids are produced and the apex of the embryo becomes flattened and one cell produces a hair with a basal intercalary meristem. Excessive growth of neighbouring cells results in the formation of a terminal depression and additional hairs are produced. Next all the cells of the first hair, except the basal one, disappear and the basal cell becomes the three-sided apical cell. In those Fucales, such as *Fucus*, where the adult condition of the apical cell is four-sided, this change in the apical cell takes place quite early, the new four-sided cell being cut off from the original three-sided one. The peculiar salt-marsh fucoids (see p. 373) are interesting in that the juvenile three-sided apical cell persists throughout life.

In the allied genus *Pelvetia* the fronds have no mid-rib and are linear, compressed or cylindrical with irregular dichotomous branching. Air vesicles may be present in some species but ordinarily are absent. The structure of the thallus is essentially similar to that of *Fucus*, but the Californian *Pelvetia fastigiata* also possesses a few cryptostomata which are otherwise absent from the genus. Only two ova mature in the oogonia, the remaining six nuclei being extruded from the cytoplasm, though in *Pelvetia fastigiata* there may occasionally be four ripe ova or else ova that contain two nuclei. In *P. canaliculata* the two mature eggs are arranged one above the other, while in the Japanese species, *P. wrightii*, they are placed side by side.

The thallus of the common *Ascophyllum nodosum*, which sometimes bears nodular galls caused by the eel-worm *Tylenchus fucicola*, is more or less perennial, and regenerates each year from a persistent base or from the denuded branches (Fig. 10.18). As in *Fucus* and *Pelvetia*, free-living or embedded forms have evolved in salt-marsh areas (cf. p. 373), and these differ not only vegetatively but also in the absence of reproductive organs. The normal fronds have a serrated margin but no mid-rib and commonly bear vesicles (*pneumatocysts*). When the vesicles are borne on the little side branches they are termed pneumatophores. Recent work has shown that in some regions at least one pneumatocyst and its associated axis is produced annually so the number of pneumatocysts may indicate the age of the plant. This, however, does not seem to be universal. Work in Norway (Baardseth, 1955) has shown that the pneumatocysts are only formed at a certain time of year, which ranges from the end of January in Eire to May in Canada, and formation is therefore probably related to sea temperatures. The axis is beset by simple, clavate, compressed branchlets that arise singly or in groups in the axils of the serrations. These are later converted into or are replaced by short-

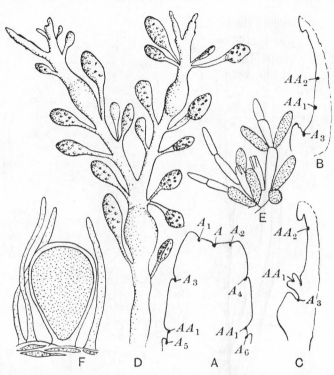

Fig. 10.18 *Ascophyllum nodosum.* A–C, diagram showing method of branching (A, apical cell; A_1–A_6, secondary initials in order of development; AA_1, AA_2, tertiary and quaternary initials); D, plant; E, microsporangia; F, megasporangium. (D × ½, E × 225, F + 2·25) (A–C, after Oltmanns; D–F, after Newton)

stalked, yellow, fertile branches which fall off after the gametes have been liberated from their conceptacles. The oogonia each give rise to four ova and the remaining four nuclei degenerate.

HIMANTHALIACEAE: *Himanthalia* (*himant,* thong; *halia,* of the sea). Fig. 10.19

The short, perennial frond or button arises from a small disc-like holdfast; the shape of the button depends on its level, the frond being short and stumpy when it grows exposed at high levels, and more elongate at the lower levels where the plants are submerged for longer periods. From March to July of each year new receptacles grow out from the centre of the buttons and form very long strap-shaped and repeatedly-forked structures filled with mucus. Although the button is regarded as the frond and the thong as the receptacle, it is possible to regard the button as part of the stipe and the thong as frond plus receptacle, as in *Durvillea*. In support of the second interpretation it should be mentioned that the thong contains horizontally running hyphae which have so far only been found in *Durvillea* in receptacular regions. In the

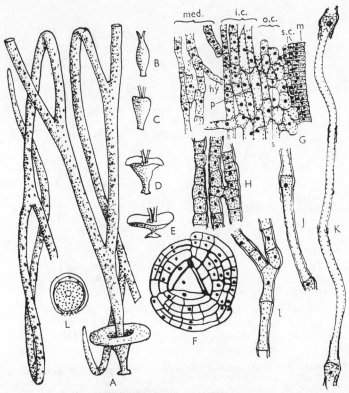

Fig. 10.19 *Himanthalia lorea.* A, fertile plant; B, C, abnormal buttons; D, button from bottom of dense zone; E, button from top of dense zone; F, t.s. apical cell; G, l.s. behind apex of young receptacle (m = meristoderm, s.c. = cells recently cut off from it, o.c. = outer cortex, i.c. = inner cortex, m = medulla, p = pit); H–K, stages in elongation of medullary cells; L, mature megasporangium (oogonium). (H–K × 125) (A–E, I, after Gibb; F–H, after Naylor)

thong these horizontal hyphae appear to be associated with the formation of air spaces. Growth curves show that the greatest length is attained by these annual thongs on plants growing in the lowest part of a dense zone, and that the shortest occur in the highest. This may be correlated with (*a*) the greater degree of desiccation at the higher levels, and (*b*) less frequent flooding which could reduce the supply of nutrient salts. The apical groove, containing the apical cell and hairs, is perpendicular to the plane of flattening. Internally it has been found (Naylor, 1951) that the extremely elongated cells of the medulla may take on an appearance resembling the trumpet hyphae of *Laminaria*. This may be a case of parallel development, but in view of other similarities to *Durvillea* it may be suggested that *Himanthalia* represents a primitive though somewhat specialized member of the Fucales. Reduction has gone so far in this genus that only one ovum matures in the ripe oogonium (macrosporangium). Liberation of gametes is controlled by the tides and exposure, and there is a definite periodicity related to these two factors.

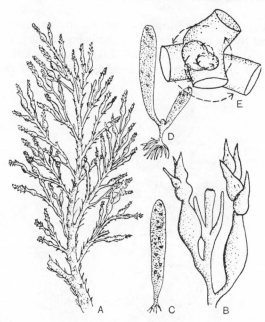

Fig. 10.20 *Cystoseira*. A, C, *ericoides* plant; B, portion of same enlarged; C, germling; D, same, rather older; E, diagram to show nature of branching in *C. abrotanifolia*. (A × ½, B × 4·5) (A, B, after Newton; C–E, after Oltmanns)

CYSTOSEIRO-SARGASSACEAE

CYSTOSEIRACEAE: *Cystoseira* (*cysto*, bladder; *seira*, chain). Fig. 10.20
    The much-branched perennial thallus is either cylindrical or compressed and arises from a fibrous woody holdfast which has more or less the structure of a conical cavern. The primary branches arise from the main stipe towards the base and divide above into filiform branches and branchlets, which sometimes do not develop very far. Seriate rows of small air vesicles are present on small branchlets. In the related *Halidrys* similar rows are also present, but these are distinctly flattened and borne in leaf-like appendages. Just before the dormant season some species shed a considerable proportion of the branch system so that the plant may appear very different. The plants are monoecious or dioecious, the conceptacles being borne in terminal or intercalary positions on the ramuli; as in some of the other genera, only one ovum develops and the remaining seven nuclei degenerate. In the sporeling the main shoot is very short and soon stops growth, and a new apical cell arises near its base

(Fig. 10.20). This new apical cell gives rise to the adult axis which is therefore
a branch of the sporeling. The later laterals develop monopodially and in
succession. The first two shoots arise opposite each other but the remainder
have a divergence of two-fifths. The genus is principally confined to the warm
sub-tropical and temperate waters; in the Mediterranean considerable
speciation and hybridization has taken place.

SARGASSACEAE: *Sargassum* (*sargasso*, Spanish for seaweed). Fig. 10.21

In the simplest forms branching is distichous but in the great majority it is
radial with a divergence of two-fifths. Like *Cystoseira* the main axis is
usually very short and the thallus is composed mainly of richly branched,

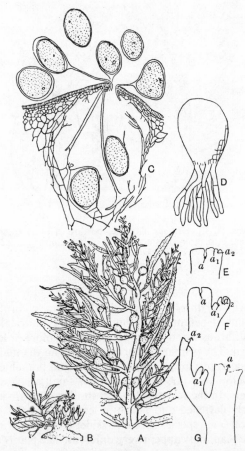

Fig. 10.21 *Sargassum.* A, *S. filipendula*; B, base of plant; C, escape of oogonia (mega-
sporangia) each with eight nuclei; D, seedling at rhizoid stage; E–G, stages in branching,
S. thunbergii (a = main initial, a_2 = branch initial, a_1 = secondary branch initial). (A × 0·45,
B, C, × 40, D × 105, E–G × 0·22) (A, B, after Taylor; C, after Kunieda; D, after Tahara;
E–G, after Oltmanns)

long laterals which produce leafy, short laterals that are built up as follows: the primary branch is a sterile phylloclade which bears cryptostomata, while the secondary branch is also sterile and is commonly reduced to an air bladder. In the simpler forms the subsequent branches are fertile and finger-like in appearance. In more advanced forms there may be further sterile phylloclades which may subtend axillary branch systems. In branching the main apical cell (a Fig. 10.21) cuts off a secondary initial (a_1), and this latter, as soon as it has emerged from the apical pit, cuts off another initial, (a_2), on the distal side. This last initial divides more rapidly than the other and grows out into the subtending leaf, which is therefore a lateral that has pushed aside the parent axis. Later the parent initial (a_1) produces further initials on a two-fifth divergence. The plants are attached by means of a more or less irregular, warty, solid, parenchymatous base or else numerous stolon-like structures grow from the main axis and anchor the plant.

The genus, which is principally confined to tropical waters, is a very large one with about 150 species, some being dioecious although others are monoecious. Morphologically the structure of the species is similar to that of *Fucus* though there is little or no medullary material, and when present it is not gelatinized. In the ripe oogonium (megasporangium) only one ovum will normally reach maturity, though occasionally eight eggs may develop. In the former case the single ovum contains all eight nuclei, but only one of these grows larger and is actually fertilized. This can be interpreted as a failure on the part of the megaspores and gametophytes to form cell walls, and is a secondary condition due to still further reduction. In *S. filipendula* there is no stalk to the young megasporangium and so it is embedded in the wall of the conceptacle. When ripe the whole oogonium, not merely the inner wall and its contents, is discharged and remains just outside the ostiole attached to the conceptacle wall by a long mucilaginous stalk which is a secondary development. In one sub-section of the genus the ostiole of the female conceptacle is closed by a disc-shaped gelatinous plug, which disappears when the oogonia are ripe.

After fertilization the first divisions take place while the zygote is still attached to the parent plant by this long stalk. The genus is very well known because of *S. natans*, the so-called Sargasso weed, which is found as large floating masses in the Sargasso Sea. At one time it was thought that plants of *S. natans*, together with one or two other pelagic species, were attached in the early stages, but there would now seem to be good evidence that they remain floating throughout the whole of their life cycle. It has been suggested that these perennial pelagic species originally arose from attached forms such as *S. filipendula* and *S. hystrix*. Pelagic species do not appear to possess the normal sexual reproduction but reproduce by vegetative fragmentation.

Anomalae

HORMOSIRACEAE: *Hormosira* (*hormo*, necklace; *sira*, chain). Figs. 10.22, 10.23

The thallus, which looks like a bead necklace, is composed of a chain of swollen vesicles (internodes) with narrow connectives (nodes). Growth takes place by means of four apical cells, which give off branches alternately in a dichotomous manner, the branches usually arising at the internodes. Apart from the discoid holdfast, there is no differentiation into appendages comparable to those of the other Fucaceae. The basal internode is solid but all the remainder are hollow; the nodes are also solid because they are composed solely of epidermis and cortex. The plants are dioecious, the conceptacles being borne on the periphery of the inflated nodes. Although eight ova are originally formed in the oogonium only four reach maturity, but in this genus, however, it is a case of degeneration of eggs and not merely of nuclei. When

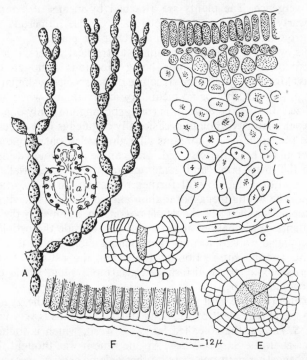

Fig. 10.22 *Hormosira banksii.* A, portion of plant; B, l.s. of apex of plant (*a* = air-filled space); C, t.s. of thallus at internode; D, l.s. of apex; E, t.s. of apex; F, cuticle being shed (semidiagrammatic). (A × ⅔, C × 150) (A, C, after Getman; B, D, E, after Oltmanns)

the ova are released from the oogonium they are still surrounded by the endochiton. This soon ruptures and releases the mature ova. After fertilization has taken place a special fertilization membrane is produced, and this is composed of two layers, the inner one being of cellulose. Another interesting feature of this genus is its capacity to form and shed a cuticle that bears the impressions of the cell outlines. The genus is monotypic, the single species, *H. banksii*, being confined to Australia and New Zealand where it grows on rocks and in tide pools of the littoral belt. Although several varieties have been described, it has recently been shown that these are merely more-distinct representatives of a range of ecological forms. Use of a statistical technique, the discriminant function, based upon measurements of bladder diameter and connective length, has differentiated these various ecological forms (see also p. 448) (Fig. 10.23).

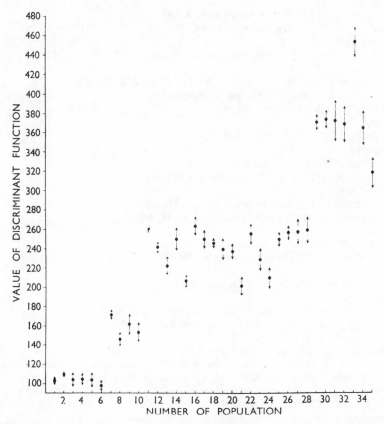

Fig. 10.23 Separation of exposed rocky coast populations of *Hormosira banksii* by use of the discriminant function. This is based upon diameter of bladder and length of connective. Group of most exposed plants on the left, successive groups of decreasing exposure to group of least exposed plants on the right. (after Bergquist)

REFERENCES

GENERAL
  Chapman, V. J. (1970).
  Fritsch, F. E. (1945). *Structure and Reproduction of the Algae*, **2**, p. 192–380 (Cambridge University Press).
ULTRASTRUCTURE
  Bouck, G. B. (1965, 1969). *J. Cell Biol.*, **26**, 523; **40**, 446.

Evans, L. V. (1962). *Ann. Bot.*, N. S., **26**, 345.
Evans, L. V. (1968). *New Phytol.*, **67**, 173.

ASCOSEIRALES
Skottsberg, C. (1907). *Erg. Schwed. Sudpol-Expd.* 1901–03, **4**, 149.

DURVILLEALES
Nizamuddin, M. (1968). *Botanica mar.*, **11**, 115.

FUCALES
Nizamuddin, M. (1962). *Botanica mar.*, **4**, 191.
Manton, I. (1964). *New Phytol.*, **63**, 244.

LAMINARIALES
Yabu, H. (1964). *Mem. Fac. Fish. Hokkaido Univ.*, **12**, 1.

Ascophyllum
Baardseth, E. (1955). *Inst. Industr. Res. and Stand. Repts.*, Eire.
Baardseth, E. (1968). *F.A.O. Special Publication.*
Gibb, D. C. (1957). *J. Ecol.*, **45**, 49.

Chorda
Roberts, M. (1967). *Br. Phycol. Bull.*, **3**, 379.

Cystoseira
Roberts, M. (1961). *J. Linn. Soc. (Bot.)*, **60**, 251.
Roberts, M. (1967). *Br. Phycol. Bull.*, **3**, 345, 367.

Cystophora
Nizamuddin, M. (1964). *Botanica mar.*, **7**, 42.
Womersley, H. B. S. (1964). *Aust. J. Bot.*, **12**, 54.

Durvillea
Naylor, M. (1949). *Ann. Bot.*, N.S., **13**, 285.
Naylor, M. (1953). *Trans. R. Soc. N.Z.*, **80**, 277.

Egregia
Chapman, V. J. (1962). *Botanica mar.*, **3**, 33, 101.

Fucus
Knight, M. and Parke, M. (1950). *J. mar. biol. Ass. U.K.*, **39**, 439.
McCully, M. E. (1966, 1968). *Protoplasma*, **62**, 287; **66**, 205.
McCully, M. E. (1968). *J. Cell Sci.*, **3**, 1.
Munda, A. I. (1964). *Nova Hedwigia*, **8**, 403.
Powell, H. T. (1963). In *Speciation in the Sea*, edit. by J. P. Harding and N. Tebble. Systematics. Association Publication No. 5, p. 63–77.

Himanthalia
Naylor, M. (1951). *Ann. Bot.*, N.S., **13**, 501.

Hormosira
Bergquist, P. L. (1959). *Botanica mar.*, **1**, 22.
Osborn, N. (1949). *Trans. R. Soc. N.Z.*, **77**, 47.

Julescrannia
Parker, B. C. and Dawson, E. Y. (1965). *Nova Hedwigia*, **10**, 273.

Laminaria
Hasegawa, Y. (1962). *Bull. Hokkaido Reg. Fish. Lab.*, **24**, p. 116–138.
Kain, J. M. (1963, 1964, 1969). *J. mar. biol. Ass. U.K.*, **43**, 129; **44**, 415; **49**, 455.
Parke, M. (1948). *J. mar. biol. Ass. U.K.*, **27**, 706.
Sundene, O. (1961, 1964). *Nytt. Mag. Bot.*, **9**, 5; **11**, 83.
Yabu, H. (1965). *Sci. Dept. Alg. Inst. Hokkaido*, **5**, 1.

Macrocystis
Clendenning, K. A. (1964). *Proc. 4th. Int. Seaweed Symp. Biarritz*, p. 55.
Neushul, M. (1963). *Am. J. Bot.*, **50**, 354.

Neushul, M. and Haxo, F. T. (1963). *Am. J. Bot.*, **50**, 349.

North, W. J. (1961). *Nature, Lond.*, **190**, 1214.

Parker, B. C. (1963). *Science, N. Y.*, **140**, 891.

Parker, B. C. (1965). *J. Phycol.*, **1**, 41.

Nereocystis

Kemp, L. and Cole, K. C. (1961). *Can. J. Bot.*, **39**, 1711.

Pelagophycus

Dawson, E. Y. (1962). *Bull. So. Calif. Acad. Sci.*, **61**, 153.

Dawson, E. Y. (1964). *Trans. S. Diego Soc. nat. Hist.*, **13**, 303.

Parker, B. C. and Bleck, J. (1965). *Trans. S. Diego Soc. nat. Hist.*, **14**, 57.

Parker, B. C. and Bleck, J. (1966). *Ann. Mo. bot. G.*, **53**, 1.

Parker, B. C. and Fu, M. (1965). *Can. J. Bot.*, **43**, 1293.

Phyllogigas

Neushul, M. (1963). *Botanica mar.*, **5**, 19.

Skottsberg, C. and Neushul, M. (1960). *Botanica mar.*, **2**, 164.

Saccorhiza

Norton, T. A. and Burrows, E. M. (1969). *Br. Phycol. J.*, **4**, 19.

I I

RHODOPHYTA

RHODOPHYCEAE

The Rhodophytes form a large but uniform group in so far as their reproductive processes are concerned. Morphologically, they vary widely in the construction of the vegetative thallus, although in all cases the thallus is fundamentally filamentous and there is no true parenchymatous construction. There are unicellular and colonial forms in the primitive Bangiophycidae. The filamentous thallus is built up essentially on one of two plans:

(a) Central filament (uniaxial) type in which there is a central corticated or uncorticated main axis which bears the branches.

(b) Fountain (multiaxial) type in which there is a mass of central filaments (rather like the cable type in the Phaeophyceae, see p. 198) which lead out like a spray to the surface.

Growth of the thallus is nearly always apical (except in the Bangiophycidae), though intercalary growth is known in the Corallinaceae (p. 281) and Delesseriacae (p. 305). Heterotrichy is a feature in many young stages and may persist in the less specialized genera. Branching is nearly always monopodial and uniformity in morphology and reproduction (see below) indicates a monophyletic origin. The chloroplasts contain, in addition to chlorophyll and carotenoids, either phycoerythrins or phycocyanins or both of these. The chloroplasts are commonly band-shaped, and parietal, though in some of the primitive Nemalionales and Bangiophycidae they are stellate. In the less-developed forms there is often only one chloroplast per cell. The chloroplasts themselves are usually considered to be of a primitive structure. The lamellae are arranged in pairs, as in the Cryptophyceae, whereas in the other divisions they are in groups of three or more and in some Chlorophyceae the lamellae are in pseudograna. The cells are uninucleate in primitive forms as in the Nemalionales, but in most orders a multinucleate condition is more frequent; this is most marked in *Griffithsia* where each cell contains a very large number of nuclei. The cell wall is double, the inner wall consisting of cellulose and the outer of pectic materials. Except for most members of the first subclass, the Bangiophycidae, the cell contents remain connected to each other by means of thin protoplasmic threads or plasmodesmata, which may be very conspicuous in the region of the fusion cell (cf. below). Electron-microscope studies have shown that in those red algae studied there are two kinds of pit: one is quite open, as in *Rhodymenia*, and the other is closed and penetrated by plasmodesmata. Some species contain special small vesicular cells rich in iodine and bromine, but the function of these is not known. Iridescence caused by special bodies within the cells is a feature of a few species, e.g. *Chylocladia reflexa*, *Chondria coerulescens*.

The characteristic reproductive bodies are usually found on separate plants, but the two sex organs may occur on the same plant, and certain abnormal cases are also known where sexual and asexual organs are present on the same thallus (see p. 305). The sexual plants are usually similar in size, but in *Martensia fragilis* and *Caloglossa leprieurii* the male plants are smaller than the female. The male structures, which are probably best termed antheridia, although they have been given other names, each give rise to a non-motile body, or spermatium, which is carried by the water to the elongated tip or trichogyne of the carpogonium or female structure. This mating system and the complete absence of flagellated reproductive cells, distinguishes the Rhodophyceae from the other algal groups, and is the basis for separating them from the remainder of the eucaryotic algae (see Chapter 1).

The carpogonium with its trichogyne is usually borne on a special lateral branch (except in the Gelidiales) which consists of a varying number of cells, the parent mother cell being known as the support cell. In many cases the cells of the carpogonial branch differ from other vegetative cells in having denser protoplasts and in lacking chloroplasts. The branch is often adventitious and the number of cells is commonly three or four.

A specialized cell, the *auxiliary cell*, is often associated with this branch, or in some cases forms a part of it; the fertilized carpogonial nucleus passes into this cell. The carpogonial branch plus the associated auxiliary cell or cells is termed the *procarp*. In some of the more primitive genera the auxiliary cell, as well as receiving the fertilized nucleus, may also have a nutritive function. Some workers argue that such cells are not truly auxiliary cells and indeed cannot be strictly auxiliary cells unless they receive the fertilized carpogonial nucleus, but this view may seem to be too extreme. Although in the past it may have appeared easy to define the carpogonial branch, more recent work has shown that the supposed three major features are by no means universal and valid (Dixon, 1961).

Carpospores are normally formed from a peculiar diploid generation that develops parasitically on the female plant. They are produced in sporangia that are terminal on gonimoblast filaments. These filaments arise usually from the auxiliary cell or fertilized carpogonium. With a few exceptions, recent work has shown that the gonimoblasts and carposporangia are diploid although previous work indicated haploidy in some cases. The only apparent exception at present is *Atractophora hypnoides* (Boillot, 1967) and this may well prove to be erroneous. It is clear that the cytological state of carpo-sporophytes, especially in primitive genera, needs exhaustive study. On germination the diploid carpospores give rise to a new independent asexual diploid generation, which reproduces by means of tetraspores formed in sporangia borne externally or sunk into the thallus. In the majority of species where there are two separate diploid generations the plants can be termed morphologically triphasic but cytologically diphasic. The tetrasporic plants are generally identical morphologically with the sexual plants, but in some cases they are very much smaller and quite different morphologically (e.g. Bonnemaisoniales). The Rhodophyceae are in fact very interesting because there is the possibility that meiosis may occur at germination of the zygote, in the carposporangia (*Liagora tetrasporifera*) or in the tetrasporangia of the tetrasporophyte. Two cases of polyploidy have been well documented in *Plumaria elegans* and *Spermothamnion repens* (Drew, 1939).

The classification of the Rhodophyceae is based primarily on the structure of the female reproductive apparatus. Apart from the Bangiophycidae, which mostly lack pit connections and have a single axile chloroplast in each cell, the remainder of the red algae, or Florideophycidae, are classified as follows:

Nemalionales

Construction is of the fountain type or is uniaxial. Carpogonia occur singly, and the so-called auxiliary cell, if present, is formed from a carpogonial branch cell or its derivative and is nutritive. When present the tetrasporangia are generally cruciate, more rarely tetrahedral.*

Bonnemaisoniales

Construction is of the uniaxial type. There are special vesicular cells and an attachment organ of modified branchlets. The asexual generation is often a dwarf, prostrate branched thallus.

Gelidiales

Construction is of the uniaxial type. The sessile carpogonia are aggregated and true auxiliary cells are lacking, the adjacent cells being only nutritive. Reduction division is delayed so that the plants are morphologically triphasic and cytologically diphasic. The tetrasporangia are cruciate, more rarely tetrahedral. Dixon (1961) believes that there is no case for the order and

---

* This order may need to be subdivided as there are 3 families with a 2-peak phycobilin and 3 families with a 3-peak phycobilin (Hirose, Kumaro and Madono, 1969).

returns the Gelidiaceae to the Nemalionales. The Nemalionales are so heterogeneous that at present it is probably better to follow Fan (1961) and retain the Gelidiales.

Cryptonemiales
Both types of thallus construction are seen. The carpogonia are borne on special accessory branches and may be aggregated into sori, nemathecia or in conceptacles. The auxiliary cell is usually borne on separate accessory branches before fertilization and is actively concerned in the post-fertilization processes. Procarps do occur in some genera. The tetrasporangia are cruciate or zonate.

Gigartinales
Both types of thallus construction exist. The support cell or a normal intercalary cell is set aside as an auxiliary cell before fertilization. The tetrasporangia are cruciate or zonate.

Rhodymeniales
Only the multiaxial type of construction occurs. The small auxiliary cells are derived from a branch formed from the supporting cell. They are cut off before fertilization but only develop after that process has taken place. The tetrasporangia are cruciate or tetrahedral.

Ceramiales
Only the Uniaxial type of construction is found here. The auxiliary cell or cells are cut off from the supporting cell or from a homologous pericentral cell after fertilization and as a direct consequence of the process. The tetraspores are usually arranged in a tetrahedral fashion, more rarely in a cruciate manner.

Some authors consider that a further two orders should be recognised, the Sphaerococcales and Nemastomales, and there is some evidence to support this. We have however, retained the genera of these two orders in the orders to which they have belonged in the past.

The antheridia are either borne scattered over the whole surface (Nemalionales), or else in localized sori. These sori are reticulate in *Rhodymenia*, band-like in *Griffithsia*, borne on special branches in *Polysiphonia*, sunk in conceptacles in the Corallinaceae, and occur on the tips of the thallus on *Chondrus*. Very little is known about the seasonal periodicity of the male plants, which are often less frequent than either the female or tetrasporic plants, but this may be due purely to lack of observation, although it is also possible that the male plants are gradually loosing their function. The antheridia often appear in an orderly sequence, being cut off usually as subterminal or lateral outgrowths from the antheridial mother cell. If they have been borne on a special part of the thallus (as in *Delesseria*) this may fall off or die away after fruiting is completed, while in other cases the mother cells simply revert to a normal vegetative state. The different types of male plant have been classified as follows:

A. The antheridial mother cell does not differ from the vegetative cells either in form or content, nor are the antheridia covered by a continuous outer envelope, e.g. *Nemalion, Batrachospermum*.

B. The antheridial mother cells are differentiated from the vegetative cells, and the antheridia are surrounded by a common outer sheath, which is later pierced by holes or else gelatinizes in order to allow the ripe spermatia to escape:

(1) The antheridia develop terminally, e.g. *Melobesia, Holmsella*.

(2) The antheridia develop subterminally:

(a) Two primary antheridia, e.g. *Delesseria sanguinea, Chondrus crispus*.

(b) Two or three primary antheridia, e.g. *Scinaia furcellata, Lomentaria clavellosa*.

(c) Three primary antheridia, e.g. *Ceramium rubrum, Griffithsia corallina*.

(d) Four primary antheridia, e.g. *Polysiphonia violacea, Callithamnion roseum*.

The primary antheridia are commonly succeeded by a second crop which arises within the sheaths of the first, but a third crop only occurs in a few genera.

In all red algae the nucleus of the spermatium is in late prophase when liberated but normally no division takes place. In *Batrachospermum* and *Nemalion*, however, a division does occur but only one of the daughter nuclei acts as the fertilizing agent. This feature has led to the suggestion that in the more advanced red algae the contents of the antheridium are equivalent to a body which formerly did divide.

The carposporophyte is the generation that arises normally after fertilization and is parasitic upon the female plants. Cytologically it is haploid in *Atractophora hypnoides* but diploid in the remaining members of the Florideophycidae. Morphologically it comprises all the structures arising from the fertilized carpogonium, fusion cell or cells and gonimoblasts. In more advanced forms it is often protected by a wall of gametophytic tissue. The carposporophyte and the wall is termed the *Cystocarp* or *Gonimocarp*. The following characteristics have been recognised in the different types of carposporophyte:

(1) Gonimoblasts arise from the carpogonium without any prior fusion with a gametophyte cell.

(2) Gonimoblasts arise from the carpogonium after fusion with a gametophyte cell but without transfer of diploid nucleus.

(3) Primary gonimoblasts transfer diploid nucleus to gametophyte auxiliary cells from which secondary gonimoblasts then arise.

(4) Gonimoblasts arise from a fusion cell, the diploid nucleus having first passed into a special gametophyte (auxiliary) cell where

(a) The gametophyte is a cell or cells of the carpogonial branch.

(b) The gametophyte cell is the support cell of the carpogonial branch or an auxiliary cell cut off before or after fertilization.

The tetraspores are formed either in superficial tetrasporangia that terminate assimilatory filaments or short laterals, or in sporangia and short laterals that are sunk into the thallus, in which case the fertile branch often becomes swollen and irregular in outline. In some cases e.g. *Plocamium*, *Dasya*, they are borne on special lateral branches or stichidia. The division of the sporangia is cruciate, zonate or tetrahedral. Experimental cultures have demonstrated that there are good grounds for believing that of the four spores in a tetrad, two will give rise to female plants and two to male plants. In some cases only two spores (bi-spores) are formed, e.g. certain Corallinaceae and Ceramiaceae. Meiosis normally occurs at the formation of the tetraspores, but when the spores develop on sexual, haploid plants, as sometimes happens, there is no meiotic division and the products function as monospores. In *Agardhiella tenera* apospory is sometimes found and again there is no meiosis so that a succession of asexual plants can occur. In the Nemalionales reproduction by means of monospores is quite common though the homology of these bodies is somewhat uncertain. In some of the Florideophycidae (*Plumaria*, *Spermothamnion*) polyspores or paraspores develop on the diploid plants, but it has recently been shown that these are in some cases morphologically equivalent to tetraspores, while in others, e.g. *Plumaria*, they form the reproductive organs of a triploid generation. Monospores, carpospores and tetraspores of some Rhodophyceae appear capable of a small degree of motion, the spores of the Bangiaceae being the most active of those investigated. The mechanism of this movement is not understood, and it is doubtful whether it is sufficient to be of significance in the reproductive processes of the plants.

The Rhodophyceae are characterized by a high degree of epiphytism and parasitism in which the parasites and many of the epiphytes demonstrate a considerable specificity. There have been few physiological studies and the distinction between epiphytes and parasites tends to be one of convenience rather than one based on experimental evidence. The epiphytes usually possess the normal morphology and pigmentation of the genus. The parasites exhibit great reduction in morphological form and often appear as whitish pads due to negligible pigmentation (*Erythrocystis* is an exception). Furthermore the parasite and host are often members of the same family or are closely related phylogenetically (adelphoparasites), as opposed to those that are not (alloparasites). A listing of some species-specific epiphytes and parasites is given in Table 11.1.

The division is principally marine, but there are a few freshwater genera, comprising in all about 200 species. The most important freshwater genus is *Batrachospermum* with many species in Australasia, but others are *Lemanea* and *Hildenbrandia*. They nearly all frequent fast-flowing streams where there is an abundance of aeration. A very interesting new genus and species, *Boldia erythrosiphon*, has recently been reported from freshwaters of the U.S.A. Essentially it is the Rhodophycean analogue of the green algal genus *Monostroma* (p. 65).

There is one section, the Corallinaceae, that is characterized by lime encrustation. During past geological ages these algae have played an important part in the building up of rocks and coral reefs, a process which is still going on.

TABLE 11.1

EPIPHYTE PARASITE	HOST
Ceramium codicola	Codium fragile
Polysiphonia lanosa	Ascophyllum nodosum
Smithora naiadum	Phyllospadix and Zostera (Phanerogams)
Porphyra nereocystis	Nereocystis luetkeana
Achrochaetium desmarestiae	Desmarestia spp.
Pleurostichidium falkenbergii	Xiphophora chondrophylla
Erythrocystis saccata	Laurencia pacifica
Janczewskia gardneri	Laurencia spectabilis
Peyssonelliopsis	Peysonellia
Choreonema thuretii	Jania rubens
Schmitziella endophloea	Cladophora pellucida
Choreocolax polysiphoniae	Polysiphonia spp.
Holmsella pachyderma	Gracilaria confervoides
Harveyella mirabilis	Rhodomela spp.
Plocamiocolax pulvinata	Plocamium pacificum
Rhodymeniocolax botryoidea	Rhodymenia spp.
Faucheocolax attenuata	Fauchea fryeana
Hypneocolax stellaris	Hypnea spp.
Gelidiocolax suhriae	Gelidium spp.
Gelidiocolax margaritoides	Gelidium spp.
Gracilariophila oryzoides	Gracilariopsis sjoestedtii
Pterocladiophila hemispherica	Pterocladia lucida
Onychocolax polysiphoniae	Polysiphonia
Callocolax globulosis	Callophyllis flabellulata
Ceratocolax hartzii	Phyllophora membranifolia

BANGIOPHYCIDAE

This subdivision is sometimes known as the Bangioideae or Protoflorideae. The plants are simple, unicellular, filamentous or membranous forms. The cells contain a single axile chloroplast. Growth of the thallus is diffuse and not apical, and plasmodesmata between cells are generally lacking. Sexual reproduction is known for a number of genera, but the carpogonium exhibits very little specialization. The spermatia are formed by repeated divisions inside the mother cell (e.g. *Porphyra*) or within special spermatangia (*Erythrotrichia*). Monospores form another means of reproduction, being produced in one of three ways:

(i) Those derived from differentiated sporangia, e.g. *Erythrotrichia*, *Erythrocladia*.

(ii) Those formed from undifferentiated cells, e.g. *Bangia*.

(iii) Those formed by successive divisions of a mother cell, e.g. *Porphyra*.

There are five orders in the subdivision but only three of these contain more than one genus. The Compsopogonales contain the freshwater genus *Compsopogon* of tropical and subtropical distribution and the Rhodochaetales has the sole genus *Rhodochaete*. *Compsopogon*, which has become a noxious alga in some parts of the world, is interesting because morphologically the erect axis consists of a row of large, colourless, axial cells surrounded by a cortex of small, highly pigmented cells which may form 2–4 layers. Pit connexions between the cells have been reported but recent work (Nichols, 1964) suggests that they may not be true plasmodesmata. Reproduction is by means of monospores which may be large or small (Fig. 11.1). *Rhodochaete* is of interest because of its phylogenetic implications (see p. 341).

The Porphyridiales comprises unicellular (Fig. 11.2), terrestrial or marine types. In *Porphyridium cruentum* the single cells, with a single stellate chloroplast, are united into a gelatinous colony of blood-red colour, often forming

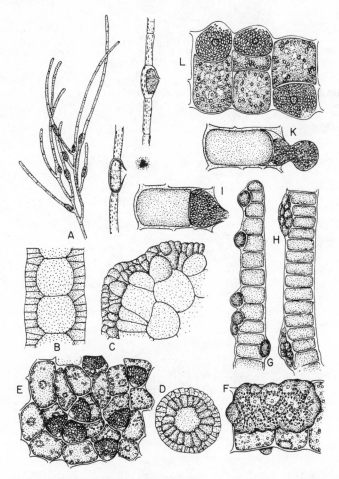

Fig. 11.1 A, *Rhodochaete pulchella* with monospores; B-L, *Compsopogon coeruleus*: B, l.s. of thallus; C-D, t.s. of thallus; E-H, development of macro-aplanosporangia; I-L, development of micro-aplanosporangia. (after Kylin)

large patches on soil or damp stones. Cell division occurs in all directions, and when a cell divides the sheath elongates to form a stalk which eventually ruptures. Monospores are the only other known means of reproduction. *P. aerugineum* is one of the few blue-coloured, red algae, which lacks com-

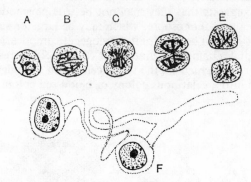

Fig. 11.2 *Porphyridium cruentum*. A-E, stages in nuclear and cell division; F, cells connected b stalks after division. (A-F × 1280) (after Zirkle and Lewis)

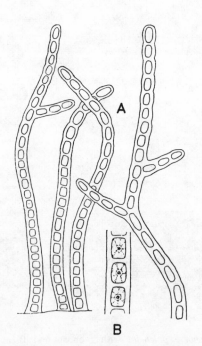

Fig. 11.3 *Asterocytis ramosa*. A, filaments and branching; B, cells. (× 220) (after Kylin)

pletely the red phycoerythrins. One of the more interesting blue Rhodo-phyceae is the thermophilic and acidophilic *Cyanidium caldarium*. This inhabitant of thermal springs reproduces by endospores (as does *Rhodospora*), possesses a single-lobed chloroplast, and a very thick cell wall which comprises up to 50 per cent protein. It has previously been assigned to the Cyanophyceae, Chlorophyceae and Cryptophyceae, but we consider it should be assigned to the Rhodophyceae as a member of the Porphyridiales (in a family Cyanidiaceae) or of a new order Cyanidiales. Because of definite physiological and biochemical differences a new order is perhaps the more logical choice. Genera such as *Asterocytis* (Fig. 11.3), which are placed in the Goniotrichales, form simple or branched filaments comprised of a single row of cells enclosed in mucilage that occur as minute epiphytes. Reproduction is by means of monospores or akinetes. *Asterocytis ornata* is one of the few blue Rhodophyceae. The Goniotrichales may be regarded as representing a more advanced condition than the Porphyridiales.

BANGIALES

These are the more advanced red algae with a heterotrichous thallus and a primitive form of sexual reproduction.

ERYTHROPELTIDACEAE: *Erythrotrichia* (*Erythro*: red; *trichia*: hair). Fig. 11.4

The plants occur as minute marine epiphytes attached to the host plant either by a one-layered base or by an irregular, lobed basal cell from which

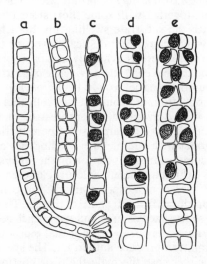

Fig. 11.4 *Erythrotrichia bertholdii.* a, lower, b, middle and c–e, upper part of the same filament. (× 390) (after Kylin)

short rhizoids radiate. The basal discs bear one or more simple, erect threads which in some species, e.g. *E. obscura*, may have additional longitudinal and transverse divisions which indicate how the *Bangia* and then the *Porphyra* (see below) type of thallus may have evolved. In some cases only the disc remains and such species are placed in the genus *Erythropeltis*. Asexual reproduction takes place by means of monospores produced in monosporangia cut off as small cells from the upper end of a vegetative cell. Antheridia are cut off in the same way as the monosporangia, while the carpogonium differs from a vegetative cell in being larger and in possessing denser contents and the surface is slightly protruded but there is no true trichogyne. After fertilization the zygote is set free, so that if it represents a carpospore only a single one arises from each act of fertilization. The carpospores apparently form a *Conchocelis*-type filament which may give rise to the parent gametophyte, or may reproduce itself by monospores.

Included in the Erythropeltidaceae is the obligate epiphyte *Smithora naiadum*. The blades arise from small discs on *Phyllospadix* or *Zostera*. Asexual reproduction is by neutral spores formed in sori. Spermatia are likewise formed in sorix, one spermatium being cut off per cell. Although carpogonia are unknown, carposporangial areas have been reported.

BANGIACEAE: *Porphyra* (purple dye). Fig. 11.5

This genus has a wide distribution extending from 15° to 71° N. and from the Cape of Good Hope to 60° S. The plant is flat and membranous, and in the common European species, *P. umbilicalis*, there are a number of growth forms; the shape, width and length of the various forms are determined by the age of the plant, the height above sea-level and the type of locality. The plants are attached by means of a minute disc which is able to produce lateral extensions from which new fronds may be proliferated. The disc is composed of long slender filaments together with some short stout ones, those near to or in actual contact with the substrate swelling up, branching and producing suckers or haptera which can penetrate dead wood or the tissue of brown fucoids. In the latter case there is evidently a capacity for epiphytism once contact is secured, and there is even some evidence of partial parasitism.

The gelatinous monostromatic fronds of *Porphyra*, which become distromatic during reproduction, are composed of cells that possess one or two stellate chloroplasts with a pyrenoid, the process of nuclear division being intermediate between mitosis and amitosis. Reproduction is by means of monospores, carpogonia with rudimentary trichogynes and antheridia. The carpogonial areas occupy a marginal position on the thallus and because of pigment differences they exhibit a different colouring to the vegetative thallus.

In the Japanese *P. tenera*, monospores are shed in considerable numbers, 10,000 spores being produced from 100 mm² of thallus. These become attached to the substrate in a short space of time and commence germination to a new plant one to two days after fertilization. In sexual reproduction all the frond, except the basal region, can produce antheridia. The species are dioecious or monoecious, but in either case the male thalli or portions of the thalli are commonly paler in colour than the female. Each antheridial mother cell gives rise to 64 or 128 antheridial cells, each of which produces one sperma-tium. The carpogonia, each derived from one cell, are borne in groups and possess a very short trichogyne. The fertilized carpogonium divides into four or eight cells that represent primitive carpospores. Fertilization has never actually been observed although there is strong evidence to suggest that it does take place. The carpospores germinate and produce prostrate, creeping filaments (usually in shells) that have been identified with the algal genus

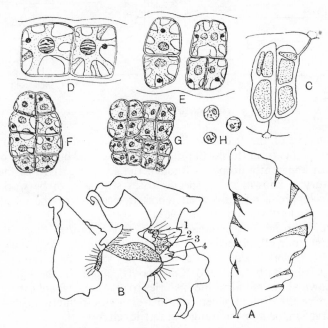

Fig. 11.5 *Porphyra*. A, thallus; B, attachment disc with three primary blades and four (1–4) secondary; C, formation of carpospores; D-H, formation of spermatia in *P. tenera*. (A × ⅓) (A, D-H, after Ishikawa; B, C, after Grubb)

Conchocelis. In the case of *P. tenera* of Japan, penetration of shells takes place within three days. This dwarf phase, which cannot tolerate any real degree of exposure, bears sporangia that produce monospores, and these on germina-tion are thought to give rise to a new adult plant, though some workers consider that they only reduplicate the '*Conchocelis*' phase. New adult plants of *P. umbilicalis* and *P. perforata* are said to arise from spores (conchospores) liberated from short swollen branches (fertile cell rows) or else arise as vegetative buds.

Japanese workers consistently report monospores from the '*Conchocelis*' phase of *P. tenera*, whereas in the case of *P. umbilicalis* only conchospores and buds are reported. The full life cycle of *P. umbilicalis* can be represented as in Fig. 11.6.

The development and growth of both adult and '*Conchocelis*' phase is related to day length and this also varies for the species. In *P. tenera* 2–12 hours of light is all that is required for production of monosporangia and liberation of spores. Short days are required for maturation of '*Conchocelis*' monosporangia in *P. yezoensis*, *P. angusta*, and *P. kuizieda*, while very short days are required for *P. pseudolinearis*.

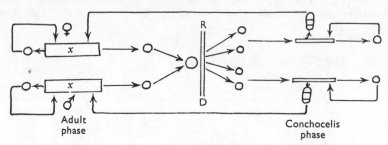

Fig. 11.6   Diagram of life cycle of *Porphyra umbilicalis*.

Production of the '*Conchocelis*' phase is not universal because in New Zealand, plants from the north produce '*Conchocelis*' but similar plants from the south apparently do not. Although the general pattern of the *Porphyra* life history is known, many specific cases still need to be worked out. A *Conchocelis* phase has also been reported recently for the allied *Smithora naiadum* (Richardson and Dixon, 1969).

Haploid chromosome numbers of two (*P. umbilicalis*), three (*P. tenera*, *P. yezoensis*), four (*P. linearis*) and five (*P. umbilicalis f. laciniata*) have been listed. In *P. umbilicalis* and its variety the same chromosome number is reported from both adult and '*Conchocelis*' phase. In the light of what is now known about red-algal life histories (p. 346) one would expect the dwarf phase to be the diploid generation in which case reduction division should be observed in the monosporangia or fertile cell rows.

REFERENCES

GENERAL

Chadefaud, M. (1963). *Revue Algol.*, **6**, 252.

Boillot, A. (1967). *C.r. hebd. Séanc. Acad. Sci., Paris*, **264**, 257.

Dixon, P. (1961). *Botanica mar.*, **3**, 1.

Dixon, P. (1963). *Oceanogr. Mar. Biol. Ann. -Rev.*, **1**, 177.

Dixon, P. (1965). Chromosomes of the Algae, edit. by M. Godward, p. 169–204, (Arnold, London).

Drew, K. M. (1939). *Ann. Bot. Lond.*, *N.S.*, **3**, 347.

Grubb, V. M. (1925). *J. Linn. Soc. Bot. Lond.*, **47**, 177.

Fritsch, F. E. (1945). *Structure and Reproduction of the Algae*, Vol. **2**, p. 397–417. (Cambridge University Press).

Kylin, H. (1956). *Die Gattungen der Rhodophyceen*, (Lund).

Papenfuss, G. F. (1966). *Phycologia*, **5**, 247.

PARASITES

Fan, K–C. (1961). *Nova Hedwigia*, **3**, 119.

Feldmann, J. and G. (1958). *Revue gén. Bot.*, **65**, 49.
Pocock, M. A. (1955). *Proc. Linn. Soc. Lond.*, **167**, 11.
PIT CONNECTIONS
Dixon, P. S. (1963). *Taxon*, **12**, 108.
BANGIOPHYCIDAE
Baker, K. M. (1956). *Bot. Rev.*, **22**, 553.
Dangeard, P. (1968). *Le Botaniste*, **51**, 5.
Bangia
Yabu, H. (1967). *Bull. Fac. Fish. Hokkaido Univ.*, **17**, 163.
Boldia
Nichols, H. W. (1964). *Am. J. Bot.*, **51**, 653.
Compsopogon
Nichols, H. W. (1964). *Am. J. Bot.*, **51**, 180.
Cyanidium
Bailey, R. W. and Staehelin, L. A. (1968). *J. gen. Microbiol.*, **54**, 269.
Chapman, D. J. (1965). Ph.D. Thesis, University of California. University Microfilms, Ann Arbor, Michigan.
Hirose, H. (1959). *Bot. Mag., Tokyo*, **71**, 347.
Mercer, F. V. Bogorad, L. and Mullens, R. (1962). *J. Cell. Biol.*, **13**, 393.
Staehelin, L. A. (1968). *Proc. R. Soc.*, **171**B, 249.
Erythrocladia
Nichols, H. W. and Lissant, E. K. (1967). *J. Phycol.*, **3**, 6.
Erythrotrichia
Heerebout, G. R. (1968). *Blumea*, **16**, 139.
Porphyra
Dring, M. J. (1967). *J. mar. biol. Ass. U.K.*, **47**, 501.
Fukuhara, E. (1968). *Bull. Hokkaido reg. Fish. Res. Lab.*, **34**, 40.
Iwasaki, H. (1965). *J. Fac. Fish. Anim. Husb. Hiroshima Univ.*, **6**, 133.
Kurogi, M. (1968). *Bull. Tohoku Reg. Fish. Res. Lab.*, **18**, 1.
Migita, S. (1967). *Bull. Fac. Fish. Nagasaki Univ.*, **24**, 55.
Migita, S. and Abé, S. (1966). *Bull. Fac. Fish. Nagasaki Univ.*, **20**, 1.
Suto, S. (1963). *Bull. Jap. Soc. Scient. Fish.*, **29**, 739.
Yabu, H. and Tokida, J. (1963). *Bull. Fac. Fish. Hokkaido Univ.*, **14**, 131.
Porphyridium
Gantt, E. and Conti, S. F. (1965, 1966). *J. Cell Biol.*, **26**, 365; **29**, 423.
Gantt, E. Edwards, M. R. and Conti, S. F. (1968). *J. Phycol.*, **4**, 65.
Pringsheim, E. G. (1968). *Arch. Mikrobiol.*, **61**, 169.
Smithora
Hollenberg, G. J. (1959). *Pac. Naturalist*, **1**, 1.
Richardson, N. and Dixon, P. S. (1969). *Br. Phycol. J.*, **4**, 181.

FLORIDEOPHYCIDAE

The plants consist of branched filaments, or are compact pseudoparenchymatous or membranous thalli with uniaxial or multiaxial type of construction. Fundamentally they are heterotrichous though this condition is lacking in many. Growth is normally from one or more apical cells and plasmodesmata are present. The carpogonium is usually highly specialized and only one spermatium is produced from each spermatangium. The zygote nucleus commonly passes into a special auxiliary cell and after fertilization there is a parasitic phase, the carposporophyte, producing carpospores. This is nearly always diploid and the carpospores give rise to an asexual generation by means of tetraspores.

NEMALIONALES

A primitive order in which the vegetative structure is both uni- and multi-axial. In many of the species the construction is loose and the main filaments can be separated when on a slide by pressure on the coverslip. There are no true auxiliary cells in many genera (*Batrachospermum*, *Nemalion*) while in others the so-called auxiliary cell is largely nutritive (*Galaxaura*).

The Nemalionales initially were made up of nine families: Chantransiaceae, Batrachospermaceae, Lemaneaceae, Naccariaceae, Achrochaetiaceae and Bonnemaisoniaceae which are all uniaxial; and Thoreaceae, Helmintho-cladiaceae, and Chaetangiaceae which are multiaxial. In the late 1930's Feldmann removed the Bonnemaisoniaceae to an order of their own, primarily on the basis of a heteromorphic alternation of generations. More recently Chihara (1962) and Levring (1954) have shown that this is not a universal feature of the family. Fan (1961) has argued that an order Bonne-maisoniales (comprising the Bonnemaisoniaceae and Chaetangiaceae) can be justified on cytological grounds by the presence of a generative (or nutritive) auxiliary cell. The Bonnemaisoniales are thus characterized by the presence of this generative (nutritive) auxiliary cell which is the hypogynous cell of the carpogonial branch. The Nemalionales which comprise the six remaining families, lack this auxiliary cell. We have followed this suggestion although it must be emphasized that many authorities do not accept the Bonnemaisoniales and that a satisfactory classification for these primitive Florideophycidae has still to be worked out.

ACROCHAETIACEAE: *Acrochaetium* (*acro*, sharp; *chaetium*, hair). Fig. 11.7

The genera of this family are characterized by a uniaxial uncorticated construction of branched filaments. Many of the species are minute epiphytes or endophytes with the monosiphonous branched threads creeping between the host cells. *Acrochaetium* is the largest genus and of world-wide distribution. In certain species there is often a basal layer that gives rise to erect monosiphonous branched threads in which each cell contains one chloroplast so that the plants are heterotrichous. Lateral branches sometimes terminate in a long cellular hair. Reproduction is primarily by monospores (occasionally bispores or tetraspores) produced on lateral or terminal sporangia. Recently it has been shown that an *Acrochaetium*-like alga is the alternate generation of *Liagora* (page 273), and many species of *Acrochaetium* for which only asexual reproduction is known may in fact be alternate generations of other Rhodophyceae. Sexual reproduction is known for a few species which may be monoecious or dioecious. Groups of two or three spermatangia are formed on small side branches and carpogonia are also formed.

After fertilization the gonimoblasts grow out from the base of the car-pogonium, the terminal cell of each row becoming a carposporangium. In allied genera alternation of the sexual and asexual generation has been reported. In *Rhodochorton floridulum* the macroscopic plant reproduces by tetraspores which give rise to a dwarf generation that may be haploid. In *R. purpureum*, West (1969) has successfully grown the slightly smaller game-tophyte generation in culture. Further work on members of the *Acrochaetium* assemblage is needed in order to establish the general character of the life histories.

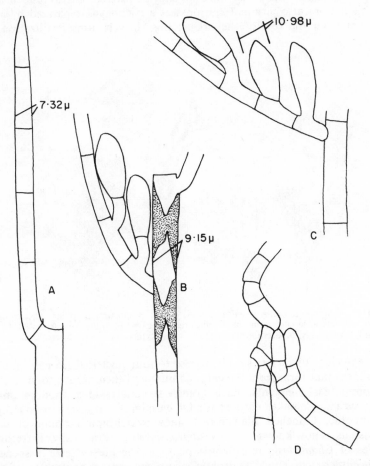

Fig. 11.7   *Acrochaetium jamaicensis*. A, branch; B, C, sporangia; D, base of plant (after Chapman)

BATRACHOSPERMACEAE: *Batrachospermum* (*batracho*, frog; *spermum*, seed). Fig. 11.8

   B. moniliforme, which is a very variable species, is found attached to stones in swift-flowing fresh waters of the tropics and temperate regions. The thallus, which is violet or blue-green in colour, is soft, thick and gelatinous, and the primary axis, which grows from an apical cell, comprises a row of large cells. Numerous branches of limited growth arise in whorls from the nodes and often terminate in long hairs. The basal cell of such a branch can give rise to a branch of unlimited growth or else to corticating cells that grow down and invest the main axis. All cells are uninucleate and contain one chloroplast and pyrenoid. Reproduction takes place by means of monospores (restricted to the juvenile phase), carpogonia and antheridia. The antheridia arise as small, round, colourless cells at the apices of short, clustered, lateral branches. The carpogonia are also terminal and possess a trichogyne separated by a constriction from the rest of the organ; this shrivels after fertilization. The

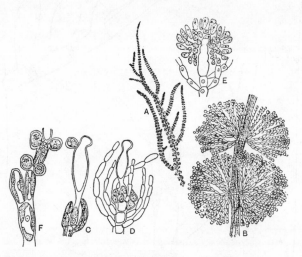

Fig. 11.8 *Batrachospermum moniliforme*. A, plant; B, portion of plant; C, carpogonial branch; D, fertilized carpogonium; E, mature cystocarp; F, antheridia. (C × 480, D × 360, E × 240, F × 640) (A, B, after Oltmanns; C–F, after Kylin)

zygote nucleus divides twice giving rise to four nuclei; these pass into protuberances that form from the carpogonium and then branch to give a mass of gonimoblast filaments. Each branch terminates in a sporangium that produces a single naked carpospore. On germination it gives rise to a diploid microscopic, branched filamentous stage, which, on account of earlier confusions, is now known as the '*Chantransia*' stage. Practically all freshwater species of *Chantransia* represent a phase in the life cycle of a species of *Batrachospermum*. The '*Chantransia*' phase can reproduce itself indefinitely, but eventually an apical cell undergoes reduction division, three nuclei abort and the new haploid cell gives rise to a new adult *Batrachospermum* plant. The gametophyte is therefore parasitic on the tetrasporophyte while the carposporophyte is parasitic on the female gametophyte.

The freshwater *Lemanea* (Lemaneaceae) has a more compact, pseudoparenchymatous thallus and much the same kind of structure is found in the marine Naccariaceae. In *Lemanea rigida*, Magne (1964) has shown that the carpospores have double the number of chromosomes (34) found in the vegetative cells (17). The nature of the sporophyte is at present unknown but one may expect it to be a dwarf filamentous alga by analogy with other Nemalionales. In *L. catenata* the carpospore germinates into a diploid protonema from which the adult arises after reduction division in the apical cell of a cladome (Magne, 1967). The peculiar freshwater genus *Thorea* (Thoreaceae) reproduces only by monospores and appears to have lost sex organs.

HELMINTHOCLADIACEAE: *Liagora* (after one of the Nereids). Fig. 11.9

The axial portion of the 'fountain'-type thallus contains large cells with which are intermingled narrower rhizoidal cells. Calcium carbonate is sometimes deposited to a greater or lesser extent on the outside of the thallus so that the plants appear to be white. One important feature is that in four species the carpospores divide to give four spores that have been regarded as

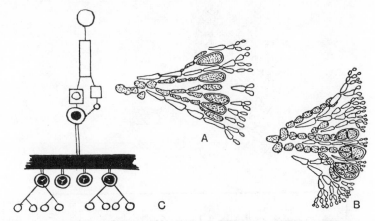

Fig. 11.9 *Liagora* A, carpospores of *L viscida*; B, carpospores in fours in *L tetrasporifera*; C, life cycle of *L. tetrasporifera*. (A, B, × 320) (A, B, after Kylin; C, after Svedelius)

tetraspores. Although no cytological evidence is available it is presumed that meiosis is delayed to the time when the carpospores divide. There is thus no independent tetrasporic diploid generation. This phenomenon was first discovered in the Atlantic *L. tetrasporifera*. A similar state of affairs has more recently been recorded for other species of *Liagora* and for *Helminthocladia agardhiana* (*H. hudsoni*).

This condition is not universal within the genus *Liagora* because recently it has been shown that the carpospores of *L. farinosa* germinate to an '*Acro-chaetium*'-like plant that produces, after meiosis, tetraspores that yield a new *Liagora* plant. The dwarf phase can be perpetuated by means of monospores (von Stosch, 1965). In this species, therefore, there is an alternation of unlike generations, as there is in *Nemalion helminthoides* and in *N. pulvinatum*.

BONNEMAISONIALES

The members of this order were formerly placed in the Nemalionales.

Thallus construction is both uniaxial and multiaxial and the life history often includes a dwarf sporophyte generation, though this is not absolutely universal. The gonimoblasts are surrounded by a wall which forms part of the cystocarp.

BONNEMAISONIACEAE: *Asparagopsis* (like asparagus fern). Fig. 11.10

The erect shoots arise from a perennial creeping thallus which possesses vesicular or iodine-containing cells whilst modified branches act as attachment organs. In the erect thallus primary and secondary laterals form a cortex which is often separated from the original axis by a space. In all Bonnemaisoniaceae except *Asparagopsis*, the thallus is flattened and the appendages arranged in two rows.

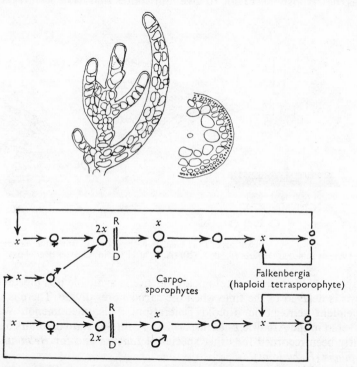

Fig. 11.10 *Asparagopsis armata*. Top left, apical portion of thallus (× 473); right, t.s. showing original main axis in centre (× 33) (after Kylin); diagram of life cycle of *Asparagopsis armata*.

In four species there is an alternation of two unlike generations, the diploid plants being small and so different from the sexual generation that they have been placed in separate genera. In *A. armata* the asexual generation is *Falkenbergia rufo-lanosa*, (Feldmann, 1966) in *A. taxiformis* it is *F. hillebrandii* (Chihara, 1962), in *Bonnemaisonia hamifera* it is *Trailliella intricata*, and in *B. asparagoides* it is *Hymenoclonium serpens*. In the last named species it is possible that meiosis occurs in the carposporangia because the dwarf *Hymenoclonium* plants give rise vegetatively to the adult gametophytes (Feldmann, 1966). In the genera *Leptophyllis* and *Delisea* isomorphic tetrasporic generations have been recorded for *Leptophyllis conferta*, *D. suhrii* and also for *D. fimbriata* (*D. pulchra*) in Australian waters, though the last species is reported as having only the haploid generation in Japanese waters. The same kind of haplophasic life history is also reported for the Japanese species *D. elegans*, *D. hypneoides* and *Ptilonia okadai* (Martin 1969). There are therefore three types of life history to be seen in the family:

(a) *Asparagopsis* type with a dwarf asexual generation, which can be either diploid (*A. armata*) or haploid (*B. asparagoides*).

(b) *Ptilonia okadai* type with only the sexual generation and carposporophyte.

(c) *Polysiphonia* type (p. 301) with isomorphic alternation.

CHAETANGIACEAE: *Scinaia* (after D. Scina) Fig. 11.11

The reasons for uniting the Chaetangiaceae with the Bonnemaisoniaceae have been outlined by Fan (1961). They are based primarily on the function of the hypogynous cell of the carpogonial branch as a nutritive auxiliary cell similar to that of *Asparagopsis*.

Scinaia is a widespread genus with most of its species in the northern hemisphere. The corresponding genera in the southern hemisphere are *Pseudoscinaia* and *Pseudogloiophloea*. The fronds, which arise from a discoid holdfast, are subgelatinous, cylindrical or compressed and dichotomously branched. The centre of the thallus is composed of both coarse and fine colourless filaments, the former arising from the apical cell and the latter from the corticating threads. There is also a peripheral zone of horizontal filaments that terminate in short corymbs of narrow assimilatory hairs interspersed with large colourless cells. These two types of epidermal cells are apparently differentiated near the apex of the thallus, the small ones giving rise to hairs, monosporangia or antheridia. The large colourless cell is thought to give protection against intense light, but it may also be a relic of a tissue which formerly had a function that has since been lost. One of two spores are formed in each monosporangium, while the spermatia arise in sori, forming bunches of cells at the ends of the small-celled assimilatory branches. The commonest species, *S. furcellata*, is monoecious, and the plants bear three-celled carpogonial branches, the reproductive cell containing two nuclei, one in the carpogonium proper and one in the trichogyne. The second cell of the carpogonial branch gives rise to a group of four large cells which are rich in protoplasm, while the sterile envelope of the cystocarp arises from the third cell.

After fertilization the zygote nucleus passes into one of the four large cells. From this large cell a single gonimoblast filament grows out which later becomes branched. Recently, (Boillot, 1968) the carpospores have been shown to give rise to a filamentous gametophyte. This represents another example of a heteromorphic alternation of generations.

In the allied genus *Galaxaura* the diploid carpospores give rise to a tetra-

K

spore generation differing morphologically in some degree from the sexual generation. There is thus an alternation of unlike generations, though the difference is in appearance rather than in size.

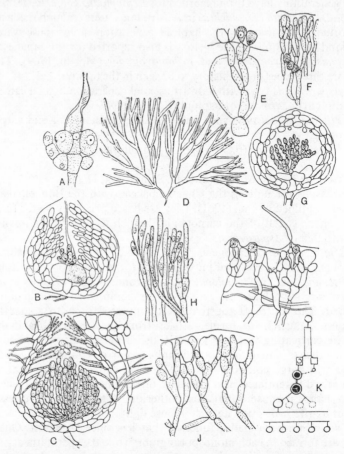

Fig. 11.11 *Scinaia furcellata*. A, carpogonial branch; B, fertilized carpogonium; C, cystocarp; D, plant; E, antheridia; F, young carpogonial branch; G, young cystocarp; H, undifferentiated threads at apex of thallus; I, monospores and a hair; J, differentiated cortex; K, life-cycle diagram. (A, E, × 700, D × ½, C × 195, F, H, × 425, G × 232, I × 340, J × 429) (C, after Setchell; A, B, E-K, after Svedelius)

GELIDIALES

The thallus is more compact than in most Nemalionales and all members are uniaxial in construction. True auxiliary cells are absent, but special nutritive cells are formed in association with the carpogonia. The order contains plants that provide the best sources of agar-agar in the world. There are three families in the order but the two largest and commonest genera, *Gelidium* and *Pterocladia* are found in the one family.

GELIDIACEAE: *Gelidium* (congealed). Fig. 11.12

In the stiff, cartilaginous thallus, which is often pinnately branched, the uniaxial construction is most clearly seen at the apex, but in the older thallus there is a central portion of large cells, derived from four primary pericentral cells, with a peripheral zone of small cells with interspersed hyphae. The

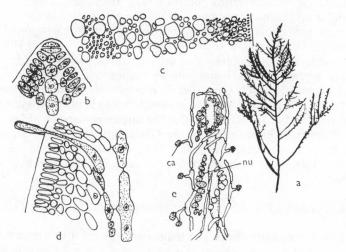

Fig. 11.12 *Gelidium corneum.* a, plant; *G. cartilagineum* b, apex; c. section of thallus; d, carpogonial branch; e, carpospores (ca) and nutritive cells (nu). (a, × ⅓; b–e, × 530) (a. after Oltmanns; b–e, after Kylin)

presence of special nutrient cells in the vicinity of the carpogonial branch results in the production of a complex structure composed of several carpogonia with small nutrient cells in short branches, and only one of these carpogonia needs to be fertilized. After fertilization a single gonimoblast filament initially grows out into the mass of nutritive cells, but it later branches so that a number of terminal carposporangia are finally formed. In *Gelidium* the cystocarp wall develops two pores or ostioles, one on either side of the supporting branch. *Pterocladia* has only one ostiole. The carpospores germinate to give a tetrasporophyte morphologically identical to the gametophyte. In the Japanese *G. amansii*, the principal source of agar, the tetraspores are shed when the water temperature rises to 20° C and the carpospores when it rises to 24° C, and this shedding occurs each afternoon. The spores become fixed to the rock surface when they have been in contact for about ten minutes. The maximum temperature for germination is about 25°–26° C. The sporelings can grow in brackish water and this markedly affects the length of the primary rhizoids. Growth is slow at first but later becomes very rapid. More than one erect shoot may arise from each basal pad or the lateral branch initials may grow horizontally giving rise to stolons from which more initials may arise.

CRYPTONEMIALES

The members of this order are very diverse in habit. A number are uniaxial and similar in structure to the Gelidiales, while others are multiaxial. There is a whole group in which there is extensive lime encrustation and considerable reduction of the thallus, e.g. *Lithothamnion*. Definite auxiliary cells are produced on accessory branches prior to fertilization and these serve not only for nutrition but also as starting points for the gonimoblast filaments.

The order divides naturally into two main subdivisions. In the first group the primary gonimoblasts fuse with nutritive cells in the carpogonial branch and secondary gonimoblasts then fuse with the auxiliary cells. In the second group the primary gonimoblasts fuse directly with auxiliary cells. In the first group are the Dumontiaceae and here the carpogonial branch and the auxiliary cell branch are widely separated. In the remaining families of this group the carpogonial branches are borne in nemathecial groups. The second group is also subdivided according to whether the carpogonia and auxiliary cells are borne on separate branches as in the Grateloupiaceae or on an adjacent branch. In certain genera, e.g. *Callophyllis*, *Weeksia*, procarps are present. There are twelve families in the order, but some of these are small and unimportant.

DUMONTIACEAE: *Dudresnaya* (after Dudresnay de St. Pol-de-Leon). Fig. 11.13

The cylindrical, highly branched thallus arises from a prostrate disc. It is soft and gelatinous and consists when young of a simple, articulated, filamentous axis with whorls of dichotomously branched ramuli of limited growth. In older plants the axis becomes polysiphonous and densely beset with whorls of branches. The polysiphonous condition is produced by enlargement of the basal cells of the primary branch whorls and of cortical threads produced from them. The plants are dioecious, the males being somewhat smaller and paler. The carpogonial branches of *D. coccinea* arise from lower cells of short side branches and when fully developed consist of seven to nine cells; they branch once or twice and may have short sterile side branches arising from the lowest cell. In the middle of the mature carpogonial branch there are two to three larger cells which function in a purely nutritive capacity, while the auxiliary cells develop in similar positions on neighbouring homologous branches. After fertilization the carpogonium sends down a protuberance containing the diploid nucleus and this cuts off two cells when it is near to the nutrient cells of the carpogonial branch. These all fuse together and sporogenous threads, each carrying a diploid nucleus, then grow out towards the auxiliary cells on the other branches. When these filaments, or connecting threads, fuse cytoplasmically with an auxiliary cell, the latter forms a protuberance into which the diploid nucleus of the connecting thread passes. When the nucleus has divided the protuberance containing one of the daughter nuclei is cut off by a wall. The gonimoblast filaments then grow out as a branched mass from this protuberance. Each sporogenous thread sent out from the original cell may unite with more than one auxiliary cell in the course of its wanderings through the thallus, so that one fertilization may result in the production of a number of carposporophytes.

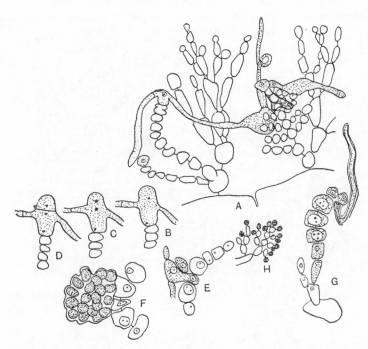

Fig. 11.13 *Dudresnaya.* A-D, stages in development of cystocarp, *D. purpurifera*; E, F, stages in development of cystocarp in *D. coccinea* after fertilization; G, *D. coccinea*, carpogonial branch; H, *D. coccinea*, antheridia. (E-G × 486, H × 510) (A-D, after Oltmanns; E-G, H, after Kylin)

SQUAMARIACEAE: *Peyssonelia* (after S. A. Peyssonel). Fig. 11.14

The characteristic members of this family are crusts of which *Peyssonelia* is the most common. The thallus is blood red and more or less firmly attached to the substrate; in some species there is lime encrustation. It commences as a monostromatic layer that gives rise on the upper surface to sparingly branched threads that become fused to each other to give a parenchymatous-like tissue. Rhizoids develop on the lower side of the primary thallus and serve as attachment organs. The spermatangia are borne laterally on erect threads that are in soral groups or nemathecia. Pericentral cells are cut off and these then act as spermatangial mother cells. The carpogonia are also borne in nemathecia. Within the nemathecium there are six to eight celled paraphyses, the lowermost cell of which can give rise to a carpogonial branch or an auxiliary cell branch. Both of these branches contain four cells. After fertilization the carpogonium fuses with the second cell of the carpogonial branch and then from this fusion cell a primary gonimoblast thread grows out to fuse with the second cell (auxiliary cell) of an auxiliary-cell branch.

The secondary gonimoblast is then formed. It comprises 8–12 cells all of which, except the basal, become carposporangia. In the tetrasporangiate plants the cruciate sporangia are also borne in nemathecia.

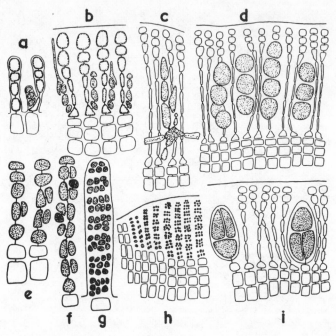

Fig. 11.14 *Peysonnelia pacifica.* a-d, development of the carpogonial branches, the auxiliary cell branches and gonimoblasts; e-h, development of spermatangia; i, tetrasporangia. (a, b, × 350, c, d, × 260, e-g × 475, h-i × 260) (after Kylin)

HILDENBRANDIACEAE : *Hildenbrandia* (after F. E. Hildenbrandt). Fig. 11.15
This genus also forms thin brownish-red crusts which adhere strongly to stones or other algae, and it is frequently difficult to distinguish from similar encrusting types such as *Peyssonelia*. The frond is horizontally expanded into a thin encrustation of several layers of cells arranged in vertical rows. The

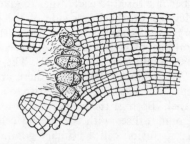

Fig. 11.15 *Hildenbrandia prototypus.* Tetraspores in conceptacles. (× 320) (after Taylor)

genus is both marine and fresh water, *H. rivularis* appearing frequently in rivers and streams. The principal mode of reproduction is by means of fragmentation, stolons (Nichols, 1965) or tetraspores produced in sporangia borne in rounded or oval conceptacles that are sunk in the thallus. The absence of sexual plants suggests specialization, though it is possible that a protonemal gametophyte exists though this has not yet been observed.

CORALLINACEAE: *Lithothamnion*; *Corallina*. (Fig. 11.16)

Heavy calcification (calcite) is the characteristic of this family which may be divided into two subfamilies:

(*a*) The Melobesioideae, which comprises the crusts and nodular Corallinaceae, e.g. *Lithothamnion, Melobesia, Heteroderma*.

(*b*) The Corallinoideae, which is composed of the jointed Corallinaceae, e.g. *Corallina, Jania, Amphiroa, Bossea*.

In *Lithothamnion* there is a multi-layered basal zone of horizontally aligned cells, the hypothallium, and a surface layer of vertically aligned cells, the perithallium. The various reproductive organs are borne in conceptacles on separate plants. The aggregation of the reproductive organs into conceptacles must be regarded as equivalent to nemathecia which subsequently become overgrown by surrounding tissue. In the male plants there are a number of two-celled filaments in the centre of each conceptacle. The basal cells of these threads cut off two elongate, antheridial mother cells, which in their turn produce two antheridia. Antheridial mother cells are also produced from the lower parts of the conceptacle walls. Since the antheridium is liberated entire it must be regarded as morphologically equivalent to a spermatium.

In the female plant the central threads of a conceptacle form three-celled carpogonial branches, the basal cell of the branch being an auxiliary cell. The outer threads of a conceptacle produce sterile two-celled filaments. Only the central carpogonia mature and usually only one of these is fertilized and produces carpospores. After fertilization the carpogonium and the cell below it fuse together and send out a filament to the lowest cell (auxiliary cell). Later all the auxiliary and nutritive sterile cells fuse to give one long fusion cell from which very short gonimoblast filaments grow out. In the tetrasporic plant there are simple filaments which give rise not only to the tetrasporangia but also to branched sterile filaments that grow out and form the conceptacle roof by division and elongation. Finally the membrane of the tetrasporangia penetrate this roof and form exit pores. The spores are arranged in a linear series.

The multiaxial type of construction can be observed extremely well in *Corallina*. The erect plants, which are jointed, cylindrical or compressed, arise from calcified, encrusting basal discs or prostrate interlaced filaments. Branching is generally pinnate, but in the closely allied genus *Jania* it is dichotomous. There is a central core of dichotomously branched filaments with oblique filaments growing out at the swollen internodes to form a cortical layer, the end cells being flattened and the whole encrusted by lime except at the joints. The plants are monoecious or dioecious, the reproductive organs being borne in terminal or lateral conceptacles. The carpogonia arise from a kind of prismatic disc formed from the terminal cells, which also function later as auxiliary cells. As a result of oblique divisions, one to three carpogonial branches are formed on each mother cell, but only one of these finally develops into the mature, two-celled carpogonial branch. After fertilization a long or rounded fusion cell is formed by the auxiliary cells,

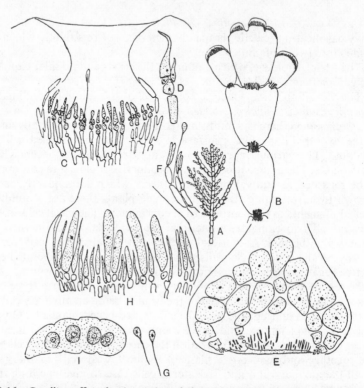

Fig. 11.16 *Corallina officinalis.* A, portion of plant; B, the same, enlarged; C, carpogonial conceptacle; D, single carpogonial branch; E, fusion cell, gonimoblasts and carpospores; F, development of antheridia; G, mature spermatia; H, young tetrasporic conceptacle; I, mature tetraspores; (C × 210, D × 342, E × 120, F × 420, G × 648, H × 240, I × 270) (A, B, after Oltmanns; C–I, after Suneson)

and this contains both fertilized and unfertilized carpogonial nuclei. The antheridia are elongated, and after liberation the spermatia round off and remain attached to the antheridial wall by means of a long, thin pedicel in *C. officinalis* and by a short stalk in *Jania rubens*. The tetrasporangia comprise a stalk cell and sporangium in which the four spores are arranged in a linear row.

GRATELOUPIACEAE: *Grateloupia* (After J. P. Grateloup). Fig. 11.17
    The plants vary in size depending on the species but they can be as long as 75 cm, with a narrow or broadly flattened axis, dichotomously or pinnately

branched. Some species bear proliferations from the margin of branches or from the surface. The thallus characteristically feels firm and gelatinous. In section there is a central region with small, elongated, anastomosing filaments which arise from star-shaped cells, while the cortex consists of rows of dichotomously branching filaments of moniliform appearance. The structure is therefore multiaxial. Primary gonimoblasts grow directly from the fertilized carpogonia to auxiliary cells borne on separate branches. The first cell of the secondary gonimoblast that arises from the auxiliary cell gives rise to numerous branches; nearly all the cells in the branches produce carpospores. The tetrasporangia are distributed throughout the cortical region in the sporophytic plants. There are a large number of species which are mostly restricted to the warmer seas.

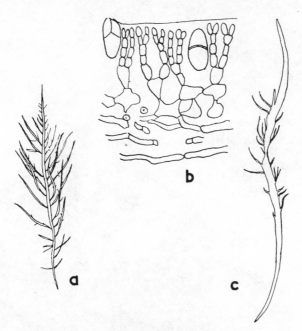

Fig. 11.17 a, *Grateloupia filicina*; b, *G. dichotoma*, t.s. of thallus with tetraspores; c, *G. prolongata*. (a × ⅔, b × 365) (a, after Chapman; b, after Boergesen; c, after Kützing)

KALLYMENIACEAE: *Callophyllis* (*callos*, beauty; *phyllis*, leaf). Fig. 11.18

This family differs from the preceding one in that the auxiliary cell is borne at the base of the carpogonial branch; it is in fact the support cell and with the carpogonium forms a procarp, an exceptional feature in this family. There are therefore no secondary gonimoblasts. The membranaceous thallus is flattened with either dichotomous or palmate branching. The central region is occupied by a mass of large cells with smaller ones intermingled, while the cortex consists of rows of short filaments arranged at right angles to the surface. The cystocarps are normally borne in outgrowths from the margin of the frond.

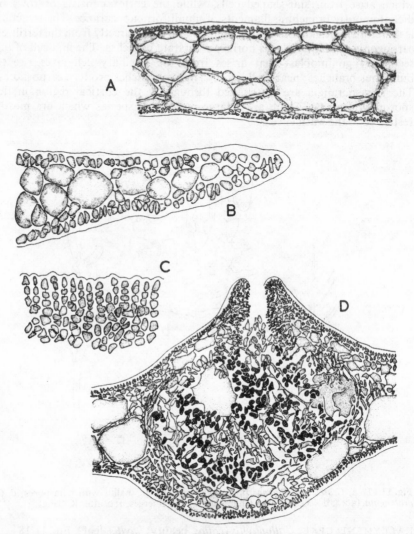

Fig. 11.18 *Camera lucida* drawings of *Callophyllis taenophylla*. A, section through mature part of blade; B, l.s. through growing margin of blade; C, surface view of growing margin of blade; D, section through young cystocarp. (A, D, × 186, B, C, × 666) (after Norris)

CHOREOCOLACACEAE: *Harveyella* (after W. H. G. Harvey). Fig. 11.19

This and the closely allied genus *Holmsella* are monotypic genera each containing a holo-parasitic species, and *Choreocolax* is another parasitic genus very nearly related to them. *Harveyella mirabilis* is parasitic on species of *Rhodomela* while *Holmsella pachyderma* is a parasite of *Gracilaria confer-voides*. They have little or no colour, as might be expected from parasites, and they send out branched filaments or haustoria into the tissues of the host. These penetrate the middle lamellae of the host cells and form secondary pit connections with them. The parasites appear as external cushions lying on the branches of the host, each cushion surrounded by an outer gelatinous

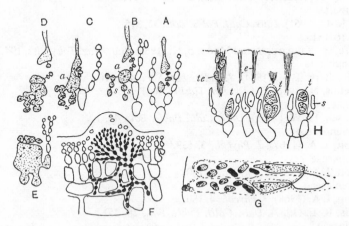

Fig. 11.19 *Harveyella* and *Holmsella*. A-E, stages in development of gonimoblasts after fertilization in *Harveyella mirabilis* (*a* = auxilliary cell, *s* = sterile filaments); F, filaments of parasite, *Holmsella pachyderma*, in host; G, antheridia of *Harveyella mirabilis*; H, tetraspores in *Holmsella pachyderma* (*e* = tracks left after tetraspores have escaped, *s* = sterile cells, *t* = tetraspores in various stages, *te* = escaping tetraspores). (after Sturch)

coat and consisting of a central area that is four to five cells thick. In *Harveyella* the female cushions bear numerous procarps with four-celled carpogonial branches, each with two sterile branches. In *Holmsella* the carpogonial branch is two-celled. The species are said to pass through the full, floridean life cycle twice every year. It is clear that their reduced morphological features are associated with their parasitic habit. In the past the family has been placed in the Gigartinales, but is now more generally placed in the Cryptonemiales.

REFERENCES

GENERAL

Dixon, P. S. (1963). *Oceanogr. Mar. Biol. Ann. Rev.*, **1**, 177.

Dixon, P. S. (1965). In *Chromosomes of the Algae*, edit. by M. Godward, p. 169–204. (Arnold, London).

Kylin, H. (1956). *Die Gattungen der Rhodophyceen*, (Lund).

Fritsch, F. E. (1945). *Structure and Reproduction of the Algae*, Vol. **2**. 418–441 (Cambridge University Press).

Magne, F. (1964). *Cahiers Biol. Marine.*, **5**, 461.

Magne, F. (1967) *Cr. hebd. Scéanc. Acad. Sci., Paris*, **264**D, 2632.

Martin, M. T. (1969). *Br. Phycol. J.*, **4** (2), 145.
ULTRASTRUCTURE
Giraud, G. (1963a). *Phys. végét.*, **1**, 203.
Giraud, G. (1963b). *J. Microscopie*, **1**, 251.
BONNEMAISONIALES
Chihara, M. (1961, 1962). *Sci. Rep. Tokyo Kyoiku Daig.*, **10**, 121; **11**, 127.
Feldmann, G. (1965). *Revue gén. Bot.*, **72**, 621.
Levring, T. (1953, 1955). *Ark. Bot., Ser.*, 2, **2**, 457; **3**, 407.
GELIDIALES
Fan, K–C. (1961). *Univ. Calif. Publs. Bot.*, **32**, 315.
NEMALIONALES
Hirose, H., Kumaro, S. and Madono, K. (1969). *Bot. Mag. Tokyo*, **82**, 197.
Chaetangiaceae
Desikachary, T. V. (1964). *J. Indian bot. Soc.*, **42**A, 16.
Svedelius, N. (1956). *Svensk. bot. Tidskr.*, **50**, 1.
Corallinaceae
Johansen, H. W. (1969). *Univ. Calif. Publs. Bot.*, **49**, 1.
Cryptonemiaceae
Abbott, I. A. (1967). *J. Phycol.*, **3**, 139.
Dumontiaceae
Abbott, I. A. (1968). *J. Phycol.*, **4**, 180.
Kallymeniaceae
Abbott, I. A. (1968). *J. Phycol.*, **4**, 180.
Norris, R. E. (1957). *Univ. Calif. Publs. Bot.*, **28**, 251.
Naccariaceae
Womersley, H. B. S. and Abbot, I. A. (1968). *J. Phycol.*, **4**, 173.
Achrochaetium
von Stosch, H. A. (1965). *Br. phycol. Bull.*, **2**, 486.
West, J. A. (1968). *J. Phycol.*, **4**, 89.
Asparagopsis
Feldmann, G. (1966). *C. r. hebd. Séanc. Acad. Sci., Paris*, **262**, 1695.
Batrachospermum
Brown, D. L. and Weier, T. E. (1968). *J. Phycol.*, **4**, 199.
Dixon, P. A. (1958). *Bot. Notiser*, **111**, 645.
Bonnemaisonea
Magne, F. (1960). *C. r. hebd. Séanc. Acad. Sci., Paris*, **250**, 2742.
Callophyllis
Bert, J. J. (1967). *Revue gén. Bot.*, **74**, 5.
Chen, L. C–M. Edelstein, T. and MacLachlin, J. (1969). *J. Phycol.*, **5**, 211.
Galaxaura
Svedelius, N. (1953). *Nova Acta R. Soc. Scient. Upsal.*, iv, **15**, 1.
Gelidium & Pterocladia
Boillot, A. (1963). *Revue gén. Bot.*, **70**, 130.
Dixon, P. S. (1958). *Ann. Bot. N. S.*, **22**, 353.
Dixon, P. S. (1959). *Ann. Bot. N. S.*, **23**, 397.
Grateloupia
Balakrishnan, M. S. (1961). *J. Madras Univ.*, **31**, 11.
Hildenbranchia
Nichols, H. W. (1965). *Am. J. Bot.*, **52**, 9.
Lithothamnion
Adey, W. H. (1966). *Hydrobiologia*, **28**, 321.

Nemalion
  Fries, L. (1968). *Svensk. bot. Trdskr.*, **61**, 457.
  Magne, F. (1961). *Cr. hebd. Séanc. Acad. Sci.*, *Paris*, **252**, 157.
  Umezaki, I. (1967). *Publs. Seto mar. Biol. Lab.*, **15**, 311.
Pseudogloiophoea
  Desikachary, T. V. and Singh, A. D. (1958). *Proc. Indian Acad. Sci.*, **47**, 163.
  Ramus, J. (1969). *Univ. Calif. Publs. Bot.*, **52**, 1.
Rhodochorton
  Knaggs, F. W. (1967). *Nova Hedwigia*, **12**, 512; **14**, 549.
  Swale, E. and Belcher, J. (1963). *Ann. Bot.*, *N.S.*, **27**, 281.
  West, J. A. (1969). *J. Phycol.*, **5**, 12.
Scinaia
  Boillot, A. (1968). *C. r. hebd. Séanc. Acad. Sci.*, *Paris*, **266**, 1831.

12

RHODOPHYTA

FLORIDEOPHYCIDEAE *(contd.)*

GIGARTINALES

Thallus construction in this order can be either of the central filament or fountain type, but it is often so modified in the adult plant that it is difficult to interpret. An ordinary intercalary cell, or occasionally the support cell, is set aside before fertilization to act as the auxiliary cell.

The major subdivision within the order depends on whether the gonimoblasts that grow out from the auxiliary cell are directed to the interior or to the exterior of the thallus. The sole exception is the family Cruoriaceae, which is regarded as the most primitive, and here both types of growth direction are to be found; also, the gonimoblasts grow out directly from the primary gonimoblast or connecting hypha. One of the two major subdivisions is divided again depending on whether the sporangia are linearly or cruciately divided; the groups of the other major subdivision depend on the number of gonimoblasts and the sporangial division. The principal features of the classification based on Kylin's scheme are shown in Fig. 12.1.

CRUORIACEAE: *Cruoria* (*cruor*, blood). Fig. 12.2

The thallus is dark red and superficially looks very like *Peyssonelia* (p. 279), but it differs in that the sparingly-branched erect threads are only held together by a gelatinous substance so that under pressure they can be forced

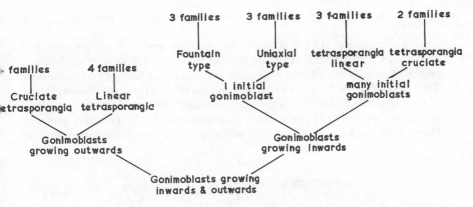

Fig. 12.1 Interrelations of Gigartinales

apart. Structurally the thallus consists of a primary, horizontal, single layer of cells attached to the substrate by short rhizoids. The erect rows of filaments that arise on the upper surface of the horizontal layer commence in the same plane, thus giving a false appearance of a multilayered basal portion. The carpogonial branches are two or three celled and arise about half way up the erect filaments. The primary gonimoblast, which is thick, fuses with cells in the middle or other threads, but the secondary gonimoblasts grow out directly from the connecting hypha and each cell functions as a carposporangium. Spermatangia are borne at the ends of the erect threads from laterally produced mother cells. The tetrasporangia are cruciately or linearly divided and arise as side branches about half-way up the assimilatory threads.

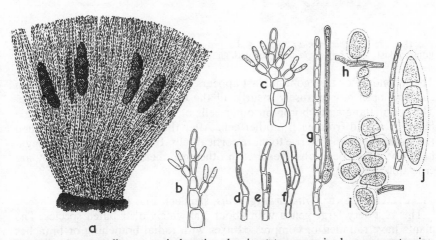

Fig. 12.2 *Cruoria pellita*. a, vertical section showing tetrasporangia; b, c, spermatangia; d–f, degenerating cells; g, carpogonial branch; h, gonimoblasts; i, gonimoblasts and carpospores; j, tetrasporangium. (a, × 75; b, c, × 470; d–g, × 292; h–j, × 165) (after Kylin)

289

NEMASTOMACEAE: *Schizymenia* (*schizo*, divide; *hymen*, membrane). Fig.
12.3

In this genus the thallus is 'leafy' but in other genera it can be branched
and is not leaf-like. The cells of the outer cortex are closely packed but
nevertheless it is possible to make out their original filamentous condition.
The cells of the medulla are loose and elongated. Interspersed in outer layers
of the cortex are gland cells, which can sometimes be seen if the thallus is held
up to the light. After fertilization the connecting hypha (primary gomino-
blast) fuses with an auxiliary cell borne on a nearby assimilatory thread.

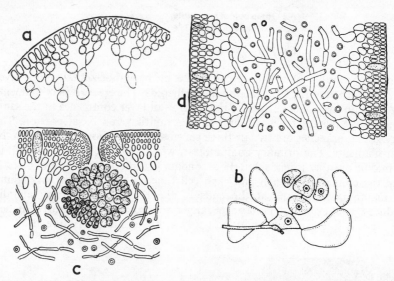

Fig. 12.3 A–C, *Schizymenia pacifica*: A, l.s. growth zone; B, young gonimoblast; C, riper
gonimoblast; D, *Schizymenia dubyi*, t.s. with gland cells. (after Kylin)

The secondary or true gonimoblast appears on the exterior side of the fusion
cell and branches profusely, nearly all the cells acting as carposporangia.
The spermatangia are borne on outer cells of the cortex, two spermatangia
being produced from each mother cell. After the spermatia have been shed
the mother cell can give rise to more cortical cells. In this genus tetrasporangia
have not so far been observed but in other members of the family they are
cruciately divided.

GRACILARIACEAE: *Gracilaria* (*gracilis*, slender). Fig. 12.4

This is a very large genus with about 100 widely distributed species. The
plants have rounded or compressed axes with radial branching or branches
arising from the sides of the flattened axes. Although an apical cell can be
recognised it is not possible to make out a primary axial thread. The thallus
is essentially composed of a mass of large central cells with smaller cells on

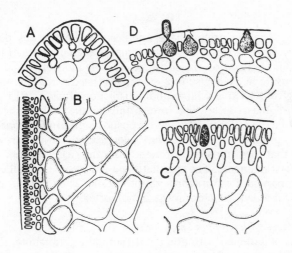

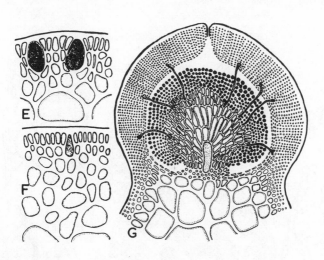

Fig. 12.4 A, B, *Gracilaria confervoides*: A, shoot apex; B, l.s. thallus; C, t.s. thallus with basal hair cell, *G. compressa*; D–F, *Gracilariopsis sjostedtii*: D, t.s. procarp; E, tetrasporangia; F, t.s. and basal cell of a hair; G, cystocarp, *Gracilaria verrucosa*. (A × 500, B–F × 275, G × 60) (after Kylin)

the periphery. The sex organs have not been studied in any detail. In the allied *Gracilariopsis sjöstedtii* the carpogonial branch consists of two plasma-rich cells. After fertilization the surrounding cells multiply rapidly to form a cystocarpic envelope. At the same time the fertilized carpogonium fuses with neighbouring cells to give a large and conspicuous fusion cell from which a mass of gonimoblasts arise. The lowest and middle cells of these filaments are sterile while the outer function as carposporangia. In *Gracilaria verrucosa* there are long filamentous cells that penetrate the cystocarpic wall and it is possible that they transfer nutriment from the richly packed wall cells to the gonimoblast filaments. The spermatangia are borne in small conceptacles and the ripe spermatia escape through the ostiole. The tetrasporangia, which are cruciately divided, develop in the outer cortex.

PLOCAMIACEAE: *Plocamium* (*plokamos*, a curl). Fig. 12.5
   The thallus is flattened with abundant sympodial branching, the branches being ultimately beset with alternating groups of two to five secund branchlets. Construction is of the central filament type and this is clearly recognisable at the apices and in transverse and longitudinal sections of the thallus. There is a three-celled carpogonial branch; the support cell serves as the auxiliary cell. The gonimoblasts develop as in *Gracilaria* and the tetrasporangia are borne in groups in special branchlets.

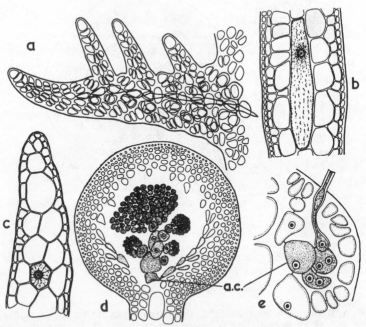

Fig. 12.5 *Plocamium coccineum*. a, side shoot; b, l.s. thallus; c, t.s. thallus; d, cystocarp; e, carpogonium. (a × 450, b × 67, c × 250, d × 165, e × 530) (after Kylin)

FURCELLARIACEAE: *Furcellaria* (*furcula*, a little fork). Figs. 12.6, 12.7

The anatomical structure is of the 'fountain' type, the central hyphal threads forming a cortex of successively smaller cells, towards the surface. The carpogonial branches develop from the innermost cortical cells. The plasma-rich auxiliary cells are borne independently of the carpogonial branch so that there is no procarp (see p. 257). After fertilization a connecting hypha grows out and fuses with an auxiliary cell from which a gonimoblast

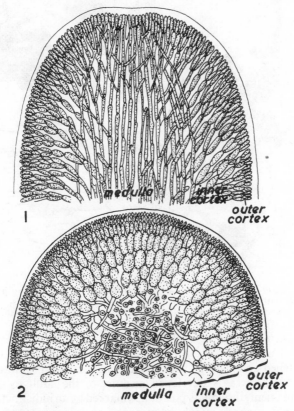

Fig. 12.6 *Furcellaria fastigiata*. 1. l.s. through apex; 2, t.s. through shoot. (after Oltmanns)

cell grows out on the interior side. This branches profusely, all cells of the branches acting as carposporangia. The mature carpospores are therefore produced in the interior of the thallus. The tetrasporangia, which are linearly divided, are found in the ultimate branchlets where they develop from the cortical filaments. In Britain, the sex organs mature in January and the carpospores and tetraspores in November, with April–May as the period of maximum vegetative growth. The single species occurs in the northern part of the Atlantic Ocean and the Baltic Sea. In recent years in Denmark and Russia it has become an important source of the material furcelleran.

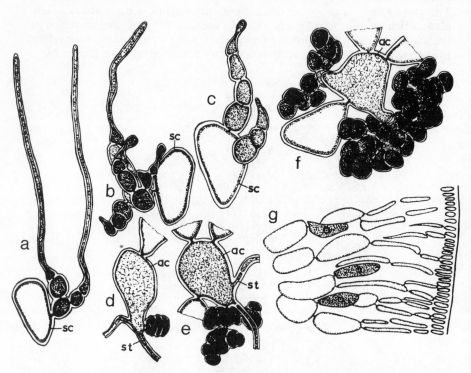

Fig. 12.7 *Furcellaria fastigiata.* a–c. carpogonial branches; d–e, young gonimoblasts; f, auxiliary cells with two gonimoblasts; g, t.s. thallus with young tetrasporangia (ac = auxiliary cell, st =sporogenous thread, sc =support cell). (a–f × 290, g × 165) (after Kylin)

RHODOPHYLLIDACEAE: *Rhodophyllis* (*rhodo*, red; *phyllis*, foliage). Fig. 12.8
The basic structure is of the central filament type, though the mature thallus is flattened and leaf-like with branches and lateral proliferations. It is not possible in the weakly developed medulla of the adult to discern any central axial filament, though its position is marked by an indistinct, branched, veining system. Growth is mainly by means of an epidermal meristem which is evidently derived from the main axial filament. The auxiliary cell is very distinct before fertilization and is derived from the support cell so that a procarp is present. There is a very large and pronounced fusion cell and in the mature cystocarp, sterile gonimoblast threads unite with the cystocarp wall to provide a passage for nutriment. The spermatangia develop from epidermal cells but are not arranged in sori. Tetrasporangia are divided linearly.

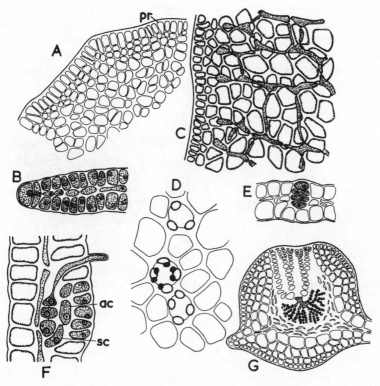

Fig. 12.8 *Rhodophyllis bifida*. A, apex of thallus from above; B, t.s. thallus; C, edge of thallus; D, thallus with antheridia seen from above; E, t.s. thallus with tetrasporangium; F, mature carpogonium; G, young cystocarp (ac = auxiliary cell, sc = support cell, pr = primary apical cell). (A, ×450; B, ×530; C, ×315; D, F, ×710; E, ×213; G, ×160) (after Kylin)

PHYLLOPHORACEAE: *Phyllophora* (*phyllo*, leaf; *phora*, bear). Fig. 12.9

The stipitate fronds expand upwards into a rigid or membranous flat lamina which is either simple or divided, and proliferations may also arise from the margin or basal disc. Morphologically the thallus is multiaxial and composed of oblong polygonal cells in the central medullary region, while the outer cortical layers consist of minute, vertically seriate, assimilatory cells. The thalli often last for many years and successive increments near branch axils or at the base of the frond are recognisable macroscopically. The plants are dioecious and the sex organs are borne in nemathecial cavities in small fertile leaflets attached to the main thallus. The carpogonial leaflets, which are sessile or have a short stalk, arise laterally from the stipitate part of the main blade. Only one cystocarp is produced from each carpogonial nemathecium. In *P. membranifolia* the carpogonial branch is three-celled and after fertilization gonimoblast filaments are formed which ramify in the tissues, finally producing pedicellate or sessile cystocarps. In *P. brodiaei* the spermatia seem to have no function and the carpogonium fuses directly with the auxiliary cell and one diploid generation is omitted. This must be regarded as a reduction phenomenon in so far as the usual rhodophycean life cycle is concerned. The resultant spores are borne in moniliform chains packed into wart-like excrescences or nemathecia which are carried on the female sexual plant. These were formerly regarded as a parasitic plant, *Actinococcus subcutaneus*. The cytology of this species is not well known.

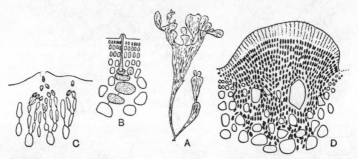

Fig. 12.9 *Phyllophora brodiaei.* A, plant; B, carpogonial branch; C, t.s. of antheridial thallus; D, nemathecia with tetraspores. (A × ⅓, B × 250, C × 450, D × 125) (B–D, after Kylin)

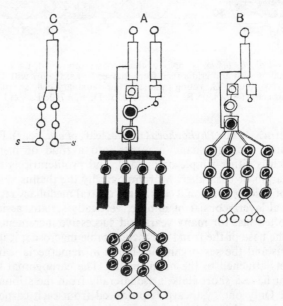

Fig. 12.10 Life cycles. A, *Phyllophora membranifolia*. B, *P. brodiaei*; C, *Ahnfeldtia plicata* (*s* = monospores) (after Svedelius)

In the Black Sea there is a huge bed of free-living species of *Phyllophora* that are used for the commercial production of agar.

The allied genus *Ahnfeldtia* with a wiry, horny thallus is a perennial in which annual growth rings can be recognised in transverse section. Reduction of the life cycle in *A. plicata* has gone still further because there is neither fertilization nor meiosis and only degenerate procarps and functionless spermatangia are formed. The nemathecia, formerly regarded as a parasite, *Sterrocolax decipiens*, contain monospores that develop as follows: the warts, which arise as small cushions from superficial cells of the thallus, contain some cells that become flask-shaped together with other cells possessing denser contents that arise in groups at the upper ends of the filaments. These latter, which probably represent degenerate carpogonia, form the generative cells and give rise to secondary nemathecial filaments in which the apical cells act as mono-sporangia, whose spores give rise to prostrate discs. In *Ahnfeldtia plicata*, therefore, there is no diploid generation (Fig. 12.10).

GIGARTINACEAE: *Gigartina* (*Gigarton*, grape stone). Fig. 12.11

This is a widespread genus, the species of which are often difficult to determine taxonomically. The plants, several of which arise from a perennial prostrate disc, vary greatly in habit from large, flat, little-branched, foliose expansions (*G. atropurpurea*) to terete fronds that are irregularly pinnate or

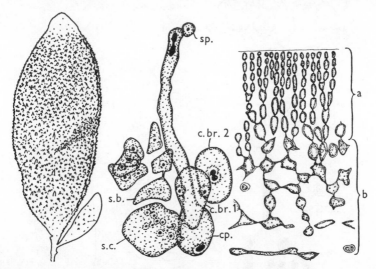

Fig. 12.11 *Gigartina*. Left, plant of *G. californica*; Centre, carpogonial branch, *G. stellata* (cp. = carpogonium, s.b. = vegetative lateral, s.c. = support cell); Right, t.s. thallus, *G. stellata* (a = outer cells, b = inner cells). (Left × ¼, Centre × 1500, Right × 160) (Centre, Right, after Newton)

dichotomously branched (*G. macrocarpa*). The female plants have papillose projections (the cystocarps) growing from the surface of the thallus. The internal construction consists of a mass of large central cells (multiaxial construction) with smaller cells towards the periphery. In some species, e.g. *G. stellata*, the cells of inner cortex and medulla produce septate hyphae which grow downwards. The carpogonial branch is three-celled and is derived from a large support cell of the inner cortex, which also functions as the auxiliary cell. In addition to the carpogonial branch the support cell either produces a vegetative lateral branch or a second carpogonial branch. Tetrasporic plants, which are far more abundant than the gametophytes, produce tetrasporangia in sori at the base of small papillate outgrowths.

RHODYMENIALES

The thallus of all species is based upon the multiaxial type of construction. In some genera the central portion becomes hollow and may be divided by septa, e.g. *Champia*. There is considerable variation in habit from flattened, membranous thalli to softer, terete and more gelatinous structures. Before fertilization small auxiliary cells are cut off from a branch derived from the support cell, and they develop only after fertilization has taken place. The order comprises two families, the Rhodymeniaceae and Lomentariaceae. In

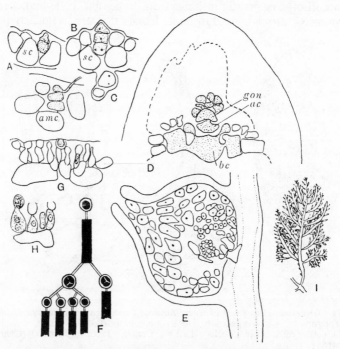

Fig. 12.12 *Lomentaria clavellosa.* A–C, development of carpogonial branch (*amc* = accessory mother cell, *sc* = support cell); D, young cystocarp (*ac* = accessory cell, *c* = support cells, *gon* = gonimoblast); E, mature cystocarp; F, *L. rosea*, life cycle; G, H, *L. clavellosa*, antheridia; I, *L. clavellosa*, plant. (A–C, G, H, × 660, D × 312, E × 90, I × 3/5) (A–C, F–H, after Svedelius; D, E, after Kylin)

the Rhodymeniaceae the gonimoblasts become profusely branched and nearly every cell gives rise to a carpospore. The thallus is occasionally hollow (without longitudinal cells bordering the cavity) but normally solid. In the Lomentariaceae each gonimoblast gives rise to a single terminal carpo-sporangium. The thallus is predominantly hollow and longitudinal cells border the cavity.

LOMENTARIACEAE (CHAMPIACEAE): *Lomentaria* (pod with constricted joints). Fig. 12.12

The mature fronds are hollow, with or without constrictions at the nodes. Branching is irregular or unilateral. The hollow central region originates from a branching structure which later on separates in order to form the outer cell layers, although a few longitudinal filaments of the original multiaxial structure remain in the centre. The plant, which is enclosed in a thick gela-tinous cuticle, may bear unicellular hairs that have arisen from the epidermal layer. The adult thallus has a group of eight to twelve apical cells, each of which produces a longitudinal filament; the corticating threads develop from lateral cells which are cut off from each filament segment just behind the apex. The male plants, which are rarely found, bear whitish antheridial sori towards branch apices. A system of branching threads, which appears as a preliminary to sorus formation, arises from a single central cell, and from each of these threads, two to three antheridial mother cells grow out and increase in length. Depending on the species, one, two or three primary antheridia arise from each mother cell and they may be followed by a crop of secondary antheridia.

The procarp consists of a support cell with a three-celled carpogonial branch. There are one or two auxiliary cells, and after fertilization only one of these receives a process and the diploid nucleus from the carpogonium. This auxiliary cell cuts off a segment on the outer side, and from this a group of cells develops that ultimately gives rise to the gonimoblasts, most of the cells forming carposporangia.

In European waters *L. rosea*, which has a diploid chromosome number of 20, is only known to produce tetraspores that apparently arise without undergoing meiosis. Individual spores germinate to give a new plant or else a whole tetrad may germinate to give a new plant. In such plants, therefore, the gametophytic generation is wholly suppressed and we have a diplont which behaves as a haplobiont in respect of its cycle. Elsewhere, records suggest that the species behaves normally.

RHODYMENIACEAE: *Rhodymenia* (*rhodo*, red; *hymen*, membrane). Fig. 12.13

The membranaceous fronds are flattened with dichotomous or palmate branching or with proliferations. The central tissue of closely packed cells is bounded by one or more dense layers of assimilatory cells. Growth of the thallus, which is derived from the 'fountain' type of construction, is related to a group of apical cells. The auxiliary cell, although cut off before fertiliza-tion does not become evident until after that process has occurred. The fusion cell is large and distinct and gives rise to a single gonimoblast that branches repeatedly, almost all cells then acting as carposporangia. The tetrasporangia, which are cruciate, are borne in the assimilatory layer. *Rhodymenia palmata* of the Atlantic and North Sea is known as dulse and is one of the edible red seaweeds.

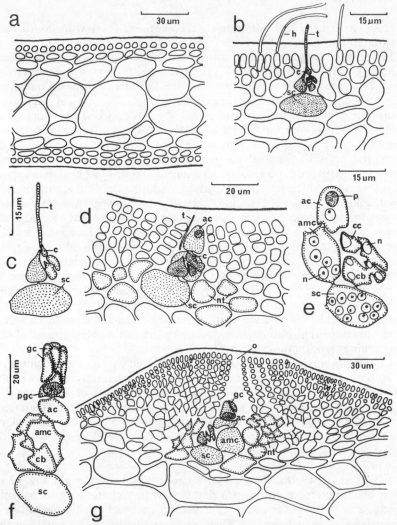

Fig. 12.13   *Rhodymenia pseudopalmata* (Lamouroux) Silva. A, t.s. of thallus; B, supporting cell and carpogonial branch with surrounding vegetative cells, some bearing hairs; C, supporting cell and carpogonial branch; D, section showing auxiliary-cell branch, third and fourth cells of carpogonial branch fusing, and early stage in development of pericarp; E, procarp showing connecting cell cut off from fused third and fourth cells of carpogonial branch; F, early stage in developing of gonimoblast; G, section through young cystocarp. (after Sparling)

CERAMIALES

The range of habit in this order is considerable even though the basic construction is uniaxial throughout. There are simple, branched filaments (e.g. *Callithamnion, Antithamnion*), leafy expansions (e.g. *Polyneura, Grinnellia, Phycodrys, Delesseria, Laingia*), nets (e.g. *Claudea, Martensia*: in *Claudea* the net form originates by branching, in *Martensia* by separation of cell rows), cartilaginous forms (*Laurencia*) and polysiphonous and corticated polysiphonous forms (*Polysiphonia, Bostrychia, Plumaria*). The order is conveniently divided into four families and of these the Ceramiaceae are presently regarded as the most primitive (Hommersand, 1963). The other three families could have evolved from different prototypes in the Ceramiaceae. The families are very large, especially the Ceramiaceae and Rhodomelaceae, and they are commonly divided into subfamilies.

RHODOMELACEAE: *Polysiphonia* (*poly*, many; *siphonia*, siphons). Fig. 12.14
The thallus in this genus generally arises from decumbent basal filaments that are attached to the substrate by means of small flattened discs at the ends of short rhizoids. These basal filaments do not represent a primary heterotrichous condition but a secondary state. Many species are epiphytic on other algae although *P. lanosa*, which is always found on the fronds of the fucoid *Ascophyllum nodosum*, is probably a hemi-parasite. The thallus is laterally or dichotomously branched and bears numerous branches which, in the perennial forms, are shed annually. The main axes and branches are corticate or ecorticate, and have a polysiphonous appearance due to the single axial cell series being surrounded by four to twenty-four pericentral cells or siphons. The primary pericentral siphons represent potential branches and are united to the parent axial cell by a pit connection. The corticating cells, when present, are always shorter and smaller and are often only found in the basal region. The ultimate branches are not polysiphonous and frequently terminate in delicate multicellular hairs.

The colourless antheridia, which are formed in clusters, are borne on a short stalk that morphologically is a rudimentary hair. In *Polysiphonia violacea* the two basal cells of the hair are sterile, the upper one giving rise to a fertile polysiphonous branch and a sterile hair. One or more mother cells are formed from all the pericentral cells on the fertile branch, and each mother cell produces four antheridia in two opposite and decussate pairs, the first and third appearing before the second and fourth. There is no secondary crop in this species. The carpogonial branches are also formed from hair rudiments, the support cell cutting off a small section from which two lateral sterile cells arise. The second 'sterile' cell is morphologically equivalent to an abortive carpogonium. Later on, a fertile pericentral cell is cut off, and this gives rise to the four-celled carpogonial branch.

After fertilization has taken place the auxiliary cell is cut off from the apex of the support cell (Fig. 12.14) and in addition two branch systems composed of nutrient cells develop from the original sterile cells. When the zygote nucleus has divided, commonly only one of the daughter nuclei passes into the auxiliary cell, which in the meantime has become fused to the carpogonium; a new wall is then laid down, cutting off the carpogonium. The auxiliary cell next fuses with the pericentral cell and after the diploid nucleus has been transferred, it unites with the other support and axial cells to give a

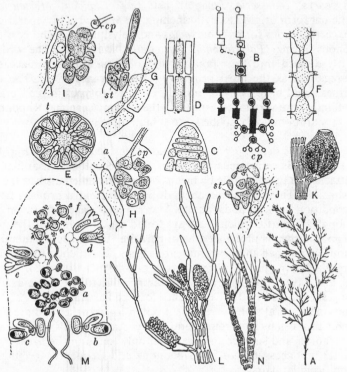

Fig. 12.14 *Polysiphonia violacea.* A, plant of *P. nigrescens*; B, life cycle; C, apex and cells cut off from central cells; D, thallus construction in l.s.; E, t.s. of thallus, *P. fastigiata* (*t* = young tetraspore); F, protoplasmic connections of axial thread; G–J, stages in development of carpospores (*cp* = carpogonium, *a* = auxiliary cell, *g* = gonimoblast, *st* = sterile cells); K, cystocarp of *P. nigrescens* with ripe carpospores; L, antheridial branch; M, *a–f*, stages in development of antheridia; N, *P. nigrescens*, tetraspores. (A × ⅓, G–I × 400, J × 260, K, N, × 33, L × 35) (A, K, N, after Newton; B, after Svedelius; C, F, schematic; D, E, after Oltmanns; G–J, after Kylin; L, after Grubb; M, after Tilden)

large fusion cell. The diploid nucleus divides many times and the daughter nuclei pass into lobes that emerge from the fusion cell. Each lobe gives rise to a short gonimoblast filament, the terminal cell of which produces a pear-shaped carpospore, while the subterminal cell gives rise to a new two-celled carpospore branch. In this way numerous carpospores are produced.

The wall of the distinctive cystocarp is two-layered, the inner wall being formed directly by cells derived from pericentral cells. The cells of the inner wall divide tangentially to give the cells that form the outer wall, so that ultimately both layers have come from the pericentral cells. The tetrasporangia develop from pericentral cells in the apical portion of branches, and as they are protected by being embedded in the thallus, the branch is usually swollen and distorted. The genus is widespread in cold and warm waters and the species is found particularly in rock pools.

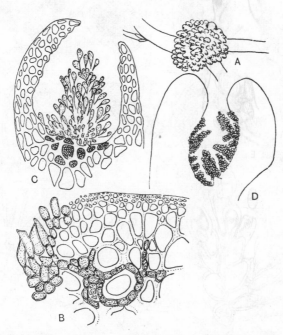

Fig. 12.15 *Janczewskia*. A, *J. moriformis* on *Chondria* sp.; B, filaments of *J. lappacea* in host, *Chondria nidifica*; C, l.s. of cystocarp of *J. moriformis*; D, antheridial conceptacle of *J. lappacea*. (A × 6, B–D × 180) (after Setchell)

Janczweskia is a remarkable hemi- or holoparasitic genus which is always to be found on other members (*Laurencia, Chondria* and *Cladhymenia*) of the same family (Fig. 12.15). All the species have organs of contact or penetration, the latter being fungal-like filaments which establish pit connections with the cells of the host. Each individual plant is a whitish, coalescent tubercular mass composed of fused branches that grow from an apical cell buried in a pit.

CERAMIACEAE: *Callithamnion* (*calli*, beauty; *thamnion*, small bush). Fig. 12.16

This is a genus of very beautiful and delicate plants that possess filamentous, radially branched fronds which are either monosiphonous or else corticated at the base, the cortication being formed by rhizoidal filaments. The plants are commonly attached by rhizoids and in many species the branchlets terminate in long slender hairs. The cells of the vegetative thallus are usually multinucleate, and in *C. byssoideum* there are protoplasmic pseudopodia

which project internally from the ends of the cells, and although these strands
are apparently capable of some movement their function is obscure.

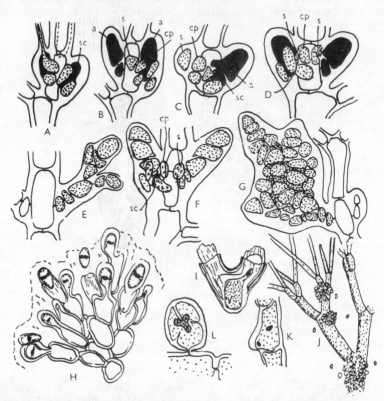

Fig. 12.16 *Callithamnion.* A–G, stages in development of cystocarp from mature carpogonium (a =auxiliary cell, cp =carpogonium, s =sporogenous cell cut off from fertilized
carpogonium, sc =support cell); H, antheridia with spermatia; I, the same enlarged; J,
antheridial sori; K, young tetrasporangium; L, mature tetrasporangium. (A–F × 520,
G × 300) (A–G, after Kylin; H–J, fifter Grubb; K, L, schematic after Mathias)

The antheridia, which form hemispherical or ellipsoidal tufts on the
branches, arise as lateral appendages, the first cell to be cut off being the stalk
cell. This stalk cell gives rise to a group of secondary cells which later on
divide to form branches composed of two to three cells, each terminating in
an antheridial mother cell. In this genus there may be two or even three crops
of antheridia arising successively in the same place, each mother cell producing about three antheridia in every crop. The cystocarps, which are
usually present in pairs enclosed in a gelatinous envelope, are formed when
two cells are cut off from a cell in the middle of a branch; these function as
the auxiliary mother cells. From one of them the four-celled carpogonial
branch is produced, while after fertilization both auxiliary mother cells
divide and cut off a small basal cell. The fertilized carpogonium also divides

into two large cells, each of which cuts off a small sporogenous cell that fuses with the adjacent auxiliary cell. As a result of this fusion each auxiliary cell can receive a diploid nucleus which soon after its entry divides into two; one daughter nucleus passes to the apex of the auxiliary cell, and the other, together with the nucleus of the auxiliary cell, is cut off by a wall. It is from the large upper cell that the gonimoblast filaments arise and so the mature cystocarp is produced.

The sessile tetrasporangia arise in acropetal succession as lateral outgrowths of the vegetative cells of young branches. In *C. brachiatum* mature tetrasporangia and antheridia have been found on the same plant, while other plants have been reported that bear both tetrasporangia and cystocarps.

In *Spermothamnion turneri* sex organs have been reported on normal tetrasporic plants, but as the procarp branch in this case develops without meiosis the carpogonium is diploid. Fusion of male and female nuclei in the carpogonium has been observed so that the carpospores are probably triploid but unfortunately their fate is not known. The type genus *Ceramium* is characterized by a large-celled central axis with much smaller cortical cells that either form bands covering the axial-cell nodes or more or less cover the cortical filament entirely. Haploid plants are also known that bear tetrasporangia but in such cases no divisions occur in the sporangia. In *S. snyderae* the tetrasporangia are replaced by polysporangia which must be regarded as homologous structures. Polyspore formation (e.g. *Pleonosporum*) may also occur.

DASYACEAE: *Dasya* (*dasus*, thick). Fig. 12.17

The much branched thallus is filiform or compressed, the branching being distichous or irregular. The single central siphon is surrounded by 4–12 pericentral siphons outside of which there is often a further band of corticating cells. The branches are like those of *Callithamnion* in that they are monosiphonous with dichotomous branching. The urn-shaped cystocarps are sessile or pedicellate while the tetrasporangia are borne in clustered rows in special lanceolate branchlets or stichidia.

DELESSERIACEAE: *Delesseria* (after Baron Delessert). Fig. 12.18

The family is divided into two subfamilies:

(a) Delesserioideae, in which the procarps arise from the central axial thread (mid-rib), e.g. *Delesseria, Membranoptera*.
(b) Nitophylloideae, in which the procarps are not restricted to the mid-rib, e.g. *Polyneura*.

The large, thin, leafy fronds possess a very conspicuous mid-rib with both macro- and microscopic veins. The complex nature of the laciniate or branched thallus can be seen from the figure. The foliar condition is clearly produced by the juxtaposition in one plane of numerous branches that arise from the axial filament. The apical cell divides to give a central cell and two pericentrals. The latter divide and give rise to the pinnate laterals of the first

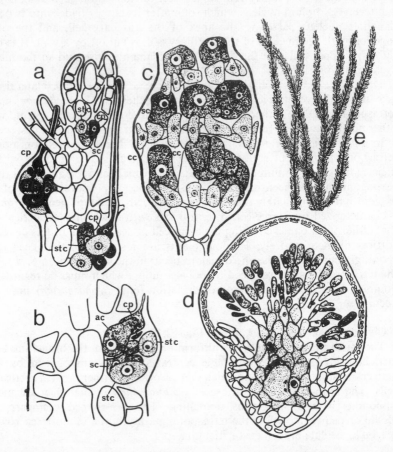

Fig. 12.17 *Dasya pedicellata*: a, b, carpogonium; c, tetrasporangial branch; d, cystocarp; e, habit (ac = auxiliary cell, cc = cover cell, cp = carpogonium, stc = sterile cell, sc = support cell). (after Kylin)

order on which secondary laterals are borne. The mid-rib is formed from the central cell and the adjacent cells of the primary laterals. The latter divide transversely so that in the mature frond they are half the length of the central cells. Growth is intercalary in the various orders of branches. The cells of the thallus also become united by means of secondary protoplasmic threads and they may also develop thin rhizoids. The reproductive organs are borne on separate adventitious leaflets. The procarps arise from the axial threads of the leaflet, each cell giving rise to two procarps. The procarp consists of a support cell which then cuts off a sterile cell and a three-celled carpogonial branch. The second cell of the fertile branch is much enlarged. As many as fifty procarps may be produced on a fertile leaflet: these ripen in acropetal

306

succession but only one is fertilized. At fertilization a second sterile cell is cut off and the other adjacent procarps and sterile cells give rise to the wall of the cystocarp.

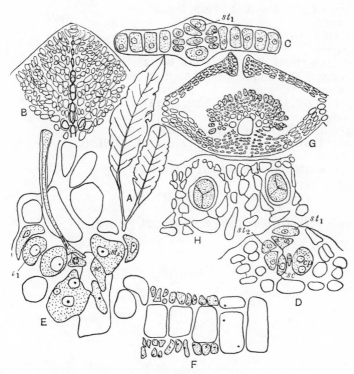

Fig. 12.18 *Delesseria sanguinea*. A, plant; B, apex of thallus to show cell arrangement; C, first stage in formation of carpogonial branch (st_1 =first, group of sterile cells); D, later stage of same (*cp* =carpogonial branch, *sc* =support cell, st_1 =first, and st_2 =second group of sterile cells); E, mature carpogonial branch (*sc* =support cell, st_1 =first sterile branch, st_2 =second sterile branch); F, formation of antheridia in related genus, *Nitophyllum*; G, t.s. of mature cystocarp in the related genus, *Nitophyllum*; H, tetraspores. (B × 258, C, D, × 404, E × 720, H × 360) (A, F, G, after Tilden; B–D, after Kylin; E, H, after Svedelius)

REFERENCES

GENERAL

   Dixon, P. S. (1964). *Proc. 4th In. Seaweed Symp. Biarritz*, p. 71, (Pergamon Press, Oxford).

   Fritsch, F. E. (1945). *Structure and Reproduction of the Algae*. Vol. 2, p. 444–747 (Cambridge University Press).

   Kylin, H. (1956). *Die Gattungen de Rhodophyceen* (Lund).

ULTRASTRUCTURE

   Bisalputra, T. (1967). *J. Ultrastruct. Res.*, **17**, 14.

   Bouck, G. B. (1962). *J. Cell Biol.*, **12**, 365.

L

CERAMIALES
  Hommersand, M. (1963). *Univ. Calif. Publs. Bot.*, **35**, 165.
GIGARTINALES
  Searles, R. B. (1968). *Univ. Calif. Publs. Bot.*, **43**, 1.
RHODYMENIALES
  Sparling, S. (1957). *Univ. Calif. Publs. Bot.*, **29**, 319.
Ceramiaceae
  Wollaston, E. M. (1968). *Aust. J. Bot.*, **16**, 217.
Delesseriaceae
  Wagner, F. S. (1954). *Univ. Calif. Publs. Bot.*, **27**, 279.
  Womersley, H. B. S. and Shepley, E. A. (1959). *Aust. J. Bot.*, **7**, 168.
Rhodomelaceae
  Scagel, R. F. (1953). *Univ. Calif. Publs. Bot.*, **27**, 1.
Callithamnion
  Boddeke, R. (1957). *Acta Bot. Neerl.*, **7**, 589.
  Hassinger-Huizenga, H. (1953). *Arch. Protistenk.*, **98**, 91.
Ceramium
  Dixon, P. S. (1960). *J. mar. biol. Ass. U.K.*, **39**, 375.
Furcellaria
  Austin, A. P. (1960). *Ann. Bot.*, *N.S.*, **24**, 255, 296.
Hypnea
  Hewitt, F. E. (1961). *Univ. Calif. Publs. Bot.*, **32**, 195.

13

FOSSIL ALGAE

Many algal fossils are known and it is therefore only possible here to outline the main types of structure represented among them. Table 13.1 outlines the principal geological strata and the generally accepted time scale.

Because the algae are generally composed of 'soft tissue', preservation, particularly in the lower geological strata, either has not occurred or else has not been very good. Those algae that characteristically deposit lime, (e.g. Codiales, Dasycladales, Corallinaceae) or silica, (e.g. diatoms) form the groups that have been most successfully preserved and therefore about which most is known. There is little doubt that even at the present time fossilization of calcareous and siliceous algae is taking place, e.g. in the Arctic benthos where crustaceous Corallinaceae are particularly abundant.

In recent years more attention has been paid to Archeozoic or pre-Cambrian rocks and the result has been the discovery of fossil organic remains of a much greater age than previously reported.

CYANOPHYCEAE

The earliest known algal fossils are those from the Gunflint Chert of

309

TABLE 13.1

GEOLOGICAL TIME SCALE
(from Kulp, 1961)

ERA	PERIOD		Beginning of interval Millions years ago
CENOZOIC	Quaternary	Pleistocene	1
		Pliocene	13
		Miocene	25
	Tertiary	Oligocene	36
	Tertiary	Eocene	58
		Pliocene	63
MESOZOIC	Cretaceous		135
	Jurassic		200
	Triassic		230
PALAEOZOIC	Permian		280
	Upper Carboniferous (Pennsylvanian)		310
	Lower Carboniferous (Mississippian)		345
	Devonian		405
	Silurian		425
	Ordovician		500
	Cambrian		600 app.
	Precambrian (Biotic era)		2,000
	Precambrian (Abiotic era)		5,000

Ontario in rocks that are two billion years old (Barghoorn and Tyler, 1965). *Animikiea septata* would appear to have been a member of the Oscillatoria-ceae. *Entosphaeroides amplus* with internal spores or endogonidia could be allied to *Chamaesiphon* (p. 21) or may have been bacterial in nature. Spore-like bodies put in the genus *Huroniospora* could have allies in either the Chroococcaceae or the Dinoflagellates. In addition there are three genera of uncertain affinities, *Archaeorestis*, *Eoastrion* and *Eosphaera*.

Other very early recognizable fossils are those described from late pre-Cambrian rocks (700–900 million years of age) near Alice Springs in Australia (Barghoorn and Schopf, 1965). The specimens tentatively ascribed to the Cyanophyceae are small septate filaments that taper from a maximum middle zone to the ends and were possibly enclosed in a sheath very like the modern *Phormidium* and *Lyngbya* species. All the above together with the following are best placed in a group called the Protophyceae.

Spongiostromata (Precambrian onwards)

Much doubt has been thrown upon the authenticity of this group, some writers regarding them as structures which originated as diffusion rings ('liesegang' phenomena) in colloidal materials or perhaps in calcareous muds.

In the original description Walcott suggested an affinity to the Cyanophyceae, but as later workers could only distinguish a purely mineral structure they suggested the idea of diffusion phenomena. Discoveries of comparable algal concretions and laminations in the Bahamas, however, have made it extremely probable that these structures were formed by algae that collected and bound the sediment. Further good examples have since been reported from the United States (see Table 13.2).*

Porostromata (e.g. *Girvanella*)

The comments concerning the taxonomic placement of the Spongio-stromata also apply here. The Porostromata includes all fossil algae of unknown position whose structure consists of a collection of well-defined tubes.

The most widely represented generā are *Girvanella* and *Rothpletzella* which date from the Silurian and Devonian of Europe, U.S.S.R., Australia and Canada. The former is known from the Cambrian to Cretaceous but was most abundant in the Carboniferous. The thallus consisted of a loose mass of small twisted tubes that compacted to form a rounded or bean-shaped mass. They were possibly formed in much the same way as the algal water-biscuits now found in South Australia. These range from tiny particles to thick bun-like forms 20 cm in diameter, whilst in them are to be found the tube-like remains of living species of *Gloeocapsa* and *Schizothrix*.

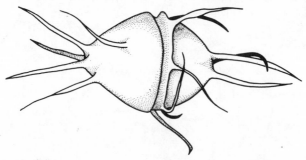

Fig. 13.1 *Hystrichodinium oligacanthum* (after Deflandre and Cookson)

DINOPHYCEAE

A large number of fossil Dinoflagellates have been recorded. The earliest are reported from the Jurassic, but even then many of the fossil forms could be ascribed to such extant genera as *Gonyaulax, Gymnodinium, Peridinium,* and *Ceratium*. Among extinct genera in the Gymnodinioideae are such forms as *Diconodinium* from the Australian Cretaceous. The genus showed the characteristic girdle, and the epitheca terminated at the apex in a short stumpy projection.

* Columnar stromatolites in which Cyanophyta were associated from precambrian Australian rocks (Glaessner, Breiss and Walter, 1969).

Among the Peridinioideae, extinct genera are represented from the Jurassic to Tertiary by such forms as *Palaeoperidinium*. A fossil member of the Gonyaulacaceae is the Cretaceous *Rhynchodiniopsis*, while *Raphiododinium* is peridinioid with long spiny protuberances. All of these forms, like many other of the fossil genera, possess the characteristic dinoflagellate orientation (see p. 149) with a distinct girdle, and in many cases there is a sulcus or sulcal region (see p. 153).

Apart from those members that can be ascribed to extinct or extant families within the Peridinioideae or Gymnodinioideae there are some forms that are ascribed to families of uncertain position, e.g. *Hystrichodinium*, *Peridinites* (Fig. 13.1). Again there is no doubt that these are dinoflagellates because of their quite distinct dinoflagellate orientation.

Associated with these planktonic algae but with a more ancient history, since they extend down at least into the Cambrian, are the wholly fossil Hystrichosphaeridae (Fig. 13.2). The exact relationships of this group are not clear but they can at least be classed as Protists, and in some cases seem to have dinoflagellate organization,

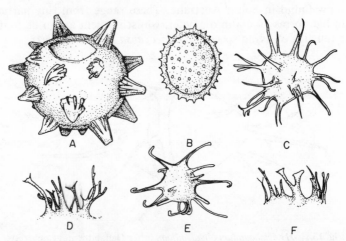

Fig. 13.2　A, *Hystrichosphaeridium striatoconus*; B, *Micrhystridium pachydermum*; C, D, *M. stellatum;* E, F, *H. heteracanthum*. (A, E, F × 622, B × 1545, C, D, × 982) (after Deflandre and Cookson)

By far the most common genus is *Hystrichosphaeridium* (Fig. 13.2) with representatives from the Silurian to the Tertiary. In *Micrhystridium* (Fig. 13.2), known from the Silurian to the Tertiary, the characteristic spiny projections are not nearly so ornate or complex. In Kansas there is evidence of two successive Cretaceous blooms which consisted of a dinoflagellate-hystrichosphaerid phytoplankton (Tasch, McClure and Oftedahl, 1964). Such a bloom must have involved adequate supplies of nutrients, including Vitamin B_{12}, and there must also have been certain optimum physical factors.

BACILLARIOPHYCEAE

Recognizable fossil diatoms are known from the Upper Jurassic, but they are most abundant in the late Cretaceous and Tertiary. The majority can be ascribed to present day forms. Fossil diatom deposits are known as Kieselguhr and are valuable commercially.

Freshwater diatom deposits, especially those from the late-glacial and post-glacial, provide a means of deducing changes that may have occurred in lakes and drainage areas. Round (1957) has demonstrated how variation in diatom species and frequency can lead to interpretation of environmental changes. Thus the dominants in the early period, e.g. *Gyrosigma*, *Cymatopleura*, indicated waters of a higher alkalinity than at present. At the time of the Boreal-Atlantic transition (pollen-analysis zone — 7,000 years ago), the early alkaline species began to be replaced by more acidic species, e.g. *Eunotia*, *Anomoeoneis*, which subsequently increased greatly. As Round points out this change is correlated with increasing warmth of the climate, with a decline of the Birch/Pine forest climax and the finish of further base supplies as the rock surfaces became leached.

Another example of the use of fossil diatoms for interpretive purposes comes from the Presto fjord (Mikkelsen, 1949). In this case, changes in the proportions of freshwater, brackish and marine diatoms showed that in the pollen analytical period VIIa the Baltic was very slightly brackish, but in VIIb, after a period of reduced salt there was a rise in salinity. Finally, at the end of IXb, there was a recent reduction in salinity (Fig. 13.3).

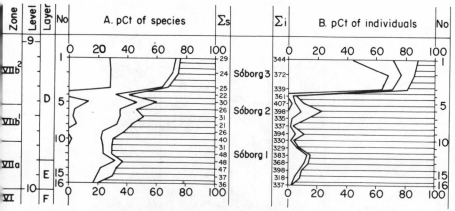

Fig. 13.3 Diatom diagram from Bp. 12 in the fiord basin of Praesto Fjord.

CHRYSOPHYCEAE, HAPTOPHYCEAE

The only other unicellular algae with any sort of fossil record are the Coccolithophoridae. The record extends back to the Jurassic, but many of the forms have been described as species of present day genera such as *Syracosphaera*, *Coccolithus*. Nearly all of the records are based upon fragments of cells, comprising the individual coccolith scales, rather than entire cells.

CHLOROPHYCEAE

It seems now that the oldest known representatives must be the large
septate filaments of *Gunflintia* from the Gunflint chert of Ontario and described by Barghoorn and Tyler (1965), and similar filaments described by
Barghoorn and Schopf (1965) from the late Precambrian of Australia. These
were multicellular and unbranched and thus very suggestive of the present day
Ulothricales.

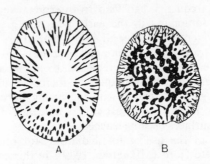

Fig. 13.4   Codiaceae. A, *Palaeoporella variabilis*; B, *Boueina hochstetteri*. (A × 12) (after
Hirmer)

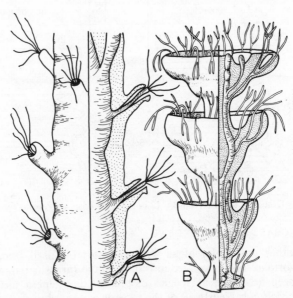

Fig. 13.5   A, *Abacella pertusa*, reconstruction; B, *Lancicula alta*, reconstruction. (after
Maslor)

CODIACEAE

This was one of the most important families of rock-building algae, especially from the Ordovician to the Pennsylvanian and from the Jurassic to the present. Examples of these algae are *Boueina* (cf. Fig. 13.4), an unbranched form from the Lower Cretaceous, and *Palaeoporella* (Fig. 13.4), which was composed of hollow cylinders of funnel-shaped bodies with slender forked branches, from the Lower Silurian. *Dimorphosiphon*, from the Ordovician, is generally regarded as the oldest-known member of the Codiaceae and has been tentatively related to *Halimeda*. It was about 10 mm long and composed of branched tubular cells without any cross walls, the cells being embedded in a calcareous matrix.

Genera that bear little resemblance to present day ones are the Devonian *Abacella* and *Lancicula* (Fig. 13.5) with their peculiar tufted branchlets. In the Permian the Codiaceae represented an important family of fossil algae with some nine genera. The generic range of this important family is shown in Table 13.2 where it will be seen that seven separate divisions of the family are represented. Of the three most important, two are still extant.

TABLE 13.2

DISTRIBUTION OF FOSSIL CODIACEAE
(after Johnson)

Genera	Cambrian	Ordovician	Silurian	Devonian	Carboniferous Miss.	Carboniferous Penn.	Permian	Triassic
GARWOODIORDEAE								
Akiyoshiphycus						—	—	
Bevocastria					—			
Garwoodia					—	– – –	– – –	– –
Hedstroemia			—	– – –	—			
Hoegiopsis		—						
Palmaphyton						—		
Ortonella			—	——	—	– – –	—	
Polymorphocodium					– – –			
HALIMEDIAE								
Calcifolium							—– – –	
Ivanovia							— –	
Anchicodium							——	
Eugonophyllum							—	
UDOTEAE								
Dimorphosiphon		—						
Hikorocodium					—	—		
Litanaia				—				
Mizziella					???	??		
Neoanchicodium							—	
Orthriosiphon					—			
Palacocodium					?—			
Palaeoporella		—	– – –	?				
Penicilloides					—			
Succodium							—	
Ulva				—				
Abacella				—				
Lancicula				—				
Ascosoma	—							
Mitscherilichia	—							
Gymnocodium							—	
Permocalculus							—	
Rothpletzella			——	———	—?		?—	—?

TABLE 13.3

DISTRIBUTION OF MAJOR FOSSIL ALGAE IN TIME
(from Johnson)

GENERA	PERMIAN	TRIASSIC	JURASSIC	CRETACEOUS	TERTIARY	QUATERNARY
CORALLINACEAE						
Melobesioideae						
Archaeolithothamnium				?		
Lithothamnium						
Lithoporella						
Corallinoideae						
Archamphiroa						
SOLENOPORACEAE						
Solenopora						
Pseudochaetetes						
Parachaetetes						
Polygonella						
Pycnoporidium						
DASYCLADACEAE						
Acicularia						
Actinoporella		?				
Clypeina					?	
Conipora						
Cylindroporella						
Goniolina						
Griphoporella					?	
Linoporella						
Macroporella						
Munieria						
Neogyroporella						
Palaeocladus						
Petrascula						
Salpingoporella						
Sestrosphaera						
Teutloporella	?					
Thyraoporella				?		
Triploporella	?					
CODIACEAE						
Cayeuxia			?	?		
Marinella						
Nipponophycus						
Boueina						
POROSTROMATA						
Girvanella						
Symploca						
ALGAE INCERTAE SEDIS						
Polygonella			?			

DASYCLADACEAE

This group contains a very large number of the fossil algae. It reached its maximum development and abundance in the upper Jurassic and lower Cretaceous, and in those days was much more important than at present. Of the 58 known genera of the Dasycladaceae some 48 are extinct (see Table 13.3 showing time distribution of the genera).

Cyclocrinus (Fig. 13.6) grew to about 70 mm and looked like a miniature golf ball borne on the end of a stalk. Narrow branches arose at the apex of the stalk and each terminated in a flattened hexagonal head; as the edges of adjoining heads were fused together to form the outer membrane, which was only weakly calcified, the cell outlines were clearly visible.

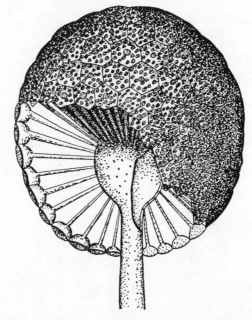

Fig. 13.6   Dasycladaceae. *Cyclocrinus porosus* (× 8) (after Hirmer)

Primicorallina (Fig. 13.7), from the Ordovician, had a stem beset with radially arranged branches, each of which branched twice into four branchlets.

Rhabdoporella (Fig. 13.8) seems to be one of the most primitive genera as it is represented by a purely cylindrical shell that is studded with pores through which the threads passed. It is known from the Ordovician and Silurian.

The type of structure found in *Diplopora* (Fig. 13.9) was also shown by many other forms from the Middle Triassic of Eurasia. It was a few centimetres long and bush-like in appearance, the main stem being covered with

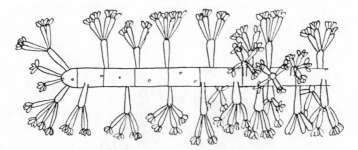

Fig. 13.7 Dasycladaceae. *Primicorallina trentonensis* (× 8·25) (after Hirmer)

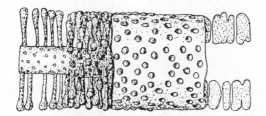

Fig. 13.8 Dasycladaceae. *Rhabdoporella pachyderma* (× 135) (after Hirmer)

whorls of branches that arose in groups of four, each bearing secondary branches that terminated in hairs. In the older thalli the outer part of the branch dropped off leaving a scar on the calcareous shell. The sporangia are reported to have been modified branches.

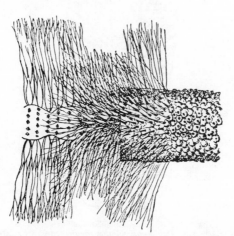

Fig. 13.9 Dasycladaceae. *Diplopora phanerospora* (× 8) (after Hirmer)

319

Palaeodasycladus (Fig. 13.10), from the Lower Jurassic, bears a resemblance to the living species of *Dasycladus*. Near the base there were only primary branches, while higher up secondary and tertiary branches were to be found.

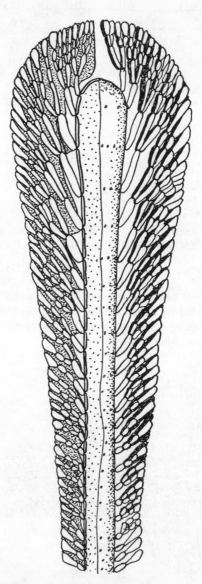

Fig. 13.10   Dasycladaceae. *Palaeodasycladus mediterraneus* (× 20) (after Hirmer)

CHAROPHYCEAE

Lagynophora, a genus from the Lower Eocene, can be ascribed to this group, while *Palaeonitella* (Fig. 13.11), from the Middle Devonian, may belong here also although its affinities are not so clear.

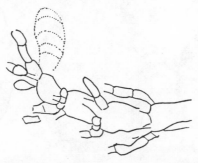

Fig. 13.11   Charales. *Palaeonitella cranii* (× 124) (after Hirmer)

Charophytes are known from much older rocks and took the form of gyrogonites. The earliest of these is *Trochiliscus podolicus* (Fig. 13.12) from beds of the Lower Devonian. In all gyrogonites it is important to note that the remains comprise only the calcareous inner part of the oogonium with the

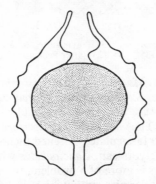

Fig. 13.12   *Trochiliscus (Eutrochitiscus) podolicus*. Gyrogonite from Lower Devonian. (after Croft)

enclosed oospore. The coronal cells, which are so important taxonomically, are invariably missing so that the relationship of the remains to living genera cannot be determined. In the Permian of North America the organ genus *Catillochara* (Fig. 13.13) includes all gyrogonites with five sinistral spirals.

PHAEOPHYCEAE

The brown algae, mainly because of their lack of calcification are very sparsely represented as fossils. If there were large forms comparable to the

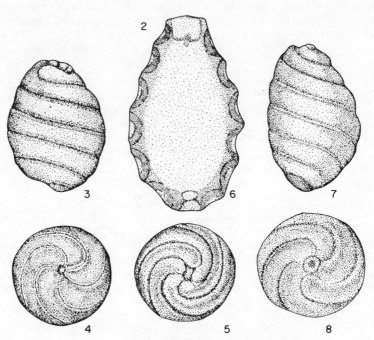

Fig. 13.13 2–8, *Catillochara moreyi*. 3–5, gyrognite; 6, wall structure; 7, lateral view; 8, basal view. (3–5 × 120, 6 × 135, 7, 8 × 125) (after Peck and Eyer)

present-day *Nereocystis, Macrocystis* and *Lessonia*, preservation does not seem to have taken place. A number of problematical fossil algae are known, especially from the Silurian and Devonian, and these may well have been members of this group. Thus some of the described species of *Buthotrephis*, which are bush-like with fasciculate, flattened dichotomising branches, look not unlike a modern Dictyotalean. *Chaetocladus* with its jointed cylindrical stems and whorls of hairs is more of a problem, though there are algae in the Chordariales and Sphacelariales with which it has at least a superficial resemblance. The Devonian genus *Drydenia*, with flattened lamina and filamentous holdfast is not unlike a laminarian. The flattened thalloid *Enfieldia* and *Hungerfordia* could be of either Phaeophycean or Rhodophycean affinities.

Parker and Dawson (1965) have recently described some interesting fossils from Californian Miocene deposits. Amongst them was one genus *Julescraneia*, which they considered as an intermediate between the two present-day Californian kelps, *Nereocystis* and *Pelagophycus*.

Also present were forms very similar to the genera *Cystoseira* and *Halidrys*, viz. *Palaeohalidrys* and *Cystoseirites* as well as algae, e.g. *Palaeocystophora*, having affinities to the present extant southern genus, *Cystophora*. This last-

322

named example raises interesting queries as to why the living forms are now
restricted to the southern oceans.

RHODOPHYCEAE

The lime-encrusted Melobesiae are represented from the Cretaceous upwards by species of *Archaeolithothamnion, Dermatolithon, Lithothamnion,
Lithophyllum* and *Goniolithon,* some of them being distinguished only with
difficulty from living forms. The Corallinaceae are also represented in the
Palaeozoic by extinct members of present-day living genera. There are a large
number of forms assigned to an extinct family, the Solenoporaceae, which
existed from the Ordovician up to the Jurassic, and recently their structure
and systematic position has been revised (Johnson, 1960). They formed
nodules which ranged in diameter from the size of peas up to several centimetres and the cells of the nodules were arranged like those of a *Lithothamnion*
although the cross walls were not well marked. In most forms the cells are
distinctly larger in size than in most present-day *Lithothamnia.* A hypothallus,
if present, appears to have been thin and poorly calcified. Conceptacles have
been described in these plants and were presumably associated with reproduction.

The so-called organs of reproduction in the Solenoporaceae have been the
subject of much discussion. There appear to be three types including the wide
dome-shaped structures regarded as conceptacles. The others are star-like
groups of extra large cells or single, elongated, spindle-shaped, cylindrical or
ovoid subjects. The three main Palaeozoic genera of this family are *Soleno-
pora, Parachaetetes* and *Ungdarella.* The relationships of the various fossil

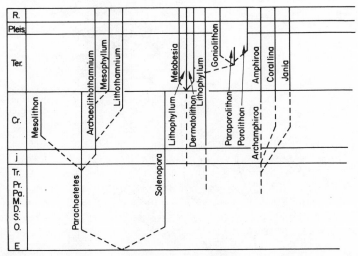

Fig. 13.14 Fossil genera of Rhodophyceae. Geological Times Span. (after J. Harlan
Johnson)

genera to the living ones have been postulated by Johnson (1963). (Fig. 13.14). Other genera that have been ascribed to the Rhodophyceae are *Palmatophycus* (Silurian) with feathery branches and fine hairs, *Thamnocladus* (Devonian) with a fleshy lamina transversèd by a central axis or strand, *Foliophycus* and *Cuneiphycus*. In the Permian the extinct Gymnocodiaceae are regarded as having been very similar to species of *Galaxaura*.

In Californian Miocene deposits a few fleshy Rhodophyceae resembling extant members of the Cryptonemiales, Gigartinales and Ceramiales have recently been discovered. These include *Paleopikea*, *Chondrides* and *Delesserites* (Parker and Dawson, 1965).

NEMATOPHYCEAE: Nematophytales. Figs. 13.15, 13.16

Two genera are now grouped in this assemblage which was established by Lang (1937), and although he regarded these forms as land plants, nevertheless they have so many features in common with the algae that it is felt proper to include them here, as does Johnson (1958, 1959). It is perhaps almost too speculative to suggest that they represent transmigrant forms, but it would appear they they must either be regarded as highly developed algae which adopted a land habitat, or else as the most primitive of all true land plants. The two genera agree closely in their morphological structure, and although they are both frequently found associated with each other in the Devonian rocks the two structures have not yet been found in organic connection.

In spite of this it is not unlikely that the leafy *Nematothallus* was the photosynthetic lamina of the stem-like *Nematophyton* and may also have functioned as the reproductive organ. In the lowest strata the plants are found

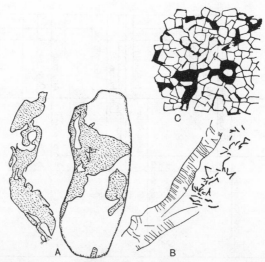

Fig. 13.15 Nematophytales. *Nematothallus*. A, specimens on rock; B, large and small tubes, the former with fine annular thickenings; C, cuticle. (A × 3/5, B, C × 150) (after Lang)

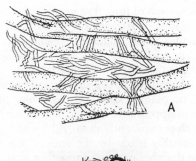

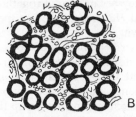

Fig. 13.16 Nematophytales. *Nematophyton*. A, longitudinal section; B, transverse section. (A, B × 120) (after Seward)

associated with remains of marine animals, which suggests that they grew under marine or brackish conditions, while in the higher strata they occur in beds, which are regarded as freshwater or continental, where they are associated with plants that were undoubtedly terrestrial. The presence of spores in *Nematothallus* has been regarded as rendering it unlikely that the plants were algal in nature, but the spores may be comparable to the hard-walled cysts such as are to be found in *Acetabularia*.

The genus *Nematophyton* is found in Silurian and Devonian rocks where it was first described under the name of *Prototaxites* and referred to the Taxaceae, but subsequently it was accepted as an alga and re-named *Nematophyton* or *Nematophycus*. Later the name *Prototaxites* was unfortunately revived and it was placed in the Phaeophyceae, while Kräusel (1936) stated that it must have had the appearance of a *Lessonia* (cf. p. 230). The valid name is therefore *Prototaxites*, but it would seem more satisfactory to retain the better name of *Nematophyton*. The largest specimen is a stem up to two feet in diameter, but whatever the size of the stem it is usually composed of two kinds of tubes, one large and the other small. The large tubes have no cross partitions, but in some species they are interrupted in places by areas which are wholly occupied by small tubes that in other parts of the thallus simply take a sinuous course between the large tubes. The wide tubes, in the specimens described by Lang (1937), show no markings indicative of definite thickening, though striations have been seen in specimens from other localities. Around the outside of the central tissue there is a cortex, or outer region, composed of the same tubes where they bend outwards towards the periphery and eventually stand at right angles to the surface. The outermost zone of all is apparently structureless and may well have been mucilaginous.

Nematothallus is composed of thin, flat, expanded incrustations of irregular shape and up to 65 mm long by 10 mm broad, and constructed of the wide and narrow tubes. The thallus is surrounded by a cuticular layer that exhibits a pseudo-cellular pattern, and which includes firm-walled spores of various sizes within the cuticle and among the peripheral tubes; in *N. pseudovasculosa* the spores have a cuticle and so the suggestion was made that these were land plants or parts of a land plant. The side tubes, which have thin, pale brown walls, are translucent in appearance and have characteristic annular thickenings. The cuticle, which is apparently readily detached, possesses distinct cell outlines that were probably made by the ends of the wide tubes from the ordinary tissue where they became fused together at the periphery, as in the living genera *Udotea* and *Halimeda*. Another species, *Nematothallus radiata*, is more imperfectly known.

From the structure described above it can be seen that the members of this group are strongly reminiscent of the Laminariales and Fucales, and it is tempting to suppose that they represent land migrants from one of these groups. There is no evidence, however, of any parenchymatous structure, so that if they were indeed members of that group they would be allied to forms with a multiaxial type of construction. Problems that have to be solved are: (1) The cuticularized spores; while no such spores with hard outer walls are known from the brown algae they are recorded from the Chlorophyceae, e.g. *Acetabularia*. The suggestion, however, that the spores may have developed in tetrads adds a further complication, at any rate so far as an algal ancestry is concerned, because the Dictyotales and tetrasporic Rhodophyceae do not show the state of differentiation found in these fossil plants. (2) The presence of a deciduous cuticle. In this connection one or two Laminariales are known to shed cuticles during reproduction, and the present authors have found a deciduous cuticle on plants of *Hormosira*, a member of the Fucales. It may be suggested that the plants perhaps had the appearance of a *Lessonia* or even of a *Durvillea*, and a stem diameter of a metre does not preclude them from being algal in character because several of the large Pacific forms may have stipes of almost this size (cf. p. 230). It has also been suggested that these forms are related to the Codiaceae, especially *Udotea*, and in certain respects it is true that they have the structure of a siphonaceous plant. Here again there are several phenomena that need to be explained: (*a*) the presence of two sizes of tubes, (*b*) the presence of a cuticle, (*c*) the presence of cuticularized spores, (*d*) the large size of stem.

The answer to the last problem has already been suggested (see above), but cuticles in the Codiaceae have not been recorded, although it is possible to detect a structure something like a cuticle in *Halimeda*; nor have any species been reported that possess two distinct sizes of tubes, although gradations in size occur in both *Udotea* and *Halimeda*. It must be admitted that there are no living members of the Codiaceae with stems that approach anywhere near the size of those of *Nematophyton*. This, however, is not an insuperable objection as the Nematophytales may bear the same relation to the living Codiaceae that the fossil Lepidodendrons bear to the living Lycopodiales. For the present, however, the problem must be left in the hope that further evidence will be found.

REFERENCES

GENERAL
Johnson, J. H. (1963). *Limestone Building Algae and Algal Limestones* (Golden, Colorado).
Kulp, J. K. (1961). *Science, N. Y.*, **133**, 1105–1114.
Pia, J. (1927). In *Hirmer's Hanbuch der Palaebotanik*, Munich and Berlin.

CARBONIFEROUS (Pennsylvanian and Mississippian)
Johnson, J. H. (1956, 1963). *Quant. Colo. Sch. Mines.*, **51**, 1; **58**, (3), 1.

DEVONIAN
Johnson, J. H. (1958). *Quant. Colo. Sch. Mines.*, **53**, 1.

JURASSIC
Johnson, J. H. (1964). *Quant. Colo. Sch. Mines.*, **59**, 1.

MIOCENE
Parker, B. C. and Dawson, E. Y. (1965). *Nova Hedwigia*, **10**, 273.

PERMIAN
Johnson, J. H. (1963). *Quant. Colo. Sch. Mines.*, **58**, 1.

PRECAMBRIAN
Barghoorn, E. S., Meinschein, W. G. and Schopf, J. W. (1965). *Science, N. Y.*, **148**, 461.
Barghoorn, E. S. and Schopf, J. W. (1965). *Science, N. Y.*, **150**, 337.
Barghoorn, E. S. and Tyler, S. A. (1965). *Science, N. Y.*, **147**, 563.
Glaessner, M. F., Preiss, W. V. and Walter, M. R. (1969). *Science*, **164**, 1056.

ORDOVICIAN
Johnson, J. H. (1961). *Quant. Colo. Sch. Mines.*, **56**, 1.

SILURIAN
Johnson, J. H. (1959). *Quant. Colo. Sch. Mines.*, **54**, 1.

BACILLARIOPHYCEAE
Round, F. E. (1957). *New Phytol.*, **56**, 98.
Krausel, R. (1936). *Natur Volk*, **66**, 13.
Lang, W. H. (1937). *Phil. Trans. R. Soc., Ser. B.*, **227**, 245.
Mikkelsen, V. M. (1949). *Dansk bot. Ark.*, **13**.

CODIACEAE
Konishi, K. (1961). *Sci. Rep. Kanazawa Univ.*, **7**, 159.

DINOFLAGELLATES
Deflandre, G. (1952). In *Traité de Paleontologie*, Ed. J. Piveteau, Tome, **1**, 89, (Paris).
Eisenack, A. (1964). *Katalog der fossilen Dinoflagellaten, Hystrichospharen und verwandten Mikrofossilien.* Stuttgart.
Tasch, P., McClure, K. and Oftedahl, O. (1964). *Micropalaeontology*, **10**, 189.

LITHOTHAMNION (Fossil)
Johnson, J. H. (1962). *Quart. Colo. Sch. Mines*, **57**, 1.

SOLENOPORACEAE
Johnson, J. H. (1960). *Quart. Colo. Sch. Mines*, **55**, 1.

NEMATOPHYCEAE
Krausel, R. (1936). *Natur Volk.*, **66**, 13.
Lang, W. H. (1937). *Phil. Trans. R. Soc. Ser. B.*, **227**, 245.

14

REPRODUCTION AND EVOLUTION

As a result of recent work on pigments in simpler members of the algae, together with electron microscope studies of cell structure, we are in a better position to assess the inter-relations of the classes than hitherto. Any evolutionary scheme, however, can only be regarded as tentative. The first photosynthetic organisms would have been those with such pigments as the bacteriochlorophylls or chlorobium chlorophylls that acquired the ability to synthesize magnesium tetrapyrroles. Such organisms would also have been anaerobes because oxygen would not have been evolved and the electron donor would have been a sulphur compound or an organic compound (e.g. organic acids).

With the development of chlorophyll and the evolution of Photosystem II (involving such pigments as biliproteins and secondary chlorophylls), the ubiquitous water became the electron donor and consequently oxygen was produced. With the production of oxygen came aerobiosis.

Probably the simplest of these plants were the blue-green algae, a stock that contained chlorophyll and biliproteins. The transition from these organisms to the simplest eucaryotic type is obscure, although it would appear that something like *Cyanidium caldarium* or the Cyanellae may represent such a transition form. The whole problem involving the origin of chloroplasts and the relationship between Cyanophyceae, Rhodophyceae, Cryptophyceae and the Cyanellae has recently been discussed by Allsopp (1969) and Carr and Craig (1970). The Rhodophyceae and Cryptophyceae are considered as primitive eucaryotes that were evolutionary dead ends. From the simple Eucaryotic ancestors, whatever they may have been, there appears to be two predominant lines of evolution.

The main line of evolution is associated with the origin of chlorophyll *b* and starch-type, storage polysaccharides ($\alpha 1:4$ glycoside linkages); it forms the *ab* line with the Charophyceae, Chlorophyceae, Prasinophyceae and from them the land plants. This assemblage has been called by Christensen the Chlorophyta. The other main line of evolution is associated with the presence of chlorophyll *c* and laminarin-type storage polysaccharides ($\alpha 1:3$ glycoside linkages); it can be termed the *ac* line. This group has been called the Chromophyta by Christensen (Scheme A).

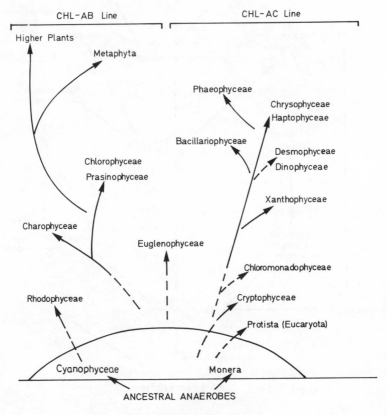

SCHEME A

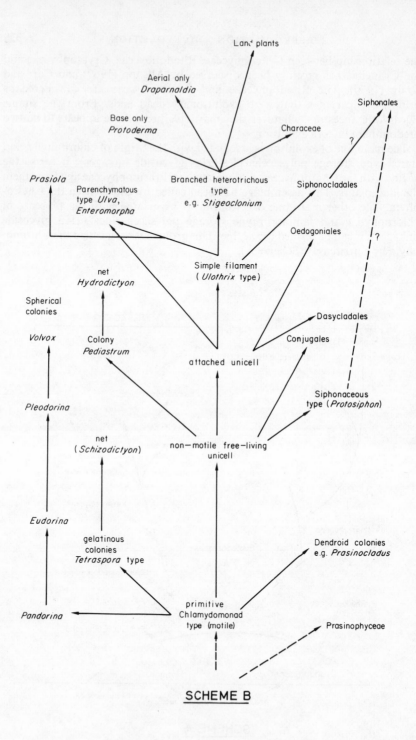

SCHEME B

CHLOROPHYCEAE

In so far as the green algae or Chlorophyceae represent the line from which it is believed land plants ultimately evolved it is convenient to consider evolution within that group first. There have been two major views about the origin of the group. According to one the Chlamydomonadaceae are regarded as the primitive source, whereas according to the other the source is the Palmellaceae. These two views are represented by schemes B and C. Of the two there would appear at present to be rather more evidence in favour of scheme B. In this scheme it will be observed that several lines of evolution divaricate from the primitive source. One line leads to the development of aggregate spherical colonies, culminating in *Volvox*; another line leads to the production of gelatinous colonies, which may become net-like (*Schizodictyon*). All these, however, would seem to represent 'blind alleys' in evolution. Studies of flagella and associated apparatus are contributing towards phylogenetic understanding (Manton, 1965) but at present they do not indicate any major deviation from the above.

The names of the genera do not necessarily imply that they formed the actual intermediate stages, but merely that forms like them were possibly involved.

In the main sequence of evolution the logical development would appear to be the non-motile unicell and then the attached unicell. From the former could have developed more colonial forms terminating in giant nets (*Hydrodictyon*), but simple siphonaceous forms may also have been produced (*Protosiphon*). The non-motile unicell would develop into various types of simple filament from whence evolution would appear to have progressed in three directions:

(*a*) To parenchymatous types, e.g. *Ulva, Enteromorpha*.

(*b*) To partially coenocytic types, e.g. Siphonocladales, and thence to the Siphonales. It has been suggested as an alternative that the Siphonales may have arisen from the siphonaceous Chlorococcales (*Protosiphon*) (Fott, 1965).

(*c*) To branched heterotrichous types (Chaetophorales). This latter line, like certain Ulotrichales would also show heteromorphy in the life history. It is the line from which it is believed the terrestrial flora arose.

The Dasycladales with the large single nuclei seem to occupy a special position (Scheme D) and it is difficult to support an argument that would locate them between the vesicular Siphonocladales and the Siphonales (Nizamuddin, 1964).

In 1951, Papenfuss propounded a third scheme for the chlorophyceae (Scheme D).

This differs from scheme B in certain important respects. The Volvocales are assigned the key position and the Ulotrichales no longer hold the central position as they do in scheme B. This latter aspect is more in line with

Kornmann's suggestion that the Ulotrichales are an isolated group (those with a heteromorphic life history). It also differs in that the Chlorococcales represent the ancestral stock of the Siphonocladales, Siphonales and Dasycladales.

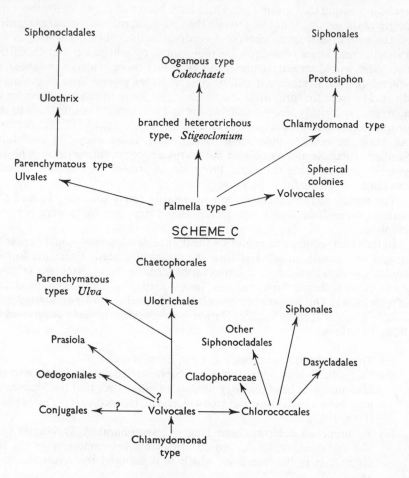

SCHEME C

SCHEME D

The ancestors of today's heteromorphic Ulotrichales probably possessed isomorphic alternation of generations from which arose the current parenchymatous members with their isomorphic alternation. It is also probable that heteromorphy arose more than once in the course of evolution, e.g. Monostromataceae and Acrosiphoniaceae.

Furthermore it seems more probable that the Siphonales and Siphonocladales arose from such stock, rather than from the Chlorococcales (e.g., *Protosiphon* — like organism), for the following reasons:

(i) There are no intermediates between the wholly haploid *Protosiphon* and the wholly diploid Siphonales although one might expect to find some forms with both generations alternating.

(ii) The Chlorococcales are nearly all freshwater whereas the Siphonales are all marine.

(iii) Intermediates can be envisaged from at least some members of the Siphonocladales and Siphonales (Scheme E).

A more detailed discussion of this problem will be found in Chapman (1954) and Nizamuddin (1964). Recently Hori and Ueda (1967) have proposed two lines of evolutionary development within the Siphonales based upon plastid structure, but evidence based on cell ultra-structure alone must be treated with caution.

Details of the suggested evolutionary development from the simple Ulothricaceous filament are shown in scheme E (Chapman, 1954). In the Siphocladalean line the position of *Urospora*, see p. 81, with macroscopic gametophyte and dwarf sporophytic, 'Codiolum' stage is of interest because it shows how forms with heteromorphic generations could have arisen, e.g. *Spongomorpha, Derbesia*. The other feature is the appearance of heterotrichy in the parenchymatous group, and also in the most advanced forms the disappearance of the haploid generation (*Entermorpha nana* and *E. procera*), The concept of the heterotrichous habit was first advanced by Fritsch in 1929, and it is clear that an appreciation of this phenomenon is important in understanding evolutionary problems among the algae. After the intercalation of a sporophytic generation there is also the disappearance of the haploid generation associated with development of the vesicular forms of the Siphonocladales.

With regard to the Conjugales two viewpoints have existed:

(*a*) The filamentous members arose from a Ulothricaceous stock and the unicellular forms developed from the filaments by reduction and fragmentation.

(*b*) The nature of Ulothricaceous reproduction makes (*a*) seem unlikely and instead the suggestions of recent workers (Randhawa, 1959; Yamagishi, 1963) have been followed, namely, that the unicellular Conjugales arose from a non-motile unicellular stock, and that these in turn gave rise to the filamentous forms.

Accepting this viewpoint the Conjugales represent two separate lines of evolution stemming from the unicellular Mesotaeniaceae (p. 114). The one line gave rise to the unicellular Desmidiaceae with exozygospores and only a hint of the filamentous condition, as seen in *Desmidium* and *Gymnozyga*, and the other line saw the evolution of the filamentous Zygnemataceae with endozygospores. The filamentous genera probably represent three separate lines of evolution based upon the nature of the chloroplast, whether stellate, plate-like or ribbon-like, and each of these lines shows the same six parallel evolutionary trends in respect of differentiation of gametes and zygospores (Scheme F).

SCHEME E

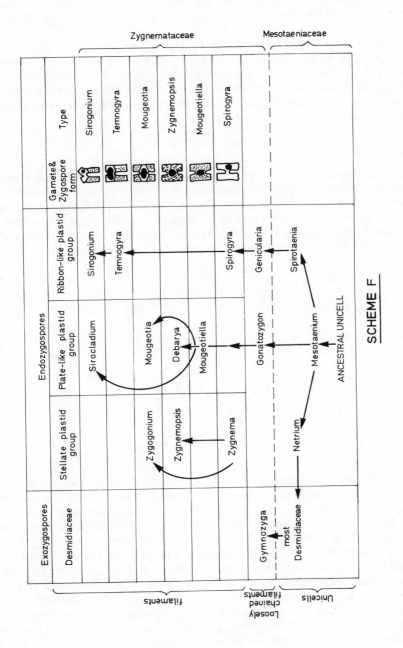

SCHEME F

With regard to the Oedogoniales, one can only assume that they arose, with much modification, from a filamentous Ulothricaceous stock (see Manton, 1965 and p. 81).

The flagellate Prasinophyceae are considered, on grounds of their ultra-structure, as being primitive — perhaps the most primitive of all green chlorophytes. At this point, however, it is perhaps too early to speculate on their relationship to the primitive Chlorophyceae. At the present time it seems that the Chaetophorales and Chlorococcales are the orders most in need of further study.

PHAEOPHYCEAE

In the Phaeophyceae, the primitive members (Ectocarpales) known today possess a thallus of branched heterotrichous filaments, with alternation of isomorphic generations. Biochemically the origin of the Phaeophyceae would appear to be with the Chrysophyceae. Both groups may have arisen from a common stock, or the Phaeophyceae may have had an origin within the Chrysophyceae via an intermediate form such as *Phaeothamnion*.

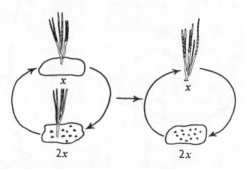

Fig. 14.1 Hypothetical life cycle of heterotrichous ancestral form (left) of *Cutleria* leading to present cycle.

Within the Phaeophyceae themselves those members with isomorphic alternation of generations probably have some relationship with one another. The apparently difficult case of *Cutleria* and *Aglaozonia* can be explained on the basis of an original heterotrichous ancestor in which the aerial part of the gametophyte persisted and only the basal part of the sporophyte (Fig. 14.1). In the case of *Microzonia* (see p. 194) one generation seems to have disappeared completely. Orders with heteromorphic alternation (Heterogeneratae) can conveniently be divided into those that are pseudoparenchymatous (Haplo-stichineae) and those which are truly parenchymatous (Polystichineae).

Schemes G and H represent two slightly different views of evolutionary development with the Phaeophyceae, in both cases starting from a branched, filamentous, heterotrichous ancestor with isomorphic alternation. Which of these two schemes is accepted depends very largely upon the view taken about the origin of the Fucales and Laminariales.

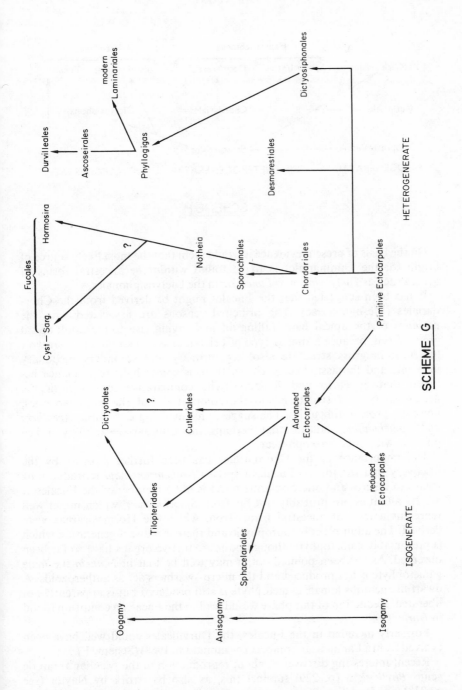

SCHEME G

modern
Laminariales

Durvilleales ← Ascoseirales ← Phyllogigas

Phyllogigas → Dictyosiphonales

Dictyosiphonales

Desmarestiales

Hormosira

Fucales { Cyst-Sarg.

Notheia

Sporochnales

Chordariales

Primitive Ectocarpales

?

HETEROGENERATE

Advanced Ectocarpales

Dictyotales ← ? ← Cuteriales

Cuteriales

Tilopteridales

Sphacelariales

reduced Ectocarpales

ISOGENERATE

Oogamy ← Anisogamy ← Isogamy

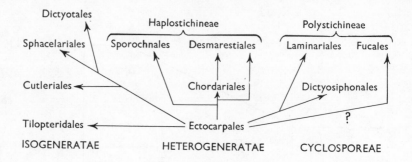

SCHEME H

On the basis of present evidence it would seem that the most likely source of origin for the Laminariales is to be found within the ancestral forms of genera such as *Dictyosiphon* or *Punctaria* in the Dictyosiphonales.

It has been suggested that the Fucales might be derived from the Chordariales (Mesogloiaceae). The principal reasons are associated with the presence of the apical hairs falling off and giving rise to the adult apical growing cell, because a similar type of behaviour is to be found in *Acrothrix* (p. 202); in gross structure also, e.g. primary and secondary medullary filaments and the assimilatory tissue, there is some slight resemblance between that of *Fucus* and *Eudesme*. The comparative looseness of the *Mesogloia* type of structure and the compactness of the Fucalean does, however, present difficulties. The suggestions involving *Chordaria*, *Acrothrix* and *Splachnidium* indicate that perhaps the Chordariales occupy a key position within the Phaeophyceae.

The importance of the Chordariales has been further stressed by the discovery that the former Fucalean species *Notheia anomala* reproduces by means of macro- and micro-swarmers and is not a member of the Fucales at all. Its affinities are probably not far from *Splachnidium*, which might well represent a relic of ancestral forms from which the Hormosiraceae were derived. The adult plant is sporophytic and there is a dwarf generation which is presumably gametophytic though no reproductive organs have so far been observed. As has been pointed out, it may well be that in *Notheia* the male gametophyte is not produced and the micro-swarmers act as antherozoids. A dwarf filamentous female gametophyte is still produced but is apparently not liberated. Reduction of this phase would lead to the Fucalean condition found in *Hormosira*.

Formerly included in the Fucales, the Durvilleales could well have been evolved from a Laminarian complex (Nizamuddin, 1968) (Scheme F).

Recent interesting discoveries about reproduction in the peculiar antarctic genus *Phyllogigas* (p. 229) support this, as also has work by Naylor (see p. 238) on *Durvillea*. The remaining Fucales are probably diphyletic, the

Hormosiraceae being derived from a Chordarialean source such as *Notheia*, while the Cystoseiraceae and Sargassaceae may be derived from the Chordariaceae.

If these arguments are valid then scheme F must be regarded as a more probable interpretation than scheme G. The above considerations also emphasize the great phylogenetic importance of the Dictyosiphonales and Chordariales. A detailed study of the former has still to be carried out, but since the Chordariales form a more or less homogeneous group, it is possible to suggest how evolution within the order may have taken place (Scheme I). The Chordariaceae would seem to represent the primitive stock with uniaxial plants (*Mesogloia*) preceding multiaxial plants (*Eudesme*, *Myriogloia*). At some stage reduction must have occurred leading to the reduced epiphytic Myrionemataceae and the saxicolous, crustaceous Ralfsiaceae. Another line of development is provided by the cushion forms such as *Elachista*. The final stage is probably represented by a genus such as *Corynophlaea* with *Petrospongium* (see p. 202) being treated as a reduced form.

Yet another line of evolution must be represented by the pseudoparenchymatous Spermatochnaceae with apical growth instead of the usual trichothallic. *Spermatochnus* and *Nemacystus* with a single axial thread may be regarded as more primitive than genera with several axial threads (e.g. *Stilophora*). Pseudoparenchymatous development is also found in *Chordaria*, representing the most advanced condition in the Chordariales, and in the peculiar monotypic genus *Splachnidium*. Another peculiar monotypic genus with no immediate obvious affinities is the antarctic *Caepidium*. Here there is a basal thalloid plant, reminiscent of *Ralfsia*, from which at certain times arise branched threads of *Chordaria*-like structure. The plant also produces a *Leathesia*-like cushion form.

RHODOPHYCEAE

The origin of the Rhodophyceae poses numerous problems. Among the Eucaryotic algae (p. 1) there appear to be no immediate relatives of this group. The class is completely lacking in any flagellated stage, and biochemical differences include the presence of large amounts of biliproteins, (see p. 4) a simple carotenoid distribution, Floridean starch and a tendency to form polysaccharide-sulphate esters.

It has been suggested that the Rhodophyceae may have arisen from the Cyanophyceae. The main argument for this relationship is the similarity in pigments, especially biliproteins. It should be remembered, however, that any connection between the Rhodophyceae and Cyanophyceae would transcend the demarcation between the Eucaryotic and Procaryotic organization and all the inherent differences that go with this.

Recently Chadefaud (1963) has suggested another possible line of development based upon a study of the reproductive organs. He points out that no sex organs are known in the Cyanophyceae whereas they are present in primitive Rhodophyceae (other than Porphyridiales) and that there is also evidence of sex in certain bacteria; hybrids are known and a form of meiosis

M

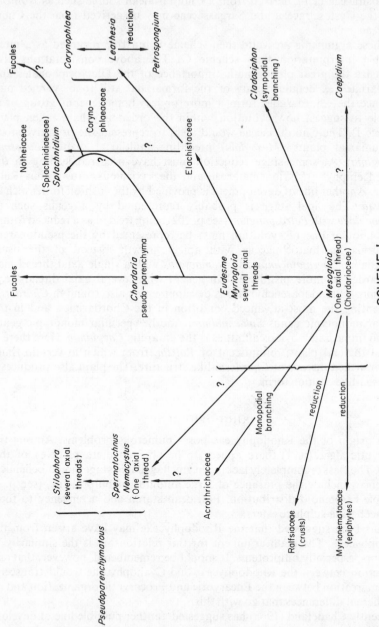

SCHEME I

occurs at endospore formation in the Sporobacteriales in which the donor cell (D) passes genetic material to the receptor cell (R) (Fig. 14.2). This process can be termed archeogamy. In *Petrocelis cruenta* a nucleus is transferred from one filament to another by a connecting thread, though there is no subsequent nuclear fusion. In *Rhodochaete parvula* a nucleus is transferred from the spermatocyst by a spermatium to a receptor cell and the two nuclei fuse. This process can be regarded as protogamy. There are two possible variants from

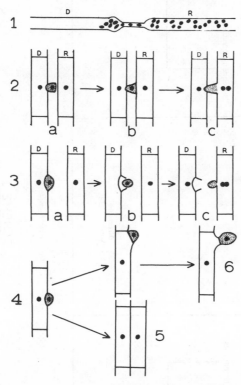

Fig. 14.2 1, *Bacillus megatherium*: donor cell (D) gives mass of chromatin to receptor cell (R) = Archeogamy; 2, *Petrocelis cruenta*: transfer of nucleus from one vegetative cell to another; non-sexual; 3, *Rhodochaete parvula*: male donor cell sends a gamete nucleus to a female receptor cell = Protogamy; 4, evolution of male donor cell from *Rhodochaete* type to Bangiales (5) and Florideophycideae (6). (after Chadefaud)

this condition. In one the spermatocyst becomes terminal and attached by a stalk (Florideophycideae) while in the other the entire cell functions as the spermatocyst (Bangiales). Chadefaud argues that the same kind of sequence can be seen in the development of the carposporophyte. If this interpretation is correct we would have to regard the bacteria as probably giving rise separately to the Cyanophyceae and Rhodophyceae, and algae, such as *Porphyridium* and *Chroothece*, would then have to be regarded as reduction forms rather than primitive genera. There is a thermophilic, acidophilic

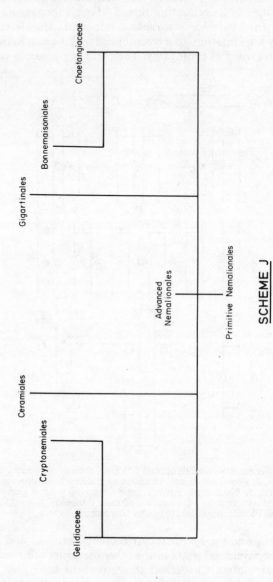

SCHEME J

organism, *Cyanidium caldarium*, which may be regarded as a very primitive Rhodophyte.

The Florideophycidae presumably arose from within the Bangiophycideae, though not necessarily from the most advanced members. Any phylogenetic treatment must take into consideration their sexual reproductive organs, and the auxiliary cell.

TABLE 14.1

REPRODUCTION	ORDER	AUXILIARY CELL
True auxiliary cells absent	Nemalionales	No procarp
Auxiliary cells present before fertilization. Morphologically simple.	Cryptonemiales	Procarp present
Auxiliary cells present before fertilization. Morphologically advanced.	Gigartinales	,,
Auxiliary cells develop after fertilization	Rhodymeniales	,,
	Ceramiales	,,

Little attention has been paid to problems of phylogeny within the Florideophycidae. Recently Fan (1961) has suggested that the Gelidiaceae and Cryptonemiales probably originated from a common ancestral stock which in its turn had been derived from members of the Nemalionales in which nutritive cells had become differentiated structures. The Bonnemaisoniales and Chaetangiaceae had a separate origin from Nemalionales in which nutritive auxiliary cells developed into generative auxiliary cells. These views are depicted in Scheme J.

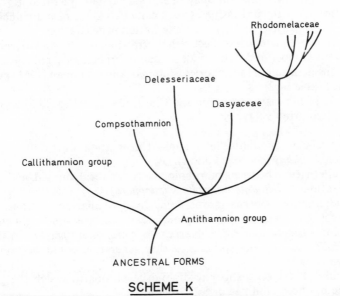

SCHEME K

Later Hommersand (1963) discussed evolution within the Ceramiales, which he regards as evolved from the Nemalionales parallel with the Cryptonemiales and Gigartinales. It seems therefore that evolution within the Florideophycideae may have progressed along a number of independent lines. Within the Ceramiales Hommersand suggests several different lines of evolution, the principal ones being shown in Scheme K.

LIFE HISTORIES

When gametes fuse to form a zygote the chromosome number is doubled and before gametes can again be produced the original haploid number must be restored. In some algae this occurs when the zygote germinates; in other algae a diploid phase is introduced and from such a phase a haploid spore is produced. In certain algae either phase is capable of reproducing itself and not giving rise to the alternate generation; hence the use of the term life history rather than life cycle. In those algae where there is only a diploid generation (Fucales, Siphonales) reduction division occurs at gamete formation and fusion of the gametes restores the double number of chromosomes.

In his studies on the Rhodophyceae, Svedelius coined a number of terms which have subsequently come into common use for the algae:

Haplont: A sexual haploid plant with only the zygote diploid (*Volvox*).

Diplont: A sexual diploid plant in which only the gametes are haploid, e.g. *Codium*.

Haplobiont: A plant possessing cytologically only one kind of individual in the life cycle: this may be either a haplont or a diplont.

Diplobiont: A plant possessing cytologically two kinds of individuals in the life cycle, and hence an alternation of generations.

Feldmann subsequently introduced the terms mono-, di- and trigenetic to describe the number of different independent morphological generations. In a later reconsideration of life histories Chapman and Chapman (1961) proposed a scheme (Table 14.2).

A study of life histories within the Chlorophyceae (Fig. 14.3) shows that there are four principal types:

(1) A unicellular or multicellular haploid generation with only the zygote diploid (*Chlamydomonas, Volvox*).

(2) An alternation between morphologically identical diploid and haploid generations (*Ulva, Enteromorpha, Cladophora*).

(3) An alternation between morphologically dissimilar generations, usually microscopic sporophytic (diploid) and macroscopic gametophytic (haploid) generation, e.g. *Ulothrix, Urospora, Acrosiphonia*. In the case of *Halicystis-Derbesia* (p. 102) neither generation could be considered microscopic).

(4) A multicellular or coenocytic diploid generation in which the gametophyte has been eliminated, e.g. Siphonales.

TABLE 14.2

	MONO-GENETIC	DIGENETIC			TRIGENETIC
		Isomorphic	Heteromorphic		
			Game-tophyte dominant	Sporo-phyte dominant	
HAPLO-PHASIC (haploid haplobiont)	Volvocales Conjugales Charales		Some *Bonnemaisoniales*		*Naccaria wiggii*
DIPLO-HAPLO-PHASIC		*Ulva Entero-morpha Dictyota*	Bangio-phycideae *Cutleria Urospora*	*Laminaria*	Most Rhodo-phyceae
DIPLO-PHASIC (diploid haplobiont	Valoniaceae Siphonales Fucales				

If the haploid plants were primitively monoecious the first stage in evolution would presumably have been the development of the dioecious condition; this could have been followed by the intercalation of a sporophyte generation through delay in occurrence of the reduction division (e.g. *Monostroma* species with 'Codiolum' zygotic phase) leading to alternation of two generations. Feldmann (1952) has put forward the view that the primitive life cycle in the Chlorophyta, Phaeophyta and Rhodophyta is the diplohaplophasic one but this is difficult to accept for the Chlorophyta because there are so many examples among the primitive orders, e.g. Volvocales, Chlorococcales.

A study of the life histories (Fig. 14.4) of the Phaeophyceae indicates that evolutionary progression in general conforms to evolutionary morphological progression. In most types the life history is not simple, because each generation is capable of repeating itself. Underlying the life history, however, is the fundamental nuclear cycle, but despite this there is no real regular alternation of generations except in the Dictyotales, Laminariales and some members of the Chordariales and Sphacelariales. It is evident that the primitive Phaeophycean was either a plant with only a haploid generation in the life history or one in which there was isomorphic alternation. Very few members of the Ectocarpaceae are known to be haplonts, and in one good case, *E. virescens*, the parthenogenetic development of the eggs suggests degeneration rather than a primitive condition. The general weight of evidence would appear to be in favour of isomorphic alternation as the primitive Phaeophycean condition (see Fig. 14.4).

The work of Chihara and Magne and the French school (see Chapt. 11) has stimulated a dramatic revision of the whole problem of Rhodophycean

life histories and their phylogeny. Although we now have a better appreciation
of the situation than previously it is certain that even the present views will
undergo further modifications as new work is carried out on the more
primitive members.

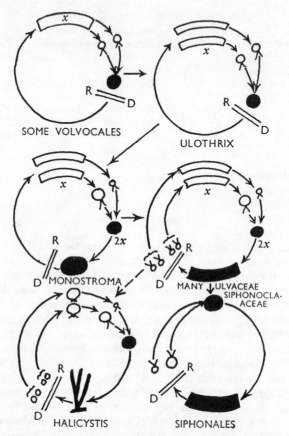

Fig. 14.3 Types of life cycle in the Chlorophyceae and their possible inter-relationships
(RD = position of reduction division in the life cycle).

The most primitive life history known at present is that of *Cyanidium
caldarium*, which reproduces by means of endospores only, so that it is
haplophasic and monogenic. A more advanced condition is to be seen in
Erythrotrichia, also haplophasic and monogenic but with male and female
plants. Within the Bangiophycideae the most advanced type of life history is
represented by *Porphyra* where it is haplophasic and heteromorphic.

Within the Florideophycidae it appears that a truly haplophasic heteromorphic life history is very rare, but it does seem to exist in *Naccaria wiggii*
and *Atractophora hypnoides*. If the higher red algae originated from the

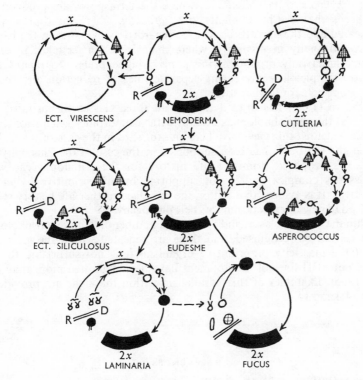

Fig. 14.4 Types of life cycle in the Phaeophyceae and their possible inter-relationships (RD = position of reduction division in the life cycle).

Bangiophycideae then these two species must be regarded as representing the primitive life history. The other alternative is to follow Feldmann and accept the diplohaplophasic life history as primitive.

The intercalation of a sporophytic generation, possibly first the diploid carposporophyte, followed by the dwarf tetrasporophyte, led to the development of the typical life history of the diplohaplophasic trigenetic type. *Liagora tetrasporifera* and other species with a similar life history (see p. 273) represent forms in which there is only the diploid carposporophyte, though Magne prefers to treat these species as reduction forms in which the tetrasporophyte has become lost.

From the basic diplohaplophasic trigenic life history various evolutionary developments have taken place. One is the continuation of the heteromorphic alternation characteristic of so many of the Bonnemaisoniales, another has resulted in the reduction of the gametophyte generation (*Rhodothamniella floridula*), while in *Lomentaria rosea* in European waters and in the genera *Hildenbrandia* and *Apophloea* the gametophyte generation has become wholly suppressed.

M2   347

Another development, which must be interpreted as retrogressive, can be seen in *Phyllophora membranifolia*, where the tetraspores are grouped into nemathecia on the diploid plant; in *P. brodiaei*, where the diploid phase disappeared and the namathecia can be regarded as parasites in the haploid thallus; and finally in *Ahnfeldtia* where meiosis no longer takes place and the nemathecia only produce monospores. Magne prefers to regard these last two examples as representing separate lines of reduction and there is clearly scope here for further study.

Feldmann (1952) has argued that in all three classes the primitive life history is the diplohaplophasic one with isomorphic generations. The arguments against this view have been put forward in the case of the Chlorophyceae and the thesis has been supported in the case of the Phaeophyceae. Feldmann's proposed primitive life history for the Rhodophyceae seems unnecessarily complex and is not supported by any primitive living representatives. It is evident that within the red algae further life history studies could lead to new interpretations of their phylogeny.

Within the various classes and divisions a number of definite morphological tendencies can be recognised, and these are repeated within the groups. In view of the plasticity of primitive organisms it is not surprising that the various potential lines of development have been exploited more than once (Fott, 1965). Examples of this parallelism within the algae are provided in Tables 14.3 and 14.4.

REFERENCES

Allsopp, A. (1969). *New Phytol.*, **68**, 591.

Carr, N. G. and Craig, I. (1970) in *Phylogenetic Aspects of Phytochemistry*, (Academic Press, New York and London).

Chadefaud, M. (1963). *Revue Algol, N.S.* **6**, 255.

Chapman, V. J. (1954). *Bull. Torrey bot. Club*, **81**, 76.

Chapman, D. J. and V. J. (1961). *Ann. Bot., N.S.*, **25**, 547.

Fan, K. C. (1961). *Univ. Calif. Publs. Bot.*, **32**, 315.

Feldmann, J. (1952). *Révue Cytol, Biol. vég.*, **13**, 4.

Fott, B. (1965). *Preslia*, **37**, 117.

Fritsch, F. E. (1945). *Ann. Bot., N.S.*, **9**, 1.

Hommersand, M. H. (1963). *Univ. Calif. Publs. Bot.*, **35**, 165.

Hori, T. and Ueda, R. (1967). *Sci. Rept. Tokyo Kyoik. Daig. B.*, **12**, 225.

Kornmann, P. (1965). *Phycologia*, **4**, 163.

Kylin, H. (1933). *Lunds univ. ärssk., N.F.*, Avd. **2**, 29.

Magne, F. (1964). *Cah. Biol. mar.*, **5**, 461.

Magne, F. (1967). *Le Botaniste*, **50**, 297.

Manton, I. (1965). *Adv. bot. Res.* **2**, 1.

Nizamuddin (1964). *Trans. Am. microsc. Soc.*, **83**, 282.

Nizamuddin (1968). *Botanica marina*, **11**, 115.

Papenfuss, E. F. (1957). *Svensk bot. Tidskr.*, **45**, 4.

Randhawa, M. S. (1959). *Zygnemaceae*. (New Delhi.)

Yamagishi, T. (1963). *Sci. Rept. Tokyo Kyo Dag.*, Sec. B. **11**, 191.

TABLE 14.3

PARALLELISM IN EVOLUTION OF THE SIMPLER TYPE OF ALGAL CONSTRUCTION

Type of Construction	Chlorophyceae	Xanthophyceae	Chrysophyceae	Cryptophyceae	Dinophyceae
(1) Motile unicell	Chlamydomonas	Heterochloris	Chromulina	Cryptomonas	Gymnodinium
(2) Encapsuled unicell	Phacotus	—	Chrysococcus	—	—
(3) Motile colony	Pandorina Volvox	—	Synura	—	Polykrikos
(4) Dendroid colony	—	Mischococcus	Dinobryon	—	Gloeodinium
(5) Palmelloid colony	Tetraspora	Botryococcus	Phaeocystis	—	Cystodinium
(6) Coccoid form	Chlorella	Chlorobotrys	Chrysosphaera	Tetragonidium	Dinothrix
(7) Simple filament	Ulothrix Urospora	Tribonema	Nematochrysis	Bjornbergiella	—
(8) Branched filament	Cladophora	—	Phaeothamnion	—	—
(9) Heterotrichous thallus	Stigeoclonium	—	—	—	—
(10) Siphonaceous filament	Codium Halimeda	Botrydium Vaucheria	—	—	—

TABLE 14.4

ANALAGOUS ORDERS IN THE ALGAE

	Chlorophyceae	Xanthophyceae	Chrysophyceae	Cryptophyceae	Dinophyceae
Flagellated single cell	Volvocales	Heterochloridales	Chromulinales	Cryptomonadales	Peridiniales
Coccoid	Chlorococcales	Mischococcales	Chrysosphaerales	Cryptococcales	Dinococcales
Palmelloid	'Tetrasporales'	Heterogloeales	Chrysocapsales	—	Dinocapsales
Filamentous	Ulotrichales	Tribonematales	Phaeothamniales	—	Dinotrichales
Siphonaceous	Siphonales	Vaucheriales	—	—	—
Amoeboid	—	Rhizochloridales	Rhizochrysidales	—	Rhizodiniales

15

MARINE ECOLOGY

ECOLOGY OF ROCKY COASTS

An observant investigator of rocky coasts cannot help but be impressed by the remarkably distinct zoning of the principal organisms; this zonation, moreover, is not confined to any one region but is found throughout the world. The vertical extent of the zones or belts depends on the tidal rise, the greater the rise the more extensive each zone. Even in the Caribbean, where the tidal range is about nine inches, it is possible to observe a zonation. Any one of the major species commonly occupies a very definite vertical range, but occasionally may be found outside this range when for example, the presence of a rock pool or cranny provides conditions favourable for its existence.

The numerous ecological studies that have now been carried out in many parts of the world have involved a great number of investigators. In any one area the number of communities recognized depends upon two factors.

The first factor to be considered is the locality.

Some localities, especially those in colder waters, are very much richer than others. Within a region itself a rocky shore is usually richer than a boulder shore, and a protected area is commonly richer than an exposed one.

The second factor is personal analysis.

Each investigator will tend to have a somewhat different concept of what comprises an algal community. The number of communities recognized will also depend upon the time and thoroughness with which the shore is examined.

The terminology that has been employed for naming the communities has led to much confusion. Algal ecology commenced after terrestrial ecology and so some investigators have attempted to apply terms used in the latter to the former, whereas others have considered that the conditions are sufficiently different to make this application impossible. Cotton (1912), for example, recognised five algal formations, subdivided into associations, at Clare Island in western Ireland:

(1) Rocky shore formation,
(2) Sand and sandy mud formation,
(3) Salt marsh formation,
(4) River mouth formation,
(5) Brackish bay formation.

At Lough Ine in south-west Ireland, Rees (1935) recognized only two formations, the exposed and sheltered coast. Rees used the term 'association' for communities where species that are associated with the dominants are controlled by the same factors, but it is not easy to prove experimentally that certain factors do control the species distribution. The term 'zone' was used for algal belts which possess horizontal continuity with well-marked upper and lower limits. The term zone has, however, a geographical connotation, e.g. temperate zone, and therefore the word 'belt' is best used for the marked horizontal communities.

In New Zealand, Cranwell and Moore (1938) used the term 'association complex' for the group of successive belts that follow one another vertically The horizontal belts are normally continuous though they can be interrupted occasionally by another community. In other words, there could be an

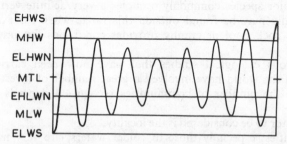

Fig. 15.1 Spring and neap tide curves. (EHWS = Extreme high water spring tides; MHW = Mean high water; ELHWN = Extreme lowest high water neap tides; MTL = Mean tide level; EHLWN = Extreme highest low water neap tides; MLW = Mean low water; ELWS = Extreme low water spring tides)

association fragment of *Durvillea* in the *Xiphophora* belt. There is still divergence of opinion about nomenclature, and even at the present time when it is desirable to secure uniformity, it is still perhaps more satisfactory to use the non-commital term 'community'.

It is important to have a satisfactory definition of the littoral. It can, for example, be described in purely physical terms and is thus regarded as extending from mean high water (M.H.W.) to mean low water mark (M.L.W.), or from extreme high water (E.H.W.) to extreme low water (E.L.W.). The littoral based upon a tide curve, could be subdivided as shown in Fig. 15.1.

Tides are by no means uniform and the various irregular tidal types would lead to some difficulties. It is now realised that the animal and plant belts of the shore vary according to exposure, latitude and topography and can be to some extent independent of the tide. The present procedure therefore is to employ a biological definition of the littoral based upon the major organisms. Ultimately, when more is known about the causal relationships between tidal phenomena and the major belt organisms it would probably be more desirable to use tidal data. The biological criteria are essentially based upon the work of the Stephensons (1949).

Basic Zonation

In 1949 the Stephensons proposed a basic grouping in the littoral which is now becoming generally accepted. Three major belts, the supra-littoral fringe, the mid-littoral and the sub (or infra-) littoral fringe, appear to be practically universal. The relationship of these belts to tide levels and the dominant organisms is as follows:

Supra-littoral	
	Upper limit for littorinids
Supra-littoral fringe	Extreme high water springs
	Upper limit for barnacles
Mid-littoral	
	Mean low water neaps
	Upper limit for laminarians or certain fucoids
Sub-littoral fringe	
	Extreme low water springs
	? Exact position not settled
Sub-littoral	

This basic zonation varies from place to place depending on the inter-relationships of tides and wave action (Fig. 15.2). Some discussion has recently taken place about the nomenclature of these belts, but so many workers have accepted the Stephenson terminology that there seems as yet no valid reason for departing from it.

Lewis (1961) proposed that the upper limit of the littoral should be set by the upper limit of either the littorinid belt or of the black lichen (*Verrucaria*) — Cyanophyceae belt that replaces littorinids in some parts of the world. He sets the lower limit as the boundary between the Laminarians (or certain

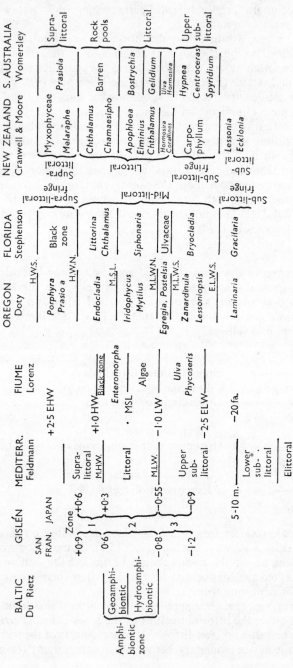

Fig. 15.2 Algal belts in different parts of the world compared on a tidal basis. (after Doty)

354

fucoids in the southern hemisphere) and the barnacle belt, and he does not recognise the sub-littoral fringe (Fig. 15.3). Other workers deny the existence of the sub-littoral fringe in their regions and therefore do not include it in a general scheme. Where it is recognised the transition is determined by the upper limit of the Laminarians, primarily in the northern hemisphere, or of fucoids, e.g. *Durvillea*, *Carpophyllum*, *Cystophora* in the southern hemisphere. The lower limit of this fringe is set a few feet below extreme low-water mark of spring tides, but at present our knowledge of this fringe is not as good as that of our knowledge of the mid-littoral and supra-littoral fringe (Chapman,

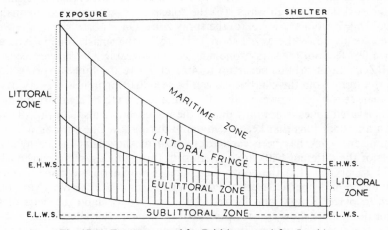

Fig. 15.3 Zones proposed for British coasts. (after Lewis)

1967). Because it is submerged for so much of the time it is not an easy zone in which to work, and it is clear that more study will be necessary before the lower limit of this fringe can be satisfactorily settled. Although contributions to its flora and fauna are made by denizens of both the eulittoral and sub-littoral there are species which appear to be restricted to the belt. It seems to present particular difficulties in tropical regions as Lawson (1966) has pointed out in his study of the West African coast.

In most regions the supra-littoral fringe is characterised by the presence of species of *Littorina* or of allied genera, i.e. *Melaraphe* in New Zealand and Australia. As well as the littorines and, in some places, in their absence, the region is commonly occupied by Cyanophyceae and marine lichens and these form a conspicuous black belt. The transition to the major belt on the shore, the eulittoral, is determined by the upper limit of the barnacles, i.e. species of *Balanus*, *Chamaesipho*, *Chthalamus*, etc. In a given region, variation of exposure may determine which particular species sets the upper limit. In most regions it is found necessary to subdivide this belt, such subdivision being based upon the dominant organisms. There is therefore considerable variation throughout the world, although some degree of uniformity is seen in the cold waters of the northern hemisphere on the one hand and the cold

waters of the southern hemisphere on the other. At least a portion of the belt
is normally dominated by the barnacles.

The levels of the upper belts in the eulittoral and of those in the supra-
littoral fringe in any locality are not entirely dependent upon the height of
the spring tides. On an exposed coast the shore is open to considerable wave
action and a heavy spray rises against the rocks to a considerable height, and
as a result both upper and lower limits are elevated several feet above normal.
This elevation is known as the 'splash zone', but its extent depends on slope
of shore and whether it is related only to storm waves or to fine-weather
waves as well. On a very exposed coast with big waves the splash zone can be
subdivided into, (a) the wash zone, which is the height the actual wave washes
up the rock face at high water, (b) the splash zone, which is represented by
actual splash from the wave, (c) the spray zone. On exposed coasts the wash
zone extends down the whole shore (Fig. 15.3); the splash zone is generally
only a few feet but the spray zone may be considerable. On the west coast of
New Zealand littorinids occur up to 40 feet above the highest spring tides,
and the spray from the giant waves can be seen drifting in nearly every day of
the year. It is likely, however, that the size and growth of algal organisms in
such elevated zones is not only due to the waves, but is also dependent upon
moist air conditions that keep them damp during low tide periods.

Sufficient work has been carried out now to show that the bathymetric
levels of the different belts are not uniform throughout a region although it
could be argued that the controlling factors would be uniform. In Great
Britain the bathymetric levels of *Fucus vesiculosus* and *F. serratus* vary from
place to place (Knight and Parke, 1950). In New Zealand the belts of the
major dominants show considerable variation, even after allowance has been
made for splash elevation. It has also been demonstrated that change of
latitude may exert an influence upon the belts. Gislen (1944) has shown that
on both sides of the Pacific the belts of brown algae tend to emerge with
increase in latitude. Knight and Parke (*loc. cit.*) have shown the same thing
to hold true for *Fucus serratus*, though work in New Zealand (Chapman and
Trevarthen, 1953) has shown that *Hormosira banksii* occupies significantly
lower levels in regions further south.

Rock pools, couloirs (or gullies) and caves, represent other features on the
rocky shore that afford rather specialized habitats. Rock pools have been
divided (Chapman, 1964) into three groups: (1) those of the upper eulittoral
and supra-littoral fringe which may remain exposed for several successive
days during the neap tide periods and where therefore conditions of temperature and salinity can become extreme, particularly in summer; (2) the pools
of the middle eulittoral, and (3) of the lower eulittoral, which may be covered
continuously for several successive days during neap tides.

Algae found in the upper pools must be capable of tolerating rapid changes
of salinity and temperature as well as being recipients of bird excrement.
Such pools tend to be populated by mats of Cyanophyceae with which are
associated diatoms and planktonic flagellates. Macroscopic algae are rare
and are generally Chlorophyceae such as *Enteromorpha* and *Chaetomorpha*.
On New Zealand shores *Pylaiella fulvescens* also occurs in such high pools.

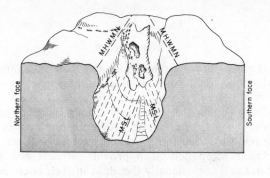

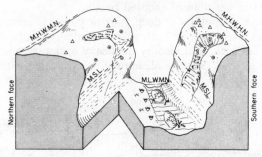

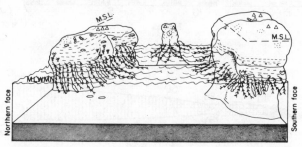

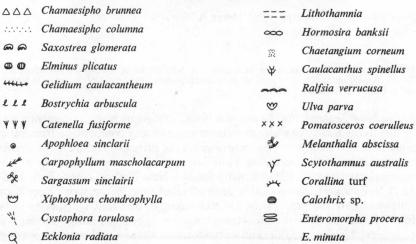

△ △ △	*Chamaesipho brunnea*	═ ═ ═	*Lithothamnia*
⠒⠒⠒	*Chamaesiphc columna*	∞∞∞	*Hormosira banksii*
⌒ ⌒	*Saxostrea glomerata*	⠿	*Chaetangium corneum*
⫧ ⫧	*Elminus plicatus*	⍦	*Caulacanthus spinellus*
⊦⊦⊦⊦	*Gelidium caulacantheum*	∿∿	*Ralfsia verrucosa*
ℓ ℓ ℓ	*Bostrychia arbuscula*	⍵	*Ulva parva*
Ɏ Ɏ Ɏ	*Catenella fusiforme*	× × ×	*Pomatosceros coerulleus*
⊙	*Apophloea sinclarii*	⇶	*Melanthalia abscissa*
⤳	*Carpophyllum mascholacarpum*	⋎	*Scytothamnus australis*
⋇	*Sargassum sinclairii*	⋏⋏	*Corallina* turf
⍵	*Xiphophora chondrophylla*	⊜	*Calothrix* sp.
⋱	*Cystophora torulosa*	⧩	*Enteromorpha procera*
Q	*Ecklonia radiata*		*E. minuta*

Fig. 15.4 Sections through a New Zealand couloir at Leigh from the mouth (lowest) to the end (top). (after Starling)

357

Studies of the flora of rock pools must therefore take into consideration the levels of the pools and also the relation of the species to their occurrence elsewhere. This is necessary because the species can be divided as follows: (a) species wholly confined to pools, (b) species characteristic of the sub-littoral, (c) adjacent eulittoral species, (d) eulittoral species that reach an upper limit in pools that are above their normal limit on the exposed rocks. On exposed coasts deep pools can provide what is essentially a highly protected habitat and consequently the algae in such pools may be those of sheltered coasts rather than of exposed coasts.

The vegetation of couloirs or deep, narrow gullies is primarily affected by light conditions, since one or both sides commonly receive much less light than adjacent flat rock, and also by the heavy backward and forward surge of the tides (Fig. 15.4). Specific algae may be associated with such couloirs,

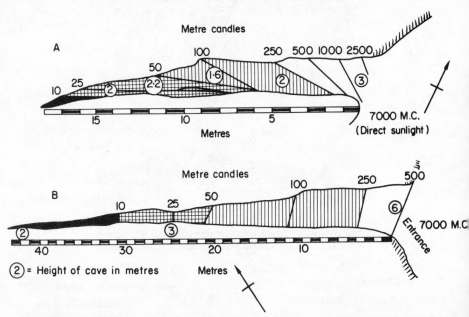

Fig. 15.5 Ground plan of caves in relation to penetration of indirect daylight: A, Red Beach; B, Stanmore Bay. The number encircled represents the height of the caves in metres. (after Dellow and Cassie, 1955)

e.g. in New Zealand *Vidalia colensoi* is most frequent here; its presence may be due to the excessive water movement or to lack of competition from other algae in such a habitat. In the case of caves the principal feature is the change in vegetation with decreasing light intensity. In a detailed study of the seasonal variations in light intensity for two New Zealand caves (Figs. 15.5–15.7) Dellow and Cassie (1955) demonstrated that the pattern of zonation follows closely the gradient of indirect daylight, and that the species to be found could be divided into stenophotic or light-demanding species, e.g. *Calothrix, Lichina, Ralfsia,* and euryphotic or shade tolerating species, e.g.

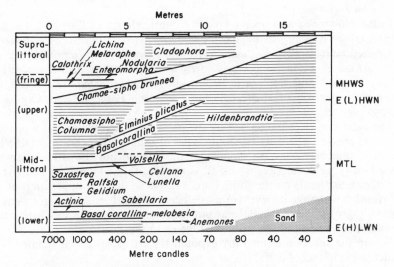

Fig. 15.6 Zonation N.W. wall Red Beach Cave, summer. (Vertical scale = 2 × horizontal scale) (after Dellow and Cassie, 1955)

Peyssonelia, Hildenbrandia, Rhodochorton. The low light intensities found in caves favour what are fundamentally 'shade' species. In New Zealand the shade-inhabiting *Codium cranwellii* is restricted to caves in the littoral region, though it also occurs in the deep sublittoral where again light is reduced.

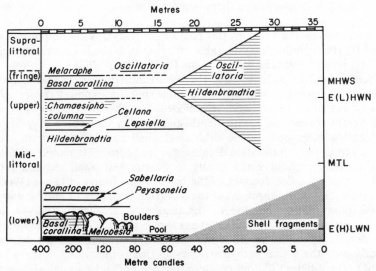

Fig. 15.7 Zoning on N. wall of Stanmore Bay Cave, winter. (Vertical scale = 4 × horizontal scale) (after Dellow and Cassie, 1955)

359

European Zonation

In the North Sea and on the colder Atlantic shores of Europe the supra-littoral fringe is occupied by species of *Littorina* with which are commonly associated belts of *Porphyra umbilicalis* and *Pelvetia canaliculata*. Where the rock surface is extensive there is often a 'black' lichen belt composed of *Lichina confinis* and *Verrucaria maura* above the *Littorina*. The *Pelvetia* is more characteristic of sheltered shores and the *Porphyra* and *Lichina* of exposed shores.

The eulittoral is characterised by the presence of barnacles and a number of fucoid species. There are two main types of zonation depending on whether the dominant barnacles are *Balanus* or *Chthalamus* but transitions between these two types occur (Lewis, 1955). It appears that *Chthalamus* can rise to slightly higher levels on the shore than *Balanus* so that the upper limit of the eulittoral can change its level from one place to another. Also, since *Pelvetia* normally grows above the *Balanus* but below the upper limit of *Chthalamus*, it may either belong to the littoral fringe or to the eulittoral.

Where the shores are not fully exposed the barnacles are associated with a succession of fucoid belts:

(a) A belt of either *Fucus spiralis* or its var. *platycarpus*.
(b) A belt of *Ascophyllum nodosum* with one of *F. vesiculosus* immediately below (or *vice versa*) with an intervening middle belt where both species occur.
(c) The lowest belt of the mid-littoral is commonly dominated by *F. serratus*, with which is often associated *Himanthalia*, or by species of *Laurencia* and *Rhodymenia* where there is shelter, or *Gigartina* or *Chondrus crispus* where there is more exposure.

One feature of the belts, in this part of the world, is the important part played by the fucoids. A comparable covering of algae is commonly absent in the corresponding cold waters of the Pacific (see p. 363).

Where the substrate is suitable, the sublittoral fringe is characterized by species of *Laminaria*, especially *L. digitata* but on exposed coasts *Alaria esculenta* is more characteristic. It is evident that the relative proportions of the different species vary with depth and locality and there may also be considerable variations. In the sublittoral when rock gives way to stones and then to shingle so the algal dominants change from *Laminaria digitata* to *L. saccharina* to *Halidrys* and finally to *Chorda* and small red and brown algae. A more detailed account of the belts can be found in Lewis (1964).

On the rocky shores throughout the world communities of marine diatoms can be found. These, because of the taxonomic problems involved, have been but little studied. Aleem (1950) has shown that the diatom communities differ depending on the nature of the substrate, and there is also a difference depending on the actual bathymetric level. This is illustrated in the scheme (p. 361). The different diatom communities are not to be found all the year round, but in general are at their best sometime between January to March or in the autumn and winter.

TABLE 15.1

DIATOM COMMUNITIES — SOUTH OF ENGLAND

	On Concrete	On Chalk
Supra-littoral	Achnanthes-Cyanophyceae Amphipleura rutilans	Amphora-Nitzschia
Eulittoral	Fragilaria-Melosira (Swanage)	Melosira-Barnacles Synedra-Pylaiella
Sub-littoral fringe		Schizonema ramosissima Schizonema grevillei Rhabdonema-Licmophora

NORTH AMERICA AND CANADA

On the North Atlantic shores, belts are found in which *Ascophyllum* and *Fucus* still occur but the algal flora is much poorer than in Europe (Stephenson, 1952). Examples of the various belts are given in Table 15.2.

TABLE 15.2

ALGAL COMMUNITIES OF NORTH ATLANTIC SHORES

	Mt. Desert Is. Maine	North Carolina
Supra-littoral fringe	Littorinids Cyanophyceae Maritime lichens	Calothrix Enteromorpha
Eulittoral	Barnacles Fucus vesiculosus Ascophyllum Rhodymenia Fucus furcatus Spongomorpha	Barnacles Modiolus Porphyra Gelidium-Polysiphonia Rhodymenia Ostrea-Padina
Sublittoral fringe	Alaria-Halosaccion	Sargassum Dictyopteris

Fucoids and laminarians do not play so important a part in this region but they are replaced by other algae.

In the far north, Nova Scotia and Prince Edward Island (Stephenson, 1954) the supra-littoral fringe is narrow and characterized by littorinids and Cyanophyceae, mainly *Rivularia*. In the eulittoral *Fucus vesiculosus* is the most plentiful fucoid but at low water it is possible to find great forests of *Chordaria flagelliformis* and *Scytosiphon lomentaria* in sheltered areas. In more exposed localities extensive areas of *Chondrus crispus* abound in the lower eulittoral and sublittoral. The Laminarian zone is not well represented on Prince Edward Island but is present on the Atlantic coast of Nova Scotia.

The belts in the North Pacific are quite different from those in the North Atlantic. The eulittoral is covered by barnacles but in the middle part there is a belt of *Fucus distichus* in one of its forms. The Laminarian zone is variably developed and in some places it is replaced by a mixed, rather impoverished algal growth (Stephenson, 1961). On the Oregon coast there is a well-marked series of belts commencing at the top of the eulittoral with *Pelvetiopsis fastigiata* or *Porphyra perforata* or *Scytosiphon lomentaria* and succeeded by *Fucus gardneri*. Species of *Cladophora* are also abundant with patches of *Cumagloia andersonii*. The next major belt is dominated by species of *Iridophycus*, *Gigartina* and *Hymenena* and towards the bottom of the eulittoral it is replaced by a belt of the laminarians, *Hedophyllum sessile* and *Alaria*. In the sublittoral the dominants vary depending on local conditions and can be species of *Laminaria*, *Lessoniopsis*, *Egregia menziesii* or *Nereocystis*. In sheltered harbours, e.g. Coos Bay, this belt carries beds of *Desmarestia* and *Laminaria saccharina*. In both littoral and sublittoral the marine phanerogam, *Phyllospadix*, occupies considerable areas.

In the southern part of California marine invertebrates are conspicuous in the upper belts of *Ralfsia* — *Prasiola*, *Endocladia* — *Gigartina* and *Halosaccion*. The lowest belts are dominated by *Laurencia*, *Sargassum* and *Cystoseira osmundacea* and at greater depths by *Egregia laevigata* and *Macrocystis*.

JAPAN

On the northern coasts the uppermost belt is dominated by *Pelvetia wrightii*, and the lower one by *Fucus evanescens* and *Laminaria longissima* while in the sublittoral there are large beds of Laminarians. In the drift-ice area on the Okhotsk coast a special belt dominated by *Heterochordaria abietina* and *Chordaria flagelliformis* occurs.* In the more temperate waters, and therefore over most of Japan, there is an upper belt of *Gloiopeltis* or, more commonly, of *Myelophycus* and *Ishige* followed by an *Hijikia* — *Eisenia bicyclis* belt or a *Gigartina intermedia* — *Sargassum* belt where salinities are lower, i.e. where river waters discharge. In the southern subtropical waters there are belts of small turf-like algae in which *Gelidium pusillum* and *Corallina pilulifera* are dominants. Excessive wave action on the open coast eliminates *Sargassum* and favours the development of *Hijikia* and *Corallina* (Fig. 15.8). The numerous inland waters of Japan contain a characteristic flora that differs from the open coast. At the higher salinities *Monostroma*, *Scytosiphon lomentaria* and *Enteromorpha compressa* predominate, whereas in less saline waters a number of associations have been recognised in which *Ulva pertusa* and *Grateloupia filicina* or *G. turutusu* are the most widespread species (Taniguti, 1961, 1962).

* In S. E. Hokkaido *Heterochordaria*, *Gloropeltis* occur with *Rhodomela larix* and *Ptilota pectinata* the order changing depending upon the factors (Saito and Atobe, 1970).

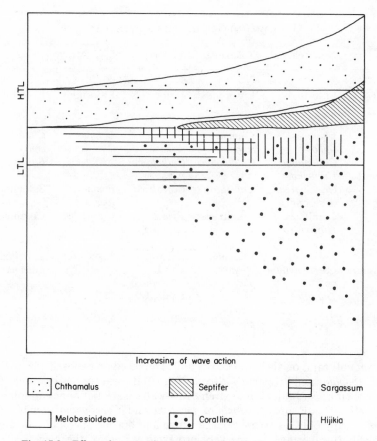

HTL

LTL

Increasing of wave action

: . : Chthamalus	//// Septifer	≡≡≡ Sargassum
Melobesioideae	: • : Corallina	|||| Hijikia

Fig. 15.8 Effect of wave action on zonation at Hotkegama. (after Taniguti)

SOUTHERN SEAS

In the South Pacific one of the principal features is that the major belts are dominated by animals. One reason for this is that, apart from *Hormosira*, the predominant fucoids are all genera that require total or almost total submergence, only being really exposed at low water of spring tides. Various examples from the South Pacific and South Atlantic are illustrated in Table 15.3.

Sarcophycus and *Durvillea* are found in the most exposed situations, while *Ecklonia*, *Hormosira* and *Cystophora* are typical of the more sheltered localities. *Xiphophora* and *Lessonia* occupy regions of moderate exposure.

In New Zealand a limited high-water belt of the tufty *Gelidium caulacantheum*, *Bostrychia arbuscula* and *Enteromorpha* occurs in sheltered areas, whilst on exposed coasts there is a seasonal belt of *Porphyra columbina* var. *laingii*. The *Hormosira — Corallina — Leathesia* belt is quite characteristic of

363

TABLE 15.3

ALGAL COMMUNITIES OF THE SOUTH PACIFIC

	New Zealand	Tasmania	Victoria	S.W. Africa	Chile
Supra-littoral fringe	Littorinids	Littorinids	Littorinids	Littorinids	Littorinids
Eulittoral	Chamae-sipho-Chthalamus	Chamae-sipho-Chthalamus	Chamae-sipho-Chthalamus	Balanus	Chamae-sipho-Balanus
	Hermella, Crassostrea or serpulids	Galeolaria	Algae or Galeolaria	Poma-toceros	Serpulids
	Corallines, Hormosira	Corallines	Pyura	Corallines	Corallines
Sublittoral	Ecklonia, Carpophyllum or Durvillea	Lessonia, Xipho-phora or Sarco-phycus (Durvillea)	Sarco-phycus (Durvillea) Cystophora	Laminaria	Durvillea Lessonia

the lower eulittoral on sheltered coasts and is replaced on exposed coasts by a belt of *Apophloea* and *Gigartina alveata*. The sublittoral species in Table 15.3 all have their upper limit above extreme low-water mark but none can tolerate any length of exposure. *Cystophora torulosa* and *Carpophyllum maschalo-carpum* occupy only a narrow vertical range and are the characteristic algae of the sublittoral fringe. In the very protected waters of the New Zealand southern fiords the uppermost belt is dominated by the lichen *Verrucaria maura* and below it is a belt of the tufty *Bostrychia vaga* that is replaced by the typical *B. arbuscula* with increasing exposure. The *Hormosira banksii* belt is replaced by one dominated by *Cladophoropsis lyallii* and associated green algae (Batham, 1965).

WARM TEMPERATE ZONATIONS
In North Carolina only the summer flora is warm temperate, the winter flora being cold temperate (Williams, 1949). A similar combination of floras exists in the Chatham Islands off New Zealand. Along the Florida-Carolina coastline the supra-littoral fringe is represented by a 'black zone' dominated by Cyanophyceae and marine lichens. The upper part of the eulittoral is dominated by barnacles (*Chthalamus*, *Tetraclita*), while the belt below by oyster and *Mytilus* (Table 15.2). In this respect the warm-water coast here is similar to the warm-water east coast of the Auckland Province in New

Zealand (see table p. 364). The lowest belt on these U.S. shores is algal and dominated by species of *Gracilaria, Ulva* and *Enteromorpha*. The sub-littoral fringe is represented by *Grateloupia* and *Sargassum*. The Stephensons have suggested that a mean winter water temperature of 10° C should form the limit between cold temperate and warm temperate shores. Bennett and Pope (1953) in Australia, however, regard the belts on the shores of Victoria (see table p. 364) as typically cold temperate, but there the mean winter temperature of the water is 11·8° C. In the event it may prove impossible to set an exact temperature at which the transition takes place.

TROPICAL ZONATIONS

Our knowledge of zonation on tropical coasts is much less extensive. In the Caribbean, algae replace animals as the major dominants in the belts (Table 15·4).

TABLE 15.4

ZONATION OF ALGAE IN THE CARIBBEAN

	Florida Keys	Jamaican Keys	Jamaica
Supra-littoral fringe	Bare white belt. Grey belt (*Bostrychia*). Black belt (Cyanophyceae).	Bare belt: Red belt (*Herposiphonia*). Black belt (Cyanophyceae).	Flat limestone or Beachrock (Cyanophyceae or *Gelidium-Bostrychia*).
Eulittoral	Yellow belt (algal turf). Lower platform (*Valonia*).	Algal turf *Cladophoropsis Herposiphonia Champia Centroceras, Cladophoropsis, Caulerpa*).	Algal turf (*Padina*). Valonia
Sublittoral	Reef flat — Coral and mixed algae (mostly calcareous red algae; Caulerpaceae Codiaceae).	Coral and mixed algae (Mostly calcareous red algae Caulerpaceae Codiaceae).	Coral and mixed algae (*Sargassum* and Caulerpaceae Codiaceae).

This may be compared with zonations recorded from a range of shores in Ghana where again algae are predominant, especially on exposed coasts (Table 15·5).

In the coral reefs of the tropical seas the calcareous red algal genera, *Lithothamnion, Lithophyllum, Goniolithon, Porolithon*,* are almost as important as the corals in reef formation. On the shoreward or lagoon side of the reef a mass of siphonaceous algae such as *Caulerpa, Valonia, Dictyosphaeria, Halimeda, Avrainvillea*, can be found while in the lagoon itself *Sargassum*,

* In the Solomon Islands *Porolithon onkodes* is the main reef rim organism, but *Neogoniolithon myriocarpon* is the principal cementing species (Womersley and Bailey, 1969).

Turbinaria, Padina, and *Laurencia* are very abundant, but in some areas *Penicillus* and *Halimeda* may be associated with them.

TABLE 15.5 (from Lawson, 1966)

ZONATION OF ALGAE ON GHANAIAN COASTS

EXPOSED⟵————————————————————————————⟶SHELTERED

Littorina				Supra-littoral fringe
Chthalamus				
Mytilus	*Littorina*			
Centroceras	*Chthalamus*	*Littorina*		Eulittoral
Ulva	*Ulva*	*Pylaiella*	*Littorina*	
Chaetomorpha	*Chaetomorpha*	*Ulva*	*Pylaiella*	
Red algal turf	Red algal turf	*Dictyopteris*	*Ulva*	
Dictyopteris	*Dictyopteris*		Red algal turf	Sublittoral
Lithothamnia			*Sargassum*	

On the Indian sub-continent the following belts appear to be generally distributed on the west coast, though the composition can vary with season (Misra, 1959).

(a) Supra-littoral fringe: a belt of black Cyanophyceae with species of *Lyngbya, Pleurocapsa* and *Calothrix* as the dominants. Shallow pools in this belt contain several species of *Ectocarpus.*

(b) Eulittoral: the upper part of this region is characterized by a bright-green belt of *Ulva* and *Enteromorpha* which may be mixed or may form a pure consociation of a single species. Shallow pools contain *Hypnea valentiae, Laurencia obtusa* and *Catenella repens,* while deep pools have *H. musciformis, Acanthophora delilei, Laurencia hypnoides* and *Caulerpa scalpelliformis.*

The lower part of the region has a belt of *Colpomenia — Iyengaria* and at the lowest levels a belt of *Gelidium — Polysiphonia — Ceramium.* The exact composition of each belt varies from place to place. In deep pools brown algae such as *Cystophyllum, Sargassum, Dictyopteris, Spatoglossum* and *Padina* abound and in shallow pools there is a green vegetation of *Caulerpa, Halimeda, Udotea* and *Cladophoropsis.*

(c) Sub-littoral — this is dominated by species of *Champia, Codium, Caulerpa,* and *Chondria* and it exists in a number of associations.

In the warm waters of the world the sub-littoral is of greater importance. On the shore, exposure even for a relatively short time involves considerable desiccation and few species seem able to tolerate it. The sub-littoral flora depends primarily upon the type of substrate, whether sand, rock or coral. Workers in the Mediterranean believe that with those clear waters a number of different belts can be recognized (Table 15.6). Even in this sea there is a pronounced difference in the summer and winter aspects of the flora (see also North Carolina). Boreal Atlantic species such as *Ulothrix flacca, U. subflaccida, Bangia fusco-purpurea* and *Porphyra* spp. dominate the flora in winter, while in summer it is the tropical and sub-tropical species such as *Siphonocladus pusillus, Acetabularia mediterranea, Pseudobryopsis myura, Liagora viscida*, which form the dominant species.

The zonations to be observed in South Africa are especially interesting (Stephenson, 1939, 1944). On the west coast the waters are relatively cold and a cold water flora and fauna predominate. The east coast is bathed by the warm waters of the Indian ocean and a very different flora and fauna exists. In the intermediate zone of the Cape itself there are transitions from the cold water zonation to the warm water zonation (Fig. 15.9). Similar changes could no doubt be observed north and south of Cape Lookout in North Carolina and in other parts of the world. Very few of these changes have, as yet, been worked out.

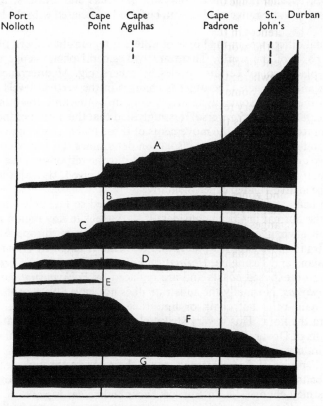

Fig. 15.9 Composition of fauna and flora of South African shores between Port Nolloth and Durban, showing the relative proportions of the different components. (A = Warm water element; B = South eastern element; C = South coast intermediate element; D = South west element; E = Western element; F = Cold water element; G = Cosmopolitan element) (after Stephenson)

In the very cold waters of the world, species are present that do not occur elsewhere. In the Arctic there are laminarians restricted to the region and in the Antarctic there are the peculiar genera *Himantothallus* and *Phyllogigas*. Nearly all the algae are sublittoral because ice action prevents littoral algae from establishing themselves. Subantarctic zonations e.g. Macquarie Island, Kerguelen, New Amsterdam (Délépine), show an uppermost *Porphyra* belt and a lower eulittoral belt of *Rhodymenia*, with *Durvillea antarctica* (with Lithothamnia) the dominant of the sublittoral fringe in which *Ascoseira* and *Phaeurus* also occur. In the sublittoral *Delesseria*, *Iridaea* and *Desmarestia* are dominants with forests of *Macrocystis* or *Lessonia* in deeper waters. Where *Phyllogigas* is present it occurs as a member of the lower sub-littoral, but even that is succeeded at lower depths by *Phycodrys* and *Phyllophora*. In the Arctic (Ellis and Wilce, 1961) the uppermost belt is a black *Verrucaria* lichen with a lower seasonal belt of stunted *Enteromorpha compressa*. This paucity is related to the winter scour by ice. In the lower eulittoral, which escapes some ice action, *Fucus distichus* is a dominant in cracks and crevices. Below low-water mark a wider range of algae can be found associated with the laminarian belt.

Seasonal differences in the floras of different regions have been observed in various parts of the world. There are two general phenomena: one is the actual replacement of certain species by others, e.g. Mediterranean, North Carolina and elsewhere; the other is changes in the vertical levels of certain algae on the shore. Some algae migrate up in winter and down in summer, while others do the very reverse. It is suggested that the nature of the response to temperature, controls the movements of those plants that migrate down in winter, and response to strong insolation determines the behaviour of those that move down in summer. It is likely that it is the response of the sporelings rather than the response of the adults that is involved, though the problem has not been investigated experimentally.

In the belts and zones that have been described so far, only what may be termed the normal has been considered. Variations in any region may occur as a result of changes in protection and exposure, e.g. in Europe *Ascophyllum* disappears in more exposed places and its niche is taken by *Himanthalia*. In New Zealand and Tasmania, *Hormosira*, *Ecklonia*, *Cystophora*, *Carpophyllum*, disappear on exposed coasts and are replaced by *Pachymenia* and *Durvillea* or *Sarcophycus*. Normally the substrate does not have a very profound effect on the main belts but chalk or limestone rocks do produce a significant change in the flora. This is very strikingly illustrated by the zonation on the chalk cliffs of Dover. Fresh water, either in the form of substantial streams or rivulets induces a change in the flora. In the case of rivulets a dense growth of *Ulva* and species of *Enteromorpha* marks their course through nearly all the normal belts on the shore. In the case of streams and rivers the dominants of the belts may change, i.e. in Europe *Fucus ceranoides* replaces *F. spiralis* in such places.

The relationship between animals and plants has not been as fully studied as it could be. There are, for instance, the algae associated with the hard shells of molluscs and barnacles (van Den Hoek, 1958). There are the browsing

TABLE 15.6

DIFFERENT SYSTEMS OF CLASSIFICATION FOR THE MARINE BELTS OF THE MEDITERRANEAN

	Forbes	Lorenz	Ardissone and Strafforello	Pruvot	Flahault	Seural	Feldmann	Molinier
		I Supra-littoral		Subterrestial zone	Subterrestial zone	Superior horizon	Supra-littoral belt	Supra-littoral
		II Region of emergent littoral	Zone I: Subzone I	LITTORAL ZONE: Superior horizon	Marine littoral zone: Superior horizon	Inter co-tidal belt: Middle horizon	Littoral belt	Meso-littoral
Mean low water-mark								
	Region I	III Region of submerged littoral	Zone I: Subzone II	LITTORAL ZONE: Middle horizon	Marine littoral zone: Inferior horizon	Inferior horizon	Upper infra-littoral	Infra-littoral (Photophilous zone
			Zone I: Subzone III					
	Region II	Region IV	Zone II					
	Region III } Region IV }	Region V }		LITTORAL ZONE: Lower horizon	Marine sublittoral zone	Littoral belt	Lower infra-littoral	
	Region V } Region VI } Region VII }	Region VI }	Zone III					Infra-littoral (Sciaphilous zone) Ellitora
				Lower limit vegetation				
	Region VIII				Abyssal zone	Coastal belt	Elittoral	

habits of limpets and sea-urchins. Abundance of the former results in the absence of fucoids in the northern hemisphere, and the abundance of the latter results commonly in a complete clearance of algal vegetation. Pools heavily populated with urchins generally possess only coralline paint.

It should be evident that any really adequate study of marine coastal zonation must involve an appreciation of the part played not only by the algae but also by the animals. The ecologist must indeed be a biologist and the communities he recognizes must be biotic communities in the strictest sense of the term. The other alternative is for teams of botanists and zoologists to work together.

REFERENCES

GENERAL

Chapman, V. J. (1964). *Bot. Rev.*, **12**, 628.
Chapman, V. J. (1957). *Bot. Rev.*, **23**, 320.
Chapman, V. J. (1964). *Coastal Vegetation* (Pergamon Press, Oxford).
Chapman, V. J. (1967). *J. Indian bot. Soc.*, **46**, 337.
Chapman, V. J. and Trevarthen, C. B. (1953). *J. Ecol.*, **41**, 198.
Cotton, A. D. (1912). *Proc. R. Ir. Acad.*, **31**, 1.
Cranwell, L. M. and Moore, L. B. (1938). *Trans. R. Soc. N.Z.*, **67**, 375.
Lewis, J. R. (1955). *J. Ecol.*, **43**, 270.
Lewis, J. R. (1961). *Oikos, Fasc.*, 11/12, 280.
Knight, M. and Parke, M. (1950). *J. mar. biol. Ass. U.K.*, **29**, 439.
Rees, T. K. (1935). *J. Ecol.*, **23**, 70.
Stephenson, T. A. and A. (1949). *J. Ecol.*, **37**, 289.

AFRICA

Lawson, G. W. (1966). *Oceanogr. Mar. Biol. Ann.-Rev.*, **4**, 405.
Stephenson, T. A. (1939). *J. Linn. Soc. (Zool)*, **40**, 487.
Stephenson, T. A. (1944). *Ann. Natal Mus.*, **10**, 261.

ARCTIC

Ellis, D. V. and Wilce, R. J. (1961). *Arctic*, **14**, 224.

AUSTRALIA

Bennett, E. and Pope, E. C. (1953). *Aust. J. mar. Freshwat. Res.*, **4**, 105.
Womersley, H. B. S. (1959). *Bot. Rev.*, **25**, 546.

INDIA

Misra, J. N. (1959). *Proc. Symp. Algol., I.C.A.R.*, **187**

JAPAN

Saito, Y. and Afobe, S. (1970). *Bull. Fac. Fish. Hokk. Univ.*, **21**, 37.
Taniguti, M. (1961, 1962). *Phytosociological Study of Marine Algae in Japan*, **2, 3**, (Tokyo).
Womersley, H. B. S. and Bailey, A. (1969). *Phil. Trans. Roy. Soc. B.*, **255**, 433.

MEDITERRANEAN

Feldmann, J. (1937). *Revue algol.*, **10**, 1.
Molinier, R. (1960). *Vegatatio*, **9**, 121.

NEW ZEALAND

Batham, E. J. (1965). *Trans. R. Soc. N.Z.*, **6**, 215.

NORTH AMERICA

Johnson, D. S. and Skutch, A. S. (1928). *Ecology*, **9**, 188.
Stephenson, T. A. and A. (1952, 1954). *J. Ecol.*, **40**, 1; **42**, 14.
Williams, L. G. (1949). *Bull. Furman Univ.*, **31**, 1.

PACIFIC

Gislén, T. (1943). *Lunds univ. ärssk.*, N.F. **39**, 1.

Stephenson, T. A. and A. (1961). *J. Ecol.*, **49**, 1, 227.

UNITED KINGDOM

Lewis, J. R. (1964). *The Ecology of Rocky Shores*. (English University Press, London.)

Diatoms

Aleem, A. A. (1950). *J. Ecol.*, **38**, 75.

Biota

Hoek, C. Van Den (1958). *Blumea*, **9**, 1.

Caves

Dellow, V. and Cassie, R. M. (1955). *Trans. R. Soc. N.Z.*, **83** (2), 321.

16

ECOLOGY OF SALT MARSHES
AND MANGROVE SWAMPS

Little is known about the ecology of algae in salt marshes in comparison with the rocky coast. This neglect has probably been due to the fact that the algae are often microscopic and more difficult to determine taxonomically. In practice, however, a detailed study of any one area often produces the some-what unexpected result of a quite extensive flora.

An investigation of any salt marsh area shows that the algal communities offer a somewhat different aspect to the algal communities of a rocky coast. Thus on a fairly low marsh the 'Autumn Cyanophyceae' appear in autumn and early winter, they disappear and are replaced in spring by the *Ulothrix* community, which in its turn is replaced during the summer months by *Enteromorpha*. Furthermore, as the continual deposition of silt gradually increases the height of the ground level in relation to the tide, the sub-mergences become fewer and the communities are replaced by others because of the modified conditions. As a result there is a definite dynamic succession of the different communities over a period of years. On a rock coast there is no succession in time and the succession in space is static.

The continual replacement of one community by another as the marsh increases in height provides changes that are more akin to those found in land habitats. A survey of the algal communities of some salt marshes is set out in

Table 16.1, and it will be observed that in the suggested nomenclature the ordinary ecological terminations for developing seres have been employed.

Table 16.1 shows that communities do not occur equally abundantly in the different areas found on a rocky coast. The reason for this probably depends on the very different types of salt marsh. For example, the Irish marshes are composed of a form of marine peat, the marshes on the west coast of England have a large sand component in the soil, the marshes on the south coast bear a tall vegetation of *Spartina* growing in a very soft mud, while the east coast marshes bear a very mixed vegetation on a mud that tends to be clay-like. In New England the marshes are formed of a marine peat while in New Zealand the soil is either muddy or sandy clay. In spite of this, however, the Sandy Chlorophyceae, Muddy Chlorophyceae, Gelatinous Cyanophyceae, *Rivularia-Phaeococcus* socies*, *Catenella-Bostrychia** and the *Fucus limicola* consocies* all have a wide distribution, though they may not necessarily appear at the same relative levels on the different marshes. On the whole, however, they are very often found in the same phanerogamic community.

On British marshes where the soil is rather sandy a *Vaucherietum** can be distinguished dominated by *V. sphaerospora*, but where the phanerogamic vegetation is very dense or heavily grazed by animals the algal vegetation is poor, e.g. south and west coast marshes. The Sandy Chlorophyceae and *Vaucheria thuretii* have a wide distribution, as also the *Catenella-Bostrychia* community, while the pan* flora appears to be richest in East Anglia and Holland.

One of the more interesting features of the algal vegetation of salt marshes is the occurrence of the marsh fucoids. These are peculiar forms which are either free living on the marsh or else embedded in the mud, and they must all at one time have been derived from the normal attached form. Sometimes they bear a fairly close resemblance to the attached form but in other cases they have been very considerably modified, and it is only the existence of intermediate forms that enables us to indicate the normal type from which they came. East Anglia is essentially the home of the marsh fucoids, although Strangford Lough in Ireland is also extremely rich. In Norfolk, for example, considerable areas can be found occupied by *Pelvetia canaliculata* ecad *libera*, while the three marsh forms of *Fucus vesiculosus*, ecads *volubilis*, *caespitosus* and *muscoides* are also abundant, the last two being embedded in the soil†.

There are three other loose lying marsh forms derived from *Fucus vesiculosus* confined to the Baltic, e.g. ecads *nanus*, *subecostatus* and *filiformis*. A small, crawling marsh form derived from *F. ceranoides* has been described from the Irish and Dovey marshes, and another larger, free living one from Strangford Lough in Ireland; like many others of this type it is profusely branched, fertile conceptacles are rare and, when present, are invariably female. *F. spiralis* vars. *nanus* and *lutarius* are other marsh derivatives, while *Pelvetia canaliculata* not only gives rise to ecad *libera* but also to a small embedded form, ecad *radicans*, which has been recorded from the Dovey

---

* The ecological terminology used in this chapter is explained in the glossary.
† These occur on Dutch marshes (Nienhuis, 1970).

TABLE 16.1

No.	Norfolk	Canvey and Dovey	Grevelingen (Nienhuis)
I	General Chlorophyceae	General Chlorophyceae	General Chlorophyceae
Ia	Low sandy Chloro- phyceae	—	—
Ib	Sandy Chlorophyceae	—	—
Ic	Muddy Chlorophyceae	—	—
II	Not investigated	Marginal diatoms	—
III	Marginal Cyanophy- ceae	Marginal Cyanophy- ceae	?
IV	*Ulothrix* community	*U. flacca* community	Vernal *Ulothrix*
V	*Enteromorpha minima* community	*E. minima–Rhizoclo- nium* community	Bliding ia minima soc.
VI	Gelatinous Cyano- phyceae	*Anabaena torulosa* community	—
VII	Not investigated	Filamentous diatoms	—
VIII	Autumn Cyanophyceae	Autumn Cyanophyceae	Cyano- phyceae* comm.
IX	*Phormidium autumnale* community	*Phormidium autumnale* community	—
X	*Rivularia–Phaeococcus*	*Rivularia–Phaeococcus*	*Rivularia–Phaeococcus*
XI	*Catenella–Bostrychia* community	*Catenella–Bostrychia*	*Scorp.* sociat– *Bostrychia*
XII	*Pelvetia–Bostrychia* community	—	—
XIII	*Enteromorpha clathrata* community	—	—
XIV	*Fucus limicola* com- munity	*Pelvetia muscoides* community	*Fucus vesiculosus f.* *volubilus* sociat.
XV	Pan community	—	—
XVI	*Vaucheria* community	?	*Vaucheria* sociat.
XVII	—	—	—

* Nienhuis (1970) considers that it is not possible to place Cyanophyceae into socio-ecological units and so on the Dutch marshes he groups them together.

SALT-MARSH ALGAL COMMUNITIES

Lough Ine	New England	New Zealand	Suggested nomenclature
—	General Chloro-phyceae	—	General Chloro-phyceae
—	—	—	Low sandy Chloro-phyceae consocies
Filamentous algae	*Rhizoclonium*	Sandy Chloro-phyceae	Sandy Chloro-phyceae consocies
—	*Cladophora-Enteromorpha*	—	Muddy Chloro-phyceae consocies
—	—	—	Marginal diatom consociation
Vertical banks assn.	—	—	Marginal Cyano-phyceae consociation
—	*Ulothrix* com-munity	?	Vernal *Ulothrix* socies
—	*Enteromorpha minima* com-munity	*Enteromorpha minima* com-munity	*Enteromorpha nana*‡ socies
Gelatinous Myx-phyceae	Gelatinous Cyano-phyceae	—	Gelatinous Cyano-phyceae socies or society depending on permanence
—	Not investigated	Not investigated	Filamentous diatom consocies
—	Autumn Cyano-phyceae	Autumn Cyano-phyceae	Autumn Cyano-phyceae consocies
—	—	—	*Phormidium autum-nale* socies
Rivularia associ-ation	*Rivularia-Phaeococcus*	*Rivularia* com-munity	*Rivularia-Phaeococcus* socies
Catenella-Bostrychia	—	*Catenella-Bostrychia*	*Catenella-Bostrychia* consocies
—	—	—	*Pelvetia limicola* consocies
—	—	—	*Enteromorpha clath-rata* socies
Limicolous Fuca-ceae association	Limicolous Fuca-ceae	Limicolous† Fucaceae	Limicolous Fuca-ceae consocies
Pan Association	Pan Association	Pan Association	Pan Association
?	*Vaucheria* com-munity	*Vaucheria* com-munity	*Vaucheria* consocies
—	—	*Gracilaria*† com-munity	*Gracilaria* com-munity

† Mangrove communities only.
‡ *E. nana* (Sommerf.) Bliding = *E. minima* of earlier authors.

marshes. There is another form, ecad *coralloides*, from Blakeney and Cumbrae marshes which requires further study of its status. *Ascophyllum nodosum* var. *minor* is a dwarf, embedded variety; ecad *mackaii* of the same species is a free living form found on American salt marshes, in Scotland and on the shores of Strangford Lough in Ireland, while ecad *scorpioides* is a partially embedded form found on the Essex marshes, on the shores of Strangford Lough and in eastern North America.* All these forms probably originated as a result of vegetative budding, although it is also possible that they have developed from fertilized oogonia that became attached to phanerogams on the marsh. There is definite evidence that *Ascophyllum nodosum* ecad *scorpioides* arises by vegetative budding from fragments of the normal plant, while it has been suggested that conditions of darkness or lowered salinity may be favourable for the development of ecad *mackii* (Gibb, 1957).

In the southern hemisphere an interesting free living form of the southern fucoid *Hormosira banksii* (limicolous Fucaceae in Table 16.1) has been recorded from mangrove swamps in New Zealand, though it does not form extensive communities.

As a group the marsh fucoids are characterized by:

(1) Vegetative reproduction as the common means of perpetuation.
(2) Absence of any definite attachment disc.
(3) Dwarf habit.
(4) Curling or spirality of the thallus.
(5) In the species derived from *Pelvetia*, *Fucus vesiculosus* and *F. spiralis* that have been investigated it has been found that the three-sided juvenile condition of the apical cell is retained throughout life, and division in the megasporangia is at best only partial.

It is suggested that these features are due to:

(*a*) exposure, which results in a dwarfing of the thallus;
(*b*) lack of nutrient salts which induces a narrow thallus;
(*c*) the procumbent habit and consequent contact with soil causes spirality because growth takes place more rapidly on the side touching the soil.
(*d*) The cause of sterility may either be a result of the high humidity (according to Baker, 1912, 1915) or, more probably, because of the persistence of the juvenile condition as represented by the apical cell and cryptostomata.

In New Zealand apart from the free living *Hormosira*, there is a *Gracilaria secundata* f. *pseudo-flagellifera* community which is widely distributed over mud flats and in mangrove swamps. It probably also occurs in Australia and would seem to be the only endemic southern hemisphere community so far recorded. The plants of *Gracilaria* are not originally free living but are attached to shells of molluscs buried in the mud.

The mangrove swamps represent an important habitat in the tropics and subtropics where they form the equivalent of the salt marshes. There are

---

* South and Hill (1970) discuss the distribution in Newfoundland.

characteristic algal genera of mangrove swamps which are to be found throughout the whole pantropical belt, though the actual species may vary. Species of *Bostrychia* and *Catenella* form a turfy covering on the mangrove trunks and pneumatophores (aerating roots). At a lower level on the same organs there is a mixture of *Caloglossa* spp, *Murrayella periclados*, *Polysiphonia*, *Bryopsis* and *Centroceras clavulatum*. Numerous epiphytic species of *Enteromorpha*, *Lyngbya*, *Rhizoclonium* and diatoms occur on these turfy algae. In the stream beds of the New Zealand mangroves, species of *Monostroma* can be found at certain seasons. A Cyanophycean community is also widespread on the pneumatophores of mangroves, and on the mud of New Zealand mangrove *Vaucheria* can form green patches. Similar green patches are formed by *Caulerpa verticillata* in tropical mangrove swamps.

In temperate regions beds of eel grass (*Zostera*) are frequently associated with salt marshes. The tropical counterpart of submerged halophytic phanerogamic meadows dominated by *Halophila*, *Posidonia* and *Thalassia* are associated with coral lagoons and sheltered waters. Very few algae are specifically associated with such meadows though a variety of species, especially small brown and red algae are abundant. There are a few examples of specific relationships, e.g. *Smithora naiadum* occurring on *Phyllospadix* on the Pacific Coast from Washington to California.

The salt-marsh pan is the marsh equivalent of the rock pool. The number, shape and size of these on different salt marshes varies very considerably, but they generally contain a certain number of algae, especially in pans which occur on the lower marshes. Although some authors deny the existence of a true persistent pan flora Chapman (1939) and Nienhuis (1970) have shown that in one area at least a definite pan flora does persist, and that many of the species comprising it reproduce during the course of their existence.

On the lower marshes the pan flora is commonly composed of Chlorophyceae, while with increasing marsh height the Chlorophyceae element decreases and the Cyanophycean element increases. A few of the constituent members, e.g. *Monostroma*, are seasonal in appearance, and on some marshes there are pans which contain algae that are normally associated with a rocky shore, e.g. *Colpomenia*, *Polysiphonia*, *Striaria*. These persist from year to year in spite of the stagnant conditions, and when compared with the habitats occupied by the same species on a rocky coast it is found that they are probably growing at an unusually high level. A comparison of the Norfolk marsh flora with that of a comparable rocky coast leads to two generalizations which are probably valid for other marsh areas:

(*a*) Some species that are littoral on a rocky coast are found growing at lower levels, usually sublittoral, on the marsh coast. This must be ascribed to the lack of a solid substrate at the normal higher levels.

(*b*) Other littoral species of the rocky coast are found growing at higher levels on the marsh coast. Some of these occur in pans or in the streams where they are continually covered by water. In the case of a few species actually growing on the marshes it is probable that they are enabled to do so because the phanerogams provide protection from desiccation.

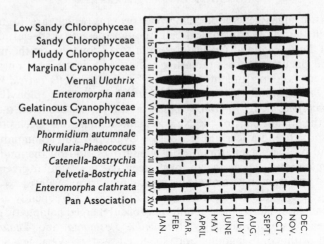

Fig. 16.1 Distribution of the algal communities throughout the year at Scolt, Norfolk. (after Chapman)

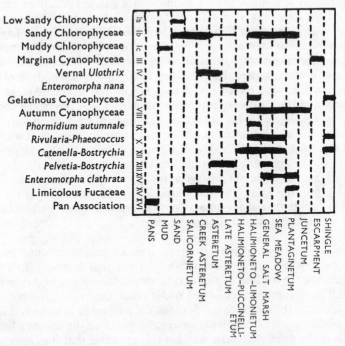

Fig. 16.2 Distribution of the algal communities in space at Scolt, Norfolk. (after Chapman)

378

One feature of salt marsh areas are the lower mud flats. These can bear an embedded carpet of *Vaucheria* but in other places, particularly estuarine areas such as the Severn in England or the Avon in New Zealand, the muddy banks contain a population of *Euglena*. This can be so dense that during exposure in daylight the banks become a deep-green colour from the emergent organisms. Before the tide returns or at night the flagellates reburrow back into the mud (Palmer and Round, 1965).

Very little work has been done on the diatom flora of salt marshes. Round (1960) has demonstrated that two distinct groups exist, one associated with high marshes and the other with low marshes, and within these groups there are communities which are related to the different types of habitat available.

It has been suggested (Carter 1932, 1933) that light and space relations, rather than factors relating to level, influence the distribution of the various species. While this may have some truth there is no doubt that the increasing height of a marsh with its consequent greater exposure does nevertheless effectively determine the upper height to which many plants can go. The species to be found on the higher marshes in Norfolk are either fucoids or gelatinous Cyanophyceae, both of which have the power of retaining moisture. The more delicate Chlorophyceae are more or less confined to the lower levels. On the other hand a dense phanerogamic vegetation, such as one finds on the south coast marshes where the tall *Spartina townsendii* must lower the light intensity considerably, does reduce the quantity of algal vegetation. A similar state of affairs has been observed on the grass covered marshes of New Zealand.

Figs. 16.1 and 16.2 show the distribution in space (e.g. among the different phanerogamic communities) and time of the marsh communities recorded from Norfolk. Some of the smaller communities, e.g. *Rivularia-Phaeococcus* and gelatinous Cyanophyceae, are apt to be overlooked in summer because the constituent species shrivel up so much or else because the colonies become covered by salt efflorescence. An examination of the distribution of the various communities on the Norfolk marshes showed that five communities are each confined to one type of habitat. This relationship may be due to:

(a) Association with a particular phanerogamic community, e.g. *Phormidium autumnale* (IX) with *Halimione portulacoides*.

(b) Dependence on certain edaphic conditions, e.g. Muddy Chlorophyceae (Ic).

(c) Dependence upon the physical character of the environment, for example slope, exposure, wave action, e.g. Marginal Cyanophyceae (III) Vernal *Ulothrix* (IV) and the Pan Association (XVI).

The factors operating on salt marshes are in many respects similar to those found on a rocky coast. The principal factor is the tidal one with all the derived factors that are associated with rise and fall of the tide (i.e. salinity changes, water loss, temperature changes, metabolic changes). Since salt marshes are restricted to protected coasts, there are not the changes that take place with transition to an exposed rocky coast.

REFERENCES

GENERAL

Chapman, V. J. (1960). *Salt Marshes and Salt Deserts of the World*, (Leon. Hill London).

Chapman, V. J. (1964). *Coastal Vegetation*. (Pergamon, London).

AMERICA

Chapman, V. J. (1940). *J. Ecol.*, **28**, 118.

ENGLAND

Baker, S. M. and Blandford, M. (1912, 1915). *J. Linn. Soc. (Bot.)*, **40**, 275; **43**: 325.

Carter, N. (1932, 1933). *J. Ecol.*, **20**, 341; **21**, 128, 385.

Chapman, V. J. (1937). *J. linn. Soc. (Bot.)*, **51**, 205.

Chapman, V. J. (1939). *J. Ecol.*, **27**, 160.

Palmer, J. D. and Round, F. E. (1965). *J. mar. biol. Assn. U.K.*, **45**, 567.

Round, F. E. (1960). *New Phytol.*, **59** (3), 332.

IRELAND

Cotton, A. D. (1912). *Clare Island Survey*, Part 15 *Scient. Proc. R. Dubl. Soc.*, **31**.

Rees, T. K. (1935). *J. Ecol.*, **23**, 69.

NETHERLANDS

Nienhuis, P. H. (1970). *Neth. J. Sea Res.*, **5**, 20.

NEW ZEALAND

Chapman, V. J. and Ronaldson, J. W. (1958). *Bull.* **125**, *D.S.I.R., N.Z.*

Ascophyllum

Gibb, D. C. (1957). *J. Ecol.*, **45**, 49.

17

SOIL ALGAE AND SYMBIOSIS

SOIL ALGAE

The study of soil algae, as such, began seriously at the commencement of the nineteenth century with the works of Vaucher, Dillwyn, Agardh and Lyngbye. In 1895 Graebner, in a study of the heaths of Northern Germany, gave the first account of soil algae as ecological constituents, and subsequently many ecologists have shown that soil algae are pioneers on bare primary or secondary soil where they prepare the ground for the higher plants that follow. An agricultural, manured soil also has a rich flora, while the same species are to be found in unmanured soils, though not in such numbers. This flora consists of about 20 species of diatoms, 24 of Cyanophyceae and about 20 species of Chlorophyceae. As might be expected, persistently damp soils have a more varied and extensive collection of algae than have generally dry soils. In the case of the diatom component of the soil flora it has been found that it is more abundant when the soil is rich in phosphates and nitrates, so that the soil salts may also be of importance.

It appears that the method of studying soil algae is of great importance. In the early years dilution techniques were employed (Roach, 1927) as was direct examination. There seems little doubt that many species can be missed by this technique (Forest, 1962, Round, 1965). Various culture methods have since been evolved, using soil samples on sand and mineral solution, on agar and mineral solution or a moist-plate culture of the soil itself. Both Round and Forest (*loc. cit.*) conclude that the moist-plate culture is the most satisfactory.

Whilst the distribution of many of the common species of soil algae is almost world wide, nevertheless some species appear typical of certain soil types. Thus peat soils are characterized by an abundance of desmid species, Cyanophyceae such as *Chroococcus* and *Gloeocapsa*, and diatoms. With a rich alkaline loam these algae are replaced by others, especially coccoid algae, Ulotrichales (*Hormidium, Ulothrix*), filamentous Cyanophyceae (*Oscillatoria, Phormidium*) and diatom genera different from those on peat soils. On the other hand, soils of one kind throughout the world may possess a similar flora, e.g. arid soils from Saudi Arabia, Mexico, Arizona, Utah and Israel all have much the same species: *Chlorosarcinopsis* was found in nearly every sample and *Chlorococcum diplobionticoideum* was almost as frequent (Chantanachat and Bold, 1962).

In temperate and monsoonal regions the soil flora may vary with the season or monsoon. Soils, however, are not permanent and are continually in a process of maturing. New soil is rapidly colonized by algae, but as colonization and succession of higher plants takes place, the amount of organic matter increases, changes occur in the nutrient status and the composition of the algal flora changes accordingly. There seems little doubt that the soil flora has an important biological significance, but at present much more experimental data is required. It may well be that the biological character of a soil will prove to be of more importance than its chemical or physical character.

SUBTERRANEAN ALGAE

There are great fluctuations in the numbers of the different species that compose the flora, but there are no species in the lower layers of the soil which do not also occur in the surface layers. The most actively growing algae are probably confined to the top inch or so of soil, where some of the species are motile, e.g. *Euglena, Chlamydomonas, Oscillatoria, Phormidium* and pennate diatoms such as *Navicula, Pinnularia, Stauroneis*. On the surface of damp soils in rain forest climates macroscopic, mucilaginous colonies of *Nostoc* can be found. The most advanced soil alga, morphologically, is *Fritschiella* (see p. 83).

In the soil there are two main groups of algae: first there are ephemerals which only appear in quantity when conditions are favourable, e.g. *Chlamydomonas, Euglena*. Secondly there are the perennials, mainly diatoms and Cyanophyceae.

Within the soil community Protozoa and rotifers may devour algae and their grazing may be selective. There are also soil fungi, especially members of the Chytridiaceae, that attack algae, but a study of both these interrelationships has scarcely begun.

Lund (1945) studied 66 different soils in Great Britain and found that, except on acid soils, the Chlorophycean component was the largest, but it was followed closely by the diatoms with the Cyanophycean element third. In the case of cultivated soils, however, the Cyanophycean element became much more important. Very little work has so far been carried out on the soils of tropical regions. A study of the soils of savanna, steppe and maxanga in the Congo yielded a flora almost entirely composed of Cyanophyceae, though in true savanna algae were rare, probably because of competition for space and light.

The growth of soil algae has been a source of interest and experiment for a number of years. In 1926 Roach found that *Scenedesmus costulatus* var. *chlorelloides* would grow in the dark when provided with mineral salts and glucose. A more recent contribution (Chantanachat and Bold, 1962) has demonstrated that at least a third of the soil algae tested were capable of fair, good or excellent heterotrophic growth in darkness in the presence of one or more carbon sources. The successful algae included species of *Neochloris*, *Spongiochloris*, *Chlorosarcinopsis* and *Friedmannia israeliensis*.

The maximum depth to which algae have been recorded in soils is 2 m. In Greenland soil algae have been found down to a depth of 400 mm and, because burrowing animals are absent, their presence there can only satisfactorily be explained by the action of water trickling down cracks. In the lower layers of the soil the algae exist in a dormant condition, either as spores or as filament fragments.

The occurrence of algae in the lower soil layers depends on (1) cultivation, (2) animals, (3) water seepage and (4) self-motility. Mechanical resistance and lack of light are said to prevent Cyanophyceae from moving down through their own locomotion, but even so some algae such as *Euglena*, may be able to move down by their own mobility. The effect of water seepage will depend on the heaviness of the rainfall, state of the soil, i.e. whether cracked or water-logged, and the nature of the algae, i.e. whether or not they possess a mucous sheath that will adhere to soil. With filamentous algae movement is helped by fragmentation or by the formation of zoospores. Many green algae are known to form zoospores when put in water after a period of drought, and hence one may presume that rain after a dry period will induce zoospore formation. Petersen (1935) has demonstrated that rain can carry algae down efficiently to depth of 200 mm, but the process is facilitated by the presence of earthworms, although these animals probably only operate indirectly in that they loosen the earth. Soil cultivation is obviously very important here.

Many soil algae, especially Cyanophyceae, can resist long spells of drought. When remoistened, bacteria usually first develop, then coccoid green algae and finally Cyanophyceae that quickly become dominant. *Nostoc muscorum* and *Nodularia harveyana* appeared from soil that had been dried for 79 years, and *Nostoc passerinianum* and *Anabaena oscillarioides* var. *terrestris* were obtained from a soil dried for 59 years.

In a study of the moisture relations of some soil algae Fritsch and Haines (1922) showed that:

(1) There is a complete absence or paucity or large vacuoles.
(2) In an open dry atmosphere nearly all the sap is retained.
(3) When the filaments dry up, contraction of the cell is such that the cell wall either remains completely investing the protoplast or else in

partial contact with it, thus ensuring that all the moisture which is imbibed will reach the protoplast.

(4) During a drought there is, as time goes on, a decreasing tendency for the cells to plasmolyse and there are also changes in the permeability of the cell wall, while the excess of moisture normally brings about changes in the reverse direction. The majority of cells which plasmolyse lack the characteristic granules, mainly of fat, that are to be found in most terrestrial algae.

(5) Those cells which survive after drought do not possess any vacuoles, and have instead a rigid, highly viscous protoplasm that is in the gel condition.

(6) If desiccation is rapid most of the cells will die but some will plasmolyse and retain their vitality in that state for weeks or months. Thus no species is likely to disappear from a flora during the rapid onset of drought.

(7) If desiccation continues the number of living resting cells will remain constant for several years.

Apart from moisture, temperature in extreme climates may be important.

Diatoms can survive very low temperatures, $-80°$ C for 8 days or $-192°$ C for 13 hours, while dry spores of *Nostoc* sp. and *Oscillatoria brevis* can survive $-80°$ C, though if they are moist temperatures below $-16°$ C are fatal. So far as the algae of tropical soils or thermal regions are concerned the dry spores of *Nostoc* sp. and *O. brevis* can tolerate 2 mins. at 100° C., the wet spores 20 mins. at 60–70° C, and the vegetative filaments 10 mins. at 40° C, this latter being a temperature that is frequently reached on open soil surfaces in the tropics. In general it seems that the distribution of soil algae is related to the physico-chemical characters of the soil, e.g., pH, texture, water capacity etc. (Gruia, 1964).

Soil algae also add certain compounds to the soil, either through death or by secretion from the cells. Cyanophyceae and Chlorophyceae have been shown to release polysaccharides in culture and this probably occurs in nature. Those Cyanophyceae that fix nitrogen play an important part (see p. 470). Certain algae also produce growth substances and antibiotics, e.g. *Chlorella*, which may affect the soil flora. It is these compounds that must play some part in the success of the soil culture (Erdschreiber) media for many algae.

REFERENCES

Chantanachat, S. and Bold, H. C. (1962). *Univ. Texas Publ.*, 6218.
Forest, H. S. (1962). *Trans. Am. Microsc. Soc.*, **81**, 189.
Fritsch, F. E. and Haines, F. M. (1922). *Ann. Bot.*, **36**.
Fritsch, F. E. and Haines, F. M. (1923). *Ann. Bot.*, **37**, 683.
Gruia, L. (1964). *Rev. Rourn. Biol. Sev. Bot.*, **9**, 295.
Lund, J. W. G. (1945). *New Phytol.*, **44**, 196; **45**, 56.
Petersen, J. B. (1935). *Dansk. bot. Ark.*, **8**, 1.

Roach, B. M. (1926). *Ann. Bot.*, **40**, 149.
Roach, B. M. (1927). *Ab. Handb. Biol. Arbeits-methoden*, Abt. XI, Pt. 3, 747.
Round, F. E. (1965). *The Biology of the Algae* (London).

SYMBIOSIS

The most striking and well known examples of symbiosis involving algae are provided in those cases where the plants are associated with animals, especially Coelenterates, or with fungi, as in the Lichens. Animal symbiotic associations are known from the Protozoa through to the Prochordates (Tunicata) (cf. McLaughlin and Zahl, 1966). Apart from these examples, however, there are other cases which are not so well known, mainly because they are not so common. *Gloeochaete*, for example, is a colourless genus of the Tetrasporaceae, the members of which possess blue-green bodies that look like chromatophores, though they are really a symbiotic member of the Cyanophyta. *Glaucocystis* is a colourless genus of the Chlorococcales in which a symbiotic member of the Cyanophyta also forms the blue-green 'chromatophores' that appear as a number of curved bands grouped in a radiating manner around the nucleus. In this case the illusion is further enhanced because the endophytes break up into short rods at cell division. It has so far proved impossible to grow the blue-green alga separately and it may thus have lost its power of independent growth.

In the flagellate *Cyanophora paradoxa* the algal symbiont also belongs to the blue-green algae, but it lacks the usual double-layered wall and has a thin single membrane instead. The cyanelle has been named *Cyanocyta korschikoffiana*. *Geosiphon*, which is variously regarded as a siphonaceous alga or as a Phycomycete, possesses small colonies of *Nostoc* enclosed in the colourless, pear-shaped vesicles that arise from an underground weft of rhizoidal threads. Reproduction by the formation of new vesicles is said to occur only in the presence of the *Nostoc*. The presence of chitinous material in the vesicular wall suggests a fungal nature for *Geosiphon*, the vesicles perhaps being galls that are formed on the threads as a result of the presence of the alga.

The principal genera taking part in lichen synthesis are *Nostoc*, *Scytonema*, *Gloeocapsa*, *Cephaleuros* and *Trentepohlia*. Under normal conditions the partnership is truly symbiotic, but under abnormal conditions the fungus may become a parasite and devour the algal component. The mycobiont cannot normally live apart from the alga, but most of the phycobionts appear capable of an independent existence.

The green bodies which are found associated with the cells of Coelentrates and Radiolarians are usually placed in 'form' genera, *Zoochlorella* and *Zooxanthella*. Most of the species are now known to be dinoflagellates, e.g. *Symbiodinium microadriaticum* in the coral *Cladocora* (cf. Fig. 17.1) and the giant Pacific clams of the genus *Tridachna* (Taylor, 1969). In the freshwater sponges *Chlorella parasitica* imparts the characteristic green colour, though *Gongrosira* and *Aphanocapsa* also occur in freshwater sponges. The non-motile cells of the coral dinoflagellates are usually found in the peripheral layers of the polyps, the host larval stages usually lacking the alga. It is the existence of the independent motile phase that has enabled the algae to be taxonomically determined. The function and relations of these symbiotic algae have been discussed over the years, and on the whole there appears to be

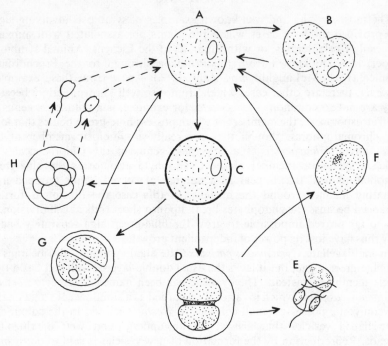

Fig. 17.1 Life cycle of *Symbiodinium adriaticum*: A, vegetative cell; B, binary fission of vegetative cell; C, vegetative cyst; D, zoosporangium with zoospore; E, zoospore; F, aplanospore; G, cyst with two autospores; H, cyst with developing gametics; I, isogametics. (after Freudenthal)

evidence for the alga obtaining food, especially nitrogenous compounds, from the animal, and the animal oxygen and carbohydrates from the alga. In the corals the algae are regarded as speeding up the rate of removal of waste products and this involves the deposition of waste calcium carbonate.

The problem of the relationship between algae and animals is by no means completely worked out, and it is not impossible that in some cases the animal is really a parasite of the alga. Some colourless Cyanophyta (*Anabaeniolum*, *Oscillospira*) typically occur in the digestive tract of mammals (including man), where they live as saprophytes. When cultured outside in the light they become autotrophic. In the worm-like *Convoluta roscoffensis* with its algal associate *Platymonas*, the animal cannot live unless infected with the alga though under certain conditions it also digests the algal cells. Exactly how the algae re-infect the host animal or are transmitted is not well known, though there are clearly various potential methods.

Examples of a looser form of symbiosis (almost a commensalism) are provided by *Anabaena cycadearum* living in the cortex of *Cycas* roots, and *Anabaena azollae* found in the leaves of the water fern, *Azolla filiculoides*. Species of *Nostoc* found in the thalli of the liverworts *Blasia* and *Anthoceros* are probably no more than space parasites obtaining shelter. This is not so, however, with *Nostoc pruniforme* that lives in the petioles and stems of large-leaved tropical *Gunnera* species, because the host probably gains from nitrogen fixation by the alga (cf. p. 470).

Epiphytism is extremely common among the algae, whilst there are also a number of epizoic forms. One may also find endophytic species, e.g. *Schmitziella endophloea* in *Cladophora pellucida*, species of *Endoderma* in various algae, and endozoic species such as *Rhodochorton endozoicum* in hydroid sheaths. *Blastodinium* is an endophyte of crustacea and worms.

The origin of the symbiotic habit among the algae is probably to be explained as cases of epiphytism or parasitism, and in the former the relationship between host and epiphytes became more intimate while in the latter case the host evolved mechanisms that enabled it to tolerate the parasite and take advantage of it.

Many examples of algal parasites are known. Among the Chlorophyta there is *Phyllosiphon arisari*, *Chlorochytrium lemnae* inhabiting the fronds of *Lemna trisulca* and species of *Cephaleuros*. *Notheia anomala* (see p. 208) is at least a hemi-parasite if not a holo-parasite in the Phaeophyta, while in the Rhodophyta some parasitic forms, e.g. *Harveyella mirabilis, Janczewskia* spp., *Faucheocolax, Colacolepis, Hypneocolax,* have all become highly modified as a result of their parasitic habit. Although these parasites have been known for a long time no intensive work has been carried out on the nature of the parasitism.

REFERENCES

Keeble, F. and Gamble, F. W. (1907). *Quart. Jl. Micro. Sci.*, **51**, 167.
McLaughlin, J. J. A. and Zahl, P. A. (1966). *Symbiosis*. Vol. I. (Academic Press, New York and London).
Smith, G. M. (1951). *Manual of Phycology*. (Chronica Botanica, Waltham).
Taylor, D. L. (1969). *J. Phycol.*, **5**, 336.
Yonge, C. M. and Nicholls, A. G. (1932). *Reports of the Great Barrier Reef Expedition*, Vol. I. (British Museum, London).

18

FRESHWATER ECOLOGY

One of the major problems in this branch of algal ecology appears to be the establishment of a successful classification on which field studies can be based. Up to 1931 the outline given by West in 1916 was in current use, but since then a scheme proposed by Fritsch (1931) has more or less taken its place. It would seem, however, that neither scheme alone is wholly satisfactory, but that a combination of the two provides a very suitable basis for workers in this field.

(a) **Subaerial communities:**

These develop at their best in the tropics although they can also be found in temperate regions. The members form communities that are either epiphytic or epilithic.

(b) **Dripping rocks and waterfalls:**

(c) **Aquatic communities:**

These vary from season to season and frequently have a marked periodicity which is controlled by diverse factors. The following five subdivisions of the aquatic communities can be recognised:

   (1) Communities of rivers and rapids.
   (2) Spring communities.
   (3) Communities of bogs and swamps.

(4) Communities of ponds and pools.
(5) Lake communities.

SUBAERIAL COMMUNITIES

Aerial epiphytic algae are most abundant in places of high humidity and are particularly prolific in rain forest, i.e. tropical jungle, moss forest of the tropics and, to a lesser extent, in temperate rain forest. Species of *Cephaleuros* and *Phycopeltis* are common on leaves (Epiphyllophytes), *C. virescens* occurring as a pest on tea bushes, while species of *Trentepohlia* are also abundant, particularly on the trunks and branches (Epiphloeophytes) of various trees as well as on rock.

Diatoms can form a flora on mosses and liverworts and if the habitat is very wet Cyanophyceae and desmids may appear. Some algae, e.g. *Stichococcus bacillaris, Chlorella ellipsoidea*, have been recorded from the caps of Agarics (Fott, 1959); *Pleurococcus* is widespread on the trunks of trees but, together with *Prasiola* and *Trentepohlia*, is also a major component of the aerial epilithic communities (Lithophytes).

Certain algae seem to demand specific habitats, some of them unusual and among such algae are species of *Basicladia* that grow on turtles, and species of *Characium* and *Characiopsis* (all Epizoophytes) that grow on the Crustacean, *Branchipus*.

DRIPPING ROCKS AND WATERFALLS

Waterfalls have been very little studied. *Lemanea* (p. 272) frequently occurs in such habitats and there is often a rich flora of Cyanophyceae and Chlorophyceae, particularly species of *Ulothrix, Lola*, and *Rhizoclonium*. Rock surfaces receiving seepage water from soils form a very suitable habitat for *Chroococcus, Gloeocapsa, Stigonema* and diatoms, including the colonial forms.

AQUATIC COMMUNITIES

Communities of Rivers and Rapids

In rapidly flowing rivers the algae are adapted to the habitat by either being encrusting, or, if they are only basally attached, the thallus is flexible and streaming. In such rapid streams attachment of motile reproductive bodies must be quick and efficient, though this aspect has not been investigated.

In rapidly flowing streams the flora is mainly composed of freshwater Rhodophyceae, e.g. *Lemanea, Compsopogon, Sacheria, Thorea, Hildenbrandia* (in acid waters, *Batrochospermum* and *Bostrychia*), *Cladophora* spp. (especially *C. glomerata*), *Vaucheria* spp., Cyanophyceae, Chrysophyceae (especially *Hydrurus*) and diatoms. Very few of the species form pure stands, though *Hildenbrandia, Heribaudiella* (*Lithoderma*), *Chaetophora* and *Rivularia* may form small compact communities. In Europe *Hildenbrandia* frequently occurs with the lichen *Verrucaria rheithropila* (Rheithropilo-Hildenbrandietum). The mountain streams feeding the Ruhr River have been divided into two regions by Budde (1928), an upper *Hildenbrandia* region and a lower *Lemanea* one. The epilithic diatom communities vary considerably. *Achnanthes*, *Gomphonema* and *Synedra* are associated with nutrient deficient acidic waters, whilst in richer, more alkaline waters *Diatoma*, *Melosira varians* and *Meridion circulare* predominate. In less swift waters of larger rivers *Tabellaria* and *Fragilaria* may be dominant.

In European streams that are more sluggish *Enteromorpha* and *Hydrodictyon* tend to replace *Cladophora*, and in very sluggish streams and also in lakes one may find temporary floating mats of *Hydrodictyon*, *Enteromorpha*, *Oedogonium*, *Spirogyra* and *Rhizoclonium*. With slower water movement sediments are deposited e.g. sand, silt, organic matter, and epipelic algal communities develop on these sediments. Such communities have not been much studied but diatoms are predominant, though one also finds Cyanophyceae, Chlorococcales, Euglenoids and Desmids. In examining such communities care has to be taken not to confuse the species with remains of the epilithic and epiphytic floras.

According to Blum (1956) the dominant benthic alga of rivers in the North Temperate zone is *Cladophora glomerata*. In the presence of certain metallic ions this species is adversely affected and may then be replaced by *Stigeoclonium tenue*, *Spirogyra fluviatilis*, *Phormidium autumnale*, etc. In Scandinavian streams it is often replaced by *Zygnema* and *Vaucheria*.

One of the most favourable locations for algal growth is immediately downstream of large rocks though the exact significance of this fact has not been worked out. As might be expected, seasonal variations occur in the vegetation. Thus in the Ruhr streams the spring period is characterized by dominance of diatoms with *Ulothrix* and *Hormidium* as subdominants. In the summer Chlorophyceae and Desmids predominate and in the winter *Ulothrix* and *Hormidium* reappear. The effect of bank vegetation (overhanging trees, reeds, etc.) may be profound, especially in the degree of light reduction.

The most important controlling factor is apparently temperature but the degree of aeration and chemistry of the water is also significant. The absence or near absence of calcium, for example, may induce a Cyanophycean rich flora. Light intensity variations and oxygen concentration changes may bring about local modifications of the water. The distribution of algae and higher aquatic plants can provide an indication of stream pollution. Highly polluted, oxygen deficient waters usually only contain bacteria and possibly Euglenoids. With slightly less pollution *Oscillatoria* and *Phormidium* will be found. In the Saline River of Michigan the entry of certain metallic ions caused the normal flora to be replaced by *Stigeoclonium tenue* and *Tetraspora* (Blum, 1957). Unpolluted oligosaprobic waters can carry a rich algal flora.

It is possible to classify the communities of rivers and streams in the following way:

(a) Communities of vertical substrates:
   (1) Those attached to boulders, rock, mud banks or roots.
   (2) Epiphytic communities on phanerogams.
   (3) Spray communities.

(b) Communities of horizontal substrates:

e.g. those attached to stones, sand or embedded in mud.

(c) Free-living communities.

The epiphytic communities are to be found attached to phanerogams, Bryophytes and larger algae, especially those such as *Cladophora* which have a rough, non-mucilaginous wall. The principal components of the epiphytic flora appear to be Cyanophyceae, diatoms and filamentous species of the Chaetophorales, such as *Aphanochaete* and *Endoderma*.

There is still considerable scope for further work on river ecology, both descriptive and experimental and the use of artificial streams with controlled conditions should be invaluable.

Spring Communities

There are three main groups of springs: cold springs, hot springs and saline springs. The first named are usually rich in calcium and bicarbonate ions and the flora is generally that of alkaliphilic diatoms. *Batrachospermum* is the only common macroscopic alga. Saline springs usually have a much richer flora because the brackish water supports both marine and freshwater species. Again, diatoms are predominant but marine species of *Enteromorpha*, *Ulothrix*, and Cyanophyceae are also recorded. Hot springs are associated with volcanic areas and the waters have a high concentration of dissolved salts.

Cyanophyceae are the major, and often the only constituents, the various species being capable of secreting carbonate of lime or silica to form brightly coloured rock masses such as travertine and sinter, the rate of deposition sometimes being as much as 1·25–1·5 mm in three days. The highest temperature recorded for water containing living plants (*Synechococcus lividus*) is 74° C (Castenholz, 1969a). The number of forms capable of living in hot springs is considerable, no fewer than 53 genera and 163 species being recorded from the thermal waters of Yellowstone National Park. Recently it has been shown that one species, *Mastigocladus laminosus* is capable of fixing atmospheric nitrogen.

Work on thermal algae in Japan (Yoneda, 1952) has shown that 470 species occur in hot waters distributed as in Table 18.1.

TABLE 18.1

	Number of Genera	Number of Species
Cyanophyceae	51	290
Rhodophyceae	2	2
Chryso- and Xanthophyceae	6	12
Bacillariophyceae	30	102
Phaeophyceae	1	1
Chlorophyceae	33	64
Total	123	471

The distribution of the Cyanophyceae in Japan in relation to water temperature is shown in Table 18.2.

TABLE 18.2

Temperature °C	Number of Genera	Number of Species
35–40	30	157
40–45	31	143
45–50	26	103
50–55	21	81
55–60	21	69
60–65	15	37
65–70	8	20
70–75	5	7
75–80*	4	7
80–85*	5	5
85–90*	4	4

It is evident that areas differ because Castenholz (1969b) reports only 6–8 species from Icelandic thermal waters and then in temperatures below those of other countries.

The flora of such springs varies according to the nature of the water, and in very acid springs the most characteristic species is the Rhodophycean *Cyanidium caldarium* and the diatom *Pinnularia braunii* var. *amphicephala*. The distribution of algae in different types of thermal springs in Japan is shown below (Table 18.3).

TABLE 18.3

DISTRIBUTION OF ALGAE IN THERMAL SPRINGS OF JAPAN

Spring type	Number of Springs	Number of Genera	Number of Species
Simple thermals	326	36	170
Simple carbonates	5	17	51
Earthy carbonates	13	13	38
Alkaline	68	14	46
Salt	265	37	133
Bitter	181	18	56
Iron carbonate	27	15	31
Vitriol	14	6	10
Alum	14	2	3
Acid	60	17	22
Sulphurous	90	17	46
Radioactive	1	5	7
Undetermined		30	140

* These are far too high (cf. Castenholz, 1969, with 74° C as the maximum).

Communities of Bogs and Swamps

These associations are commonly found in waters that are acid, particularly those in *Sphagnum* high moor bogs. They are very mixed with little or no periodicity, probably because of the relatively uniform conditions. Zygnemataceae, desmids and diatoms are most frequent, the desmid element changing considerably with altitude, chemistry of the water and type of substrate.

Communities of Ponds and Pools

The flora exists under very varied conditions with a regular or irregular periodicity. In the temperate regions Chlorococcales, Zygnemataceae (dominant in spring) and diatoms (dominant in winter) form the chief elements with a phytoplankton of flagellates. There is usually not enough aeration for the larger filamentous forms such as Cladophoraceae. The substrate and fauna are also important factors in determining the type of vegetation. The flora of tropical ponds for example contrasts sharply with that of temperate regions as there is:

(1) an excess of Cyanophyceae
(2) a scarcity of *Vaucheria*, *Oedogonium*, Xanthophyceae and Ulotrichales
(3) an abundance of filamentous desmids together with *Spirogyra*.

In America Transeau (1913) concluded that freshwater pond algae can be divided into seven classes based on abundance, duration and reproductive season, and these classes and their periodicity are represented in Fig. 18.1.

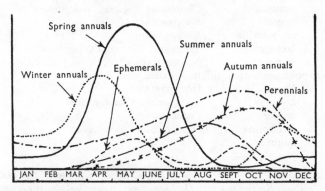

Fig. 18.1 Chart showing the estimated relative importance of the different types of algal periodicity throughout the year in the waters of E. Illinois. The irregulars are not depicted. (after Transeau from West)

Species of some groups may be present every year and there are other species that occur sporadically (Bethge, 1952–55). It is only after many years that extended cyclic periodicity becomes evident. Bethge (*loc. cit.*) recorded a ten-year cycle of growth for the production of cells of *Asterionella*.

In a study of a Harpenden pond Fritsch and Rich (1913) distinguished four seasonal phases:

(a) Winter phase with *Microspora, Eunotia* and epiphytic diatoms, and with *Ranunculus aquatilis* and *Callitriche* as the dominant phanerogams.

(b) Spring phase dominated by Conjugales, *Oedogonium* and *Conferva*, with *Ranunculus aquatilis* as the most important phanerogam.

(c) Summer phase with *Euglena*, desmids and *Anabaena* associated with a phanerogamic vegetation of *Lemna, Glyceria* and *Bidens*.

(d) Sparse autumn phase with *Lyngbya* and *Trachelomonas* but without any dominant phanerogam.

The algal periodicity is thus reflected by a similar periodicity in the phanerogamic vegetation. Compared with a Bristol pool the general trend of periodicity is very similar; there is a winter phase characterized by a hardy filamentous form (*Cladophora* or *Microspora*) and diatoms, and a spring phase with Zygnemataceae and an autumn phase with Oscillatoriaceae. The summer phase in the two pools was very different, and could be ascribed to the greater drying up of the Harpenden pool during that period.

A similar study by Brown (1908) of Indiana pools revealed similar seasonal phases (Table 18.4).

TABLE 18.4

SEASONAL PHASES OF ALGAL FLORA IN INDIANA POOLS
(Brown, 1908)

The phases in one pond (Fig. 18.2) were as follows:

Phase	Dominants
autumn	*Closterium, Euglena, Oedogonium*
winter	*Spirogyra* spp.
early spring	*Spirogyra* sp.
late spring	*Spirogyra, Euglena, Oedogonium*
summer	No one species

In another pond somewhat different phases were recorded:

Phase	Dominants
autumn	*Oedogonium, Chaetophora*
winter	*Vaucheria*
late spring	*Oedogonium, Pleurococcus*
summer	*Chaetophora*

Although the spring phases are essentially similar to those of the Harpenden pool with either *Spirogyra* or *Oedogonium*, nevertheless there are great differences. The two ponds described above also possessed floras that were essentially different and they must therefore be regarded as containing two separate associations.

The flora of pools is dependent not only upon general climatic conditions, such as rainfall and insolation, but also upon irregular microclimatic factors, e.g. aeration in the body of water itself. In many species there is a profound relationship between the meterological data and the frequency of the flora, e.g. *Microspora* and the Pleurococcaceae depend on temperature, *Oedogonium* and *Hormidium* depend on sunshine. The factors influencing the growth of

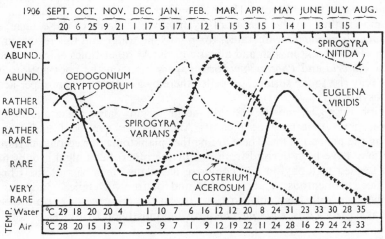

Fig. 18.2 Abundance and frequencies of the most important algae in a pond near Indiana University from 1906 to 1907. (after Brown)

aquatic algae are (1) seasonal, (2) irregular, (3) correlated. The first group which are very obvious and need not be detailed, are important in large bodies of water, but they tend to be masked by the other two groups in small bodies of water:

(2) Irregular factors:
   (a) Abnormal rainfall or drought.
   (b) Abnormal sunshine.
   (c) Abnormally high or low temperatures.

(3) Correlated factors:
   (a) Species depending on the enrichment of the water by decay of other members of the flora.
   (b) Forms influenced in their development by competition with others.
   (c) Forms influenced in their development by the presence of a suitable host, e.g. epiphytic forms.

A correlation can frequently be established between the amount of sunshine and the phenomenon of reproduction. This is in accordance with experimental work which has shown that reproduction is often initiated by the presence of bright light. An unusual concentration of the salts in the water during a period of drought may, however, counteract the influence of sunshine. There may also be a correlation with the nitrate factor for some species. Thus *Volvox* can receive a severe check when the nitrate is high and only reproduces at times of low nitrate value. *Euglena*, if present, attains to a maximum soon after the nitrate minimum unless other factors intervene.

Some pools may be permanently rich in an ion, e.g. calcium and carbonate ions in limestone areas, and they will possess a flora determined at least, in part, by this enrichment.

When conditions for growth are particularly favourable, the water becomes thick with the cells of algae, generally of a single species, and provides what is known as a water bloom. These blooms generally develop in late summer and mostly they involve species of Cyanophyceae. Some of these algae, e.g. *Microcystis aeruginosa*, liberate a toxin into the water and this results in the death of zooplankton, fish and aquatic birds. At other times epipelic species of *Oscillatoria* and *Phormidium* form papery mats on the pool bottom and these may rise to the surface, propelled there by gas derived from super-active photosynthesis.

Epiphytic algae, mostly diatoms, occur on any large algae and aquatic phanerogams that may be present. Algal masses can be found tangled among the phanerogams. These are neither planktonic nor epiphytic, though they may have been derived from one or the other. Such algal groups have been termed metaphyton and comprise flagellates, members of the Chroococcales, filamentous algae, desmids and diatoms. Detailed study of the zooplankton will, from time to time, reveal the presence of epizoic species such as *Characiopsis*, *Chlorangium*, growing mostly on Crustacea.

Lake Communities

These are large bodies of water and they provide many more well defined habitats than do ponds and pools. Because of their size, the physical and chemical environment is more constant, but because of their greater depth there is oxygen and temperature stratification during the summer. The sharp drop in temperature that occurs between the upper (Epilimnion) and lower (Hypolimnion) waters is known as the thermocline. Variations in type of rock, nature of inflowing streams, aquatic flora and climate all exert an effect upon the algae. The principal features of lake ecology are illustrated in Fig.

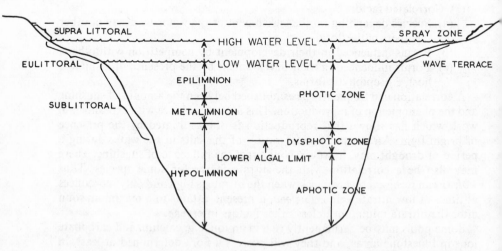

Fig. 18.3 Lake features. (modified from Round)

18.3 (After Round, 1965), which differs from the original where the eulittoral is regarded as the zone from low-water level to the top of the spray.

West was the first investigator to show that a rich desmid flora is associated with an ancient hard-rock substrate (Oligotrophic waters), but diatoms are abundant in younger, softer-rock areas or where there has been much silting with consequent solution of mineral salts (Eutrophic waters). A summary of the main features of these two types of water is set out in Table 18.5.

TABLE 18.5

EVOLUTIONARY TREND—LAKE WATERS

Characters			
Hard ancient rocks unchanged	Softer more recent rocks or some silting		Soft rocks or much silting
Generally deep	Decreasing depth		Generally shallow
No O_2 decrease with depth at thermocline			O_2 decreases with depth at thermocline
High $\dfrac{\text{Alkali}}{\text{Ca}+\text{Mg}}$			Low $\dfrac{\text{Alkali}}{\text{Ca}+\text{Mg}}$
Poor in dissolved minerals	Minerals in solution increasing		Rich in dissolved minerals
Rich in number of species	Decreasing number of species		Poor in number of species
Poor in actual numbers of individuals	Increasing number of individuals		Rich in actual numbers of individuals
Desmids abundant	Diatoms and desmids	Diatoms and *Eudorina*	*Asterionella* and Cyanophyceae

OLIGOTROPHIC waters ⟶ Over many years ⟶ EUTROPHIC waters

A third type is the Dystrophic water, which is found on moorlands, where desmids form the most abundant part of the flora. In the course of years Oligotrophic waters may also change into Dystrophic waters.

Originally there was thought to be a fairly sharp distinction between oligotrophic, eutrophic and dystrophic waters, but with the realization that one type can gradually change into another it is evident that any such distinctions cannot be sharp. A considerable body of data is now available on the physical and chemical environment in pools and lakes (Hutchinson, 1954). Among these physical factors is oxygen concentration. In sheltered lakes as compared with open lakes there is an oxygen stratification which closely follows the bottom contours, while the influence of any rivers entering the lake together with the problem of periodic floods is yet a further factor.

Where there is a shallow littoral shore the communities may be difficult to recognise unless there is a rocky substrate, in which case there may then be a zonation that is dependent upon changes of water level and wave action: this type of zonation has been observed in several continental lakes. In deeper waters the communities are more distinct because a zonation develops which is primarily maintained by the light intensity factor or by competition.

The type of eulittoral flora is dependent upon the degree of exposure to wave action, the water chemistry position, (high or low) on the eulittoral and the vertical range of the eulittoral. Such factors are indeed reminiscent of those operating on a rocky coast. In Alpine lakes, e.g. the Traunsee, the uppermost belt is dominated by the Cyanophyceae, *Gloeocapsa sanguinea* status *alpinus*, *Scytonema crustaceum*, and *Chlorogloea microcystoides* (Kann, 1958) (Fig. 18.4). Nearer the mean water level there is a belt of

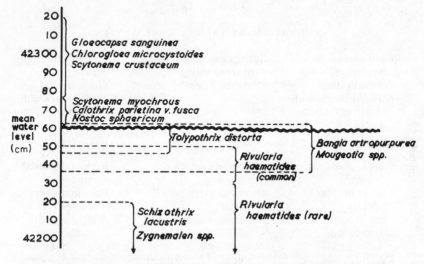

Fig. 18.4   Eulittoral zonation on shore rocks of the Traunsee. (after Kann)

Scytonema myochrous, *Calothrix parietina*, *C. fusca* and *Tolypothrix distorta* var. *penicillata* (Fig. 18.4). On Lake Maggiore, on the chalk rocks of the eastern side, the upper *Gloeocapsa* is associated with *Scytonema myochrous* and two of its forms (Kann, 1963) while on the siliceous rocky shores is a community very comparable to the *Scytonema-Calothrix* community of mean water level above. The spray zone of Lake Windermere carries a similar flora, the *Tolypothrix distorta* dominant being accompanied also by *Pleurocapsa fusca*. At 5–25 cm below mean low-water *Rivularia haematites* occurs in the Traunsee, though if the shore is bouldery (Holstein See) other species of *Rivularia* appear instead. At still lower levels *Schizothrix lacustris* and Zygnemataceae become associated with *Rivularia* (Fig. 18.4). On Windermere *Ulothrix zonata*, diatoms and Cyanophyceae occur around the low water zone (Godward, 1937).

On the clean extended beaches of Lake Maggiore the predominant species is *Coleochaete scutata*, though other Chaetophorales, e.g. *Stigeoclonium tenue*, *Chaetophora incrassata*, *Gongrosira de baryana* are very frequent. Below low-water mark the same algae are associated with *Dichothrix orginiana*, *Rivularia biasolettiana* and *Chaetophora elegans* and a very comparable community occurs on Lake Windermere at the same level. At intermediate depths in Lakes Windermere and Maggiore there are diatoms and filamentous green algae, while below 2 m Cyanophyceae once more predominate. The epilithic flora of base-poor Swedish lakes (Quennerstedt, 1955) comprises mixed Cyanophyceae above and below low-water level, but in the latter case the species present are associated with a rich diatom flora.

Round (1965) has pointed out that there are three distinct components of lake floor algae (Epipelic flora). There are the macroscopic algae, *Chara*, *Nitella*, rooted in the sediment by means of branched underground rhizoids, non-motile colonial or mucilaginous masses *Spirogyra*, *Mougeotia*, and motile microscopic algae, diatoms, Cyanophyceae, flagellates. In protected bays of large lakes, e.g. Lakes Rotorua and Rotoiti in New Zealand, Characeae may form thick underwater swards that have lasted for many years. In such places they have produced a repellant, black, reducing, organic mud of considerable depth. In some European alpine lakes rounded balls of *Aegagropila* (p. 92) lie on the bottom and in oligotrophic lakes, it is possible to find in quiet bays, loose masses of *Spirogyra* and *Mougeotia*. The composition of the flora can be altered by the nature of the substrate, peat, sand, mud etc., and also by the sand-to-silt ratio. The composition of the flora also varies with depth and season (Table 18.6).

TABLE 18.6

VARIATIONS IN EPIPELIC (BOTTOM LIVING) COMMUNITIES OF LAKE WINDERMERE WITH DEPTH AND SEASON (After Round, 1956)

Depth	Diatoms	Cyanophyceae	Diatoms and season			
			10/1/49	2/5/49	22/8/49	28/11/49
1 metre	550	175	—	773	496	24
2	243	124	307	306	376	36
3	260	180	224	460	264	26
4	145	74	243	175	19	4
5	216	67	122	136	666	12
6	144	88	204	163	399	24
8	17	0	69	6	—	—
10	5	0	24	1	—	—
12	0	0	3	0	—	—

An epiphytic flora develops on macroscopic phanerogams, bryophytes, lichens and large algae. This flora is influenced by depth, chemical composition of the water, season, nature of the host surface and life period of the host. On Lake Windermere Godward (1937) recognised the following epiphytic communities:

(1) Epiphytic communities growing on aquatic macrophytes:
  (a) On submerged plants between 0 and 0·5 m depth. This flora possesses a conspicuous Cholorophycean element, e.g. Conjugales, Chaetophorales and Ulothricales.
  (b) On submerged plants between 1 and 3 m, dominated by *Oedogonium*, *Coleochaete* and diatoms.

(c) A community on submerged plants between 3 m and 6 m which is
     comprised of *Coleochaete*, a few diatoms and some Cyanophyceae.

(2) Communities on dead leaves and organic debris:

(a) Up to 12 m — wholly diatoms.

(b) Between 2–16 m — four diatom species and *Microcoleus delicatulus*.

The depth range of the diatoms was found to be greatest at the time of
their maximum in spring and smallest in mid-winter. It was also discovered
that the diatom frequency and light intensity often show an opposite trend
in the upper layers and a similar trend in the lower layers of the lake. The
nature of the habitat, whether organic or inorganic, makes a considerable
difference to the behaviour of the different species, and each individual
species responds variously to the differences of these environments (Round,
1957) (Fig. 18.5).

The nature of the substrate was also found to be of great importance. This
is illustrated in Fig. 18.6 E, where it can be seen that so far as the tips of the

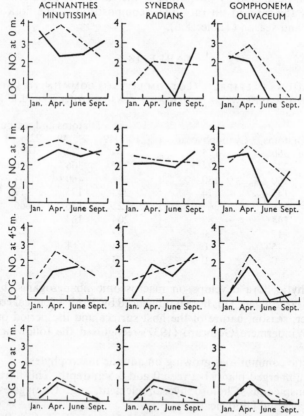

Fig. 18.5 Distribution of diatoms on slides suspended at different depths at different
seasons of the year off two types of shore. (after Godward)

leaves are concerned, the total number of epiphytes increases up to the third or fourth leaf from the apex, after which there is a decline. The diatom flora however is an exception to this behaviour, because it increases regularly with the age of the substrate so that the oldest leaves bear the greatest number of diatomaceous epiphytes. There are distinct differences in the epiphytic flora of the upper and lower surfaces of the leaves, and it was observed that in the case of the first few leaves below the apex the upper surface was infinitely superior in the number of epiphytes, probably because of the greater light intensity. In addition to distribution in relation to increasing age, there is

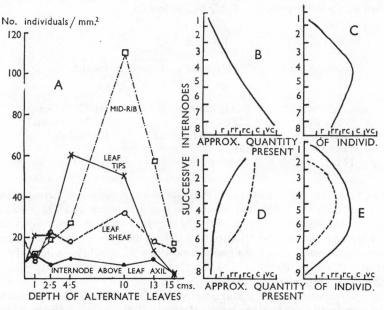

Fig. 18.6 A, distribution of total epiphytes on successive leaves of *Oenanthe fluviatilis*; B, C, distribution of *Cocconeis placentula* on successive internodes of plants of *Equisetum limosum* well separated (3 stems average) (*r* =less than five individuals per 0·1 sq mm., *rr* =about five, *rc* =about ten, *c* =about thirty, *vc* =about fifty); D, distribution of *Cocconeis placentula* (——) and *Eunotia pectinata* (– – – –) on crowded plants of *Equisetum limosum* (three stems average); E, distribution of *Stigeoclonium* sp. (——) and *Coleochaete scutata* (– – – –) on fairly crowded stems of *Equisetum limosum* (two stems average) (after Godward)

also the relation to the different parts of the phanerogamic substrate. Fig. 18.6, A illustrates the distribution of epiphytes on the different parts of a phanerogam, and it will be observed that it is only on the leaf tips that the maximum is reached at the third or fourth leaf, while the leaf sheaths show a slight maximum at about the tenth leaf with a well-marked maximum for the mid-rib at the same level. These maxima on the lower level leaves are to be associated with the diatom flora. It will also be observed that the number of epiphytes on the internodes remains more or less constant, but rapid growth of the substrate, e.g. the leaf lamina, tends to prevent colonization by epiphytes. The density of epiphytes that are attached to dead organic material is dependent upon the habitat of the substrate, e.g. if it is floating then there are few epiphytes, if it is attached or submerged the epiphytes are numerous, while if it is lying at the bottom the epiphytes will be few. The various species to be found are all a residuum from the last living state of the material. In some cases the appearance of epiphytes is due to change in the host with age, e.g. old filaments of the Zygnemataceae lose their mucilage sheath and they then become colonized by many epiphytes.

Experimental work and observation show that the greatest growth and number of epiphytes are partly related to conditions of good illumination, a feature which is illustrated by Table 18.7 below.

TABLE 18.7

TOTAL NUMBER OF EPIPHYTES COLLECTED ON SUSPENDED
SLIDES

	Sandy bottom	Muddy bottom
Water level	225 ⎫	262 (no *Eunotia*)
5 cm	176 ⎬ (no *Eunotia*)	108 (102 *Eunotia*)
12 cm	176 ⎭	0
17 cm	37 (all *Eunotia*)	0

When considering the effect of illumination not only are there problems associated with the individual plants, such as the upper and lower surfaces of leaves, but also the density of the host plants may be highly significant, e.g. Fig. 18.6 shows the distribution of various epiphytes on plants of *Equisetum limosum* under different conditions of spacing. Where there is screening of leaves, either on the same plant or by several plants, then the epiphytes develop on the unscreened portion.

The inter-relations of host and epiphyte are important, and epiphytes tend to develop in the depressions where the cells of the host adjoin each other. Results from experiments carried out with scratched slides suspended in the water make it clear that depressions in a surface increase the number of epiphytes very considerably (Table 18.8).

TABLE 18.8

EFFECT OF SURFACE UPON EPIPHYTES

Epiphyte	No. in scratches on slides	No. elsewhere
Cocconeis	517	297
Stigeoclonium sp.	665	198
Chaetopeltis	138	54
Ulvella	747	200
Coleochaete scutata	40	13

So far as the attachment organs of the epiphytes are concerned, there is no apparent relation between the nature of the substrate and the method of attachment. The differences seen above therefore must be explained by the behaviour of the motile reproductive bodies which either come to rest in the depressions or else are swept there by micro-currents in the water. Ponds with muddy bottoms have a reduced number of epiphytes probably because the pH and the gases evolved are toxic. Summing up, it can be said that the factors influencing the distribution of epiphytes are as follows:

(1) Age of substrate
(2) Rate of growth of substrate
(3) Light intensity
(4) Screening
(5) Nature of the surface
(6) Chemical surroundings.

Of these (3) is probably the most important, although it is difficult to separate the effects from those of (1) and (4).

REFERENCES

GENERAL

Fott, B. (1959). *Algen Kunde* (Fischer Verlag, Jena).

Hutchinson, G. E. (1954). In *Limnology*, Vol. 1 (Wiley, New York).

Kann, E. (1963). *Memoire Ist. ital. Idrobiol.*, **16**, 1.

Round, F. E. (1965). In *Biology of the Algae.* (London).

Tiffany, L. H. (1951). In *Manual of Phycology*, Edit. by G. M. Smith, p. 293–311. (Chronica Botanica Co., Waltham, Mass).

West, G. S. (1916). In *Algae*, p. 418. (Cambridge University Press).

LAKES

Godward, M. (1937). *J. Ecol.*, **25**, 496.

Kann, E. (1958). *Verh. int. Verein. theor. angew. Limnol.*, **19**, 433.

Quennerstedt, N. (1955). *Acta. Phytogeogr. Suec.*, **36**, 1.

Round, F. E. (1956). *Arch. Hydrobiol. Planktonk.*, **52**, 398.

Round, F. E. (1957). *J. Ecol.*, **45**, 133.

PONDS

Bethge, H. (1952–55). *Ber. dt. bot. Ges.*, **65**, 187; **66**, 93; **68**, 319.

Brown, H. B. (1908). *Bull. Torrey Bot. Club.*, **35**, 223.

Fritsch, F. E. and Rich, F. (1913). *Ann. Biol. Lac.*, **6**, 1.

Transeau, E. N. (1913). *Amer. Micr. Soc.*, **32**, 31. *Trans.*

STREAMS

Blum, J. L. (1956). *Bot. Rev.*, **22**, 291.

Blum, J. L. (1957). *Hydrobiologia*, **9**, 361.

Budde, H. (1928). *Arch. Hydrobiol. Planktonk.*, **19**, 433.

THERMAL

Castenholz, R. W. (1969a). *Bact. Rev.*, **33**, 476.

Castenholz, R. W. (1969b). *J. Phycol.*, **5**, 360.

Yoneda, Y. (1952). *Mem. Coll. Agric. Kyoto Univ.*, **62**, 1.

19

PHYTOPLANKTON

Apart from attached littoral and sublittoral algae there is in the surface waters a vast population of floating algae (phytoplankton). This consists primarily of Bacillariophyceae, Cryptophyceae, Pyrrhophyta, Cyanophyceae (particularly in fresh water), Haptophyceae, Chlorophyta (mainly in fresh waters) and Chrysophyceae, and the numbers are so enormous that it is in these waters that the rate of carbon fixation is very high. The growth of the crop depends primarily on five factors: rate of reproduction, rate of removal of individuals by death or grazing by zooplankton, photosynthesis, in relation to light intensity, availability of nutrients, and temperature. To these it has recently been suggested should be added the effect of excreted algal substances, as they may partly determine the annual sequence of the organisms. Despite what may seem to be a prolific literature, as Strickland (1965) points out, our knowledge of primary marine and freshwater production is extremely meagre.

Plankton is found in the ocean, in freshwater lakes, ponds and rivers. It also occurs in abundance in specialized bodies of water such as sewage purification lagoons (p. 418). In every case the flora is composed of larger microplankton species 50–500 μm diameter, which are readily collected in plankton nets, and smaller species 10–50 μm diameter, the *nannoplankton*, which are only collected if water samples are filtered or centrifuged. The principal component of the nannoplankton are Haptophyceae (Coccolitho-phorids) (p. 175), Silicoflagellates (p. 164), Chrysophyceae, Cryptophyceae, Dinoflagellates, small flagellates and coccoid Chlorophyta. In many parts of the world there is little or no information about the composition of the nannoplankton. Some workers also recognize the ultra-plankton, organisms of 1–10 μm diameter. There is no general agreement however about either the size divisions or the names.

Many of the species have elaborate morphological outgrowths but it is very doubtful whether these assist in flotation. Gas vacuoles and oil globules are probably the main buoyancy features, though if there is a mass of mucilage this may assist as its specific gravity is probably very close to that of water. However, the general consensus of opinion at present is that water turbulence is the main factor keeping phytoplankton afloat.

MARINE PHYTOPLANKTON

Vast areas of ocean are involved and for this reason the geographical distribution of most species is only imperfectly known. Apart from variations in composition of the flora associated with the different water masses of the world, differences are also found between the coastal or neritic phytoplankton and the open ocean or oceanic plankton. The neritic plankton is commonly richer because the coastal waters contain more nutrients, and also because there are species derived from the non-planktonic littoral communities. Data is still meagre about the seasonal distribution of species and their occurrence at different depths. Variations in distribution are related to a number of different factors. The primary factor is light, and the depth to which solar radiation can penetrate to permit positive net assimilation is known as the euphotic depth. Usually this is the depth to which 1 per cent of mid-day surface illumination penetrates in summer. This varies from one to two — in very highly productive waters rich in plankton, to 125 m in the clearest ocean waters. The general range is between 15–75 m (Strickland, 1965). Another major factor is temperature; in Monterey Bay high temperatures favour *Ceratium* species and low temperatures favour *Thalassiosira* (Bolin and Abbott, 1963). *Thalassiosira antarctica* is a particular cold water species with a temperature tolerance between − 1·8° C and 3·5° C and it also has a strict salinity tolerance being able to live in waters containing 32·6–34·5° salt. The comparable Arctic species, *T. hyalina*, is likewise restricted to the colder waters of the Arctic and is absent where the warm gulf stream waters enter (Round, 1965). In the Indian Ocean phytoplankton peak production is related to the south-west monsoon, which probably produces a great inflow of nutrients. It is also believed that the physiological state of the organisms, particularly as regards reproduction, is highly important (Subrahmanyan and Sen Gupta, 1965).

Dinoflagellates are particularly valuable as a means of distinguishing various water masses in respect of chemistry and temperature, e.g. *Ceratium* species were used successfully in the Atlantic (Graham and Bronikovsky, 1944) and Ferguson-Wood (1954) used the Dinoflagellates to delineate water masses around Australia (Fig. 19.1). The North Sea has been studied very

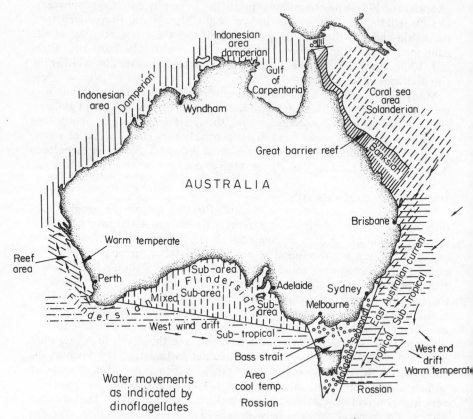

Fig. 19.1 Australian biogeographical ocean regions based upon dinoflagellate association. Note the following water masses: 1, Cool temperate Bass Strait, characterized by *Ceratium horridum*; 2, E. Australian current, characterized by *Ceratium, Heterodinium* and *Histioneis*; 3, Coral Sea characterized by *Peridinium* flora; 4, Barrier reef with fewer *Ceratium*; 5, N. and N.W. Australia characterized by *Peridinium abei, P. orum, P, ventricum* and *P. hirobis*; 6, W. Australia with sparse flora of *Ceratium karstenii, C. lunula, C. massiliense* var. *macroceroides*; 7, West wind drift area with *C. ranipes, C. ornithocerus*; 8, Gt. Australian Bight. (after Ferguson-Wood)

thoroughly with samples taken at known depths and within a very short space of time, and the various water masses delineated (Braarud *et al*, 1953) (Fig. 19.2). A comparable study of the Ligurian sea revealed variations in the three dominant phytoplankton species in a transect across the sea and these variations probably reflect the existence of two separate water masses (Leger, 1964) (Fig. 19.3).

In recent years attempts have been made to use statistical methods, to establish ecological affinities in plankton populations, but it will take time before these methods become established to the extent they are in terrestrial ecology (Cassie, 1961).

In nearly all temperate waters it has been found that there are distinct
spring and autumn maxima when some component species occur in great
abundance. In tropical waters there is very little variation, though in inshore
waters, adjacent to coasts with monsoonal climates, a variation does occur

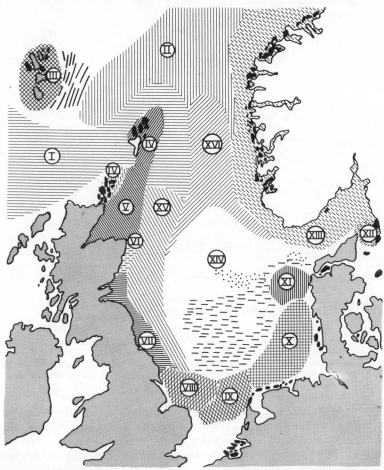

Fig. 19.2 Phytoplankton Zones of the North Sea. I = Inflowing Atlantic; II = Local
Atlantic; III = Faero water; IV = Shetland–Orkney water; V = *Thalassiosira gravida* water;
VI, X, XI = Coastal; VII = rich coastal; VIII = Humber–Wash; IX = Southern bight;
XII = Kattegat; XIII = Norway water; XIV = Central; XV = S. flowing Atlantic; XVI =
Mixed water. (from Braarud)

associated with the rains. In winter in temperate waters there is a rapid
vertical mixing of the waters bringing up fresh nutrients from the lower
layers, and this, when associated with increasing day length, more intense
insolation and higher water temperature, is responsible for the development
of the spring maximum. The spring blooms, particularly of diatoms in colder
waters and in deep water, are further related to light intensity and light

penetration (Anderson and Banse, 1961). Continual vertical movement maintains a rich medium and, as a result, cold upwellings (e.g. off S-W Africa and California) are very rich in plankton. Lack of trace elements may reduce the numbers in certain species; iron affects the abundance of *Skeletonema*

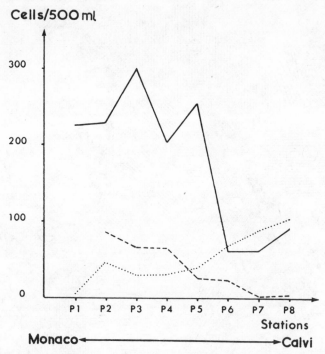

Fig. 19.3   Variation in three phytoplankton species between Monaco (P 1) and Calvi (P 8). (*Rhizosolenia hebetata* = ——————, *Rhizosolenia alata* = --------- and *Peridinium leonis* = ...). (after Léger)

costatum, and manganese and also vitamins, especially B_{12}, affects *Cyclotella* sp., as a result of their absorption by other organisms. The most spectacular of the seasonal maxima are the planktonic dinoflagellate blooms referred to as 'red tides' (see page 155), which are most commonly found in the temperate and tropical waters of California, Florida, South America and South West Africa. The two most common genera involved are *Gymnodinium* and *Gonyaulax* and sometimes *Prorocentrum*. These blooms occur during the summer months and are closely related to the warm temperatures and abundant supply of nutrients caused by the upwelling of cold nutrient-rich water which is a common feature of these regions. Dinoflagellates are not responsible for all red tides, the red coloration in the Red Sea, for example, is due to a bloom of the red Cyanophyte, *Trichodesmium*. Green discolorations of the sea off the West coast of India have been caused by the chloromonad, *Hornellia*.

The effect of upwelling water has been studied in detail in Monterey Bay

(Bolin and Abbott, 1963) where it causes high plankton volumes. The steadiness of the upwelling is a major factor influencing the magnitude of local plankton blooms. These water masses have been studied in other parts of the world, the most intensive being in Walvis Bay, S.W. Africa. All upwelling areas seem to have a complex hydrography and this complicates the interpretation of biological data (Anderson and Banse, 1961). Apart from the upwelling there is a relationship between the plankton volumes and presence of nutrients, such as phosphate and silicate (Fig. 19.4).

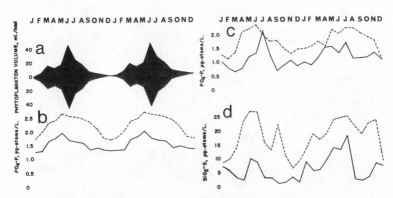

Fig. 19.4 a, annual cyclic variation in phytoplankton standing crop (ml./haul), central Californian coast, 1954–60, two cycles shown; b, average cycle PO₄-P, µg-atoms/L. (1951–55); c. average cycle PO₄-P, µg-atoms/L. (1954–55); d, average cycle SiO₂-S, µg-atoms/L. (1954–55). (————— = surface, ------ = 50 m) (after Bolin and Abbott)

An exception to the normal spring and autumn maxima is provided by the enclosed waters of the Mediterranean where the maxima occur in November–December and March, the one in March being the larger.

During the 1939–45 war, successful attempts were made to increase the plankton, and hence the fish, by adding artificial fertilizers to large, more or less enclosed, areas of water such as lochs. Generally only quantitative changes occur in the phytoplankton; qualitative changes develop only after a number of years unless there has been an algal bloom (Brook, 1965). A technique such as this can have valuable application, as, for example, in the Philippine fish ponds.

In recent years new techniques, such as the continuous plankton sampler and the electronic counter have made plankton studies slightly less laborious. Data obtained from the North Sea using the electronic counter show distinct patches of Dinoflagellates (Fig. 19.5). In the future more information is required about the seasonal stratification of phytoplankton, but this is related to the difficulties of occupying ocean stations for any great length of time, and is also related to the fact that at a fixed station the body of water itself can change. In the North Sea, where this kind of work has been done, changes only occur with depth in summer when a thermocline develops (Fig. 19.6).

A vast amount of work has been performed on the productivity of oceanic

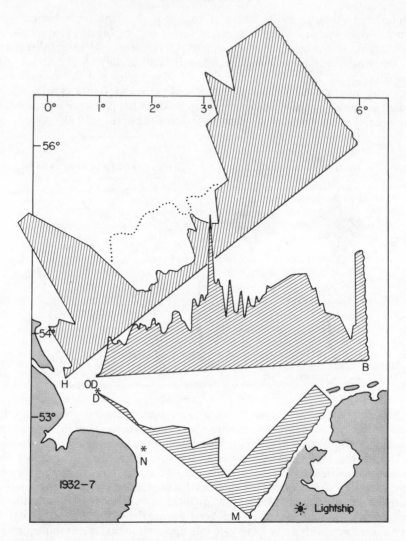

Fig. 19.5 The maximal quantities of dinoflagellates in numbers per mile recorded by continuous plankton recorder along three lines in the North Sea. (from Lucas, 1964)

waters with determinations of oxygen production using either Winkler light and dark bottles and determinations of the amount of radioactive [14]C (as HCO_3^- or CO_3^{--}) tracer absorbed by the cells under given light conditions. Discussion is still taking place about the validity of the various techniques associated with the use of ^{14}C, but the available results do give at least an indication of the productivity of various waters. Another technique is to determine the amount of chlorophyll *a* present in a sample of phytoplankton, and fair agreement has been reported between the two methods.

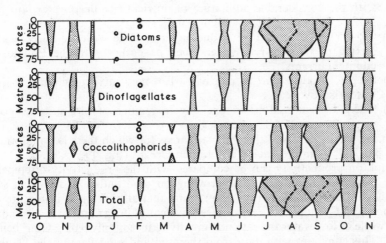

Fig. 19.6 The vertical distribution of the phytoplankton in the North Sea during one year's sampling. (from Halldal, 1953)

Danish workers have distinguished four classes of oceanic production in tropical and subtropical zones:

(1) Daily productivity 0·5–3 g C produced/m² in one day. This is associated

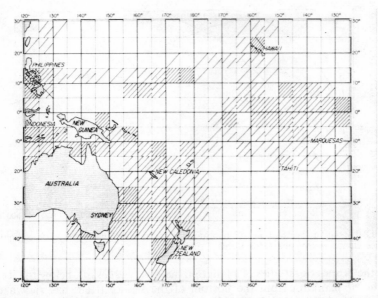

Fig. 19.7 Rate of ^{14}C in fixation in surface waters of the Pacific Ocean (1961). ($/ = 0·1$ mg C $h^{-1} m^{-3}$; $// = 0·2$ mg C $h^{-1} m^{-3}$; $/// = 1·0$ mg C $h^{-1} m^{-3}$)

with the considerable admixture of nutrient-rich deep water into the surface waters, e.g. Benguela current.

(2) 0·2–0·5 g C produced/m² in one day associated with the fairly steady mixing of nutrient-rich waters such as occurs with the Equatorial counter currents.

(3) 0·1 g C produced/m² in one day where any mixing of nutrient-rich waters is slow.

(4) 0·05 g C produced/m² in one day in regions of old, nutrient-poor waters, e.g. central Sargasso Sea.

On the whole, however, there are still areas for which there is no information, particularly in the Pacific and Indian Oceans (Fig. 19.7).

Plankton productivity can be compared with the productivity of epipelic algae which are mainly diatoms growing on mud and sand flats. On salt marshes in Georgia, Pomeroy (1959) found that at high water a mixture of diatoms and Cyanophyceae fixed the equivalent of 0·2 g C/m² over a 10 hour day, while at low water the rate was greatly reduced, only 0·02 g C/m² being fixed. This compares with data from the mud and sand flats of the Danish Wadden sea where values ranging from 0·038 to 0·047 g C produced/m² over a 10 hour day have been recorded (Grøntved, 1965). Variations in productivity as between groups of algae can be very considerable: thus Cyanophyceae can fix 35–390 mg C/kg/h., a mixture of diatoms and flagellates produce 2·4–757 mg C/kg/h and pure diatom communities produce 4–454 mg C/kg/h.

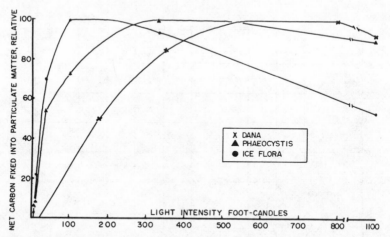

Fig. 19.8 Relative levels of net C fixation into particulate matter at various light intensities. (Dana = gross phs) (after Bunt)

In the Antarctic, studies of plankton, mainly *Phaeocystis*, below the ice, as well as of plankton from the bottom of the sea-ice itself, has shown that both the ice flora and the plankton have a degree of 'shade' adaptation well in excess of that of Arctic plankton (Dana station 10531) (Fig. 19.8). The ice organisms show an increasing inhibition of photosynthesis above 100 ft-candles, and light saturation with the antarctic plankton was reached at 300 ft-candles. Temperatures below 5° C brought a very great drop in the carbon fixation, and this is very significant as temperatures in the experimental waters (McMurdo Sound) rarely rise above $-1.5°$ C. (Bunt, 1964). The whole problem of nutrition in antarctic waters demands further study (see p. 434).

In recent years some very interesting work has been done on the productivity of artificial marine systems (Odum *et al*, 1963). Artificial reef ponds were set up in Texas with suitable animals and plants and these developed

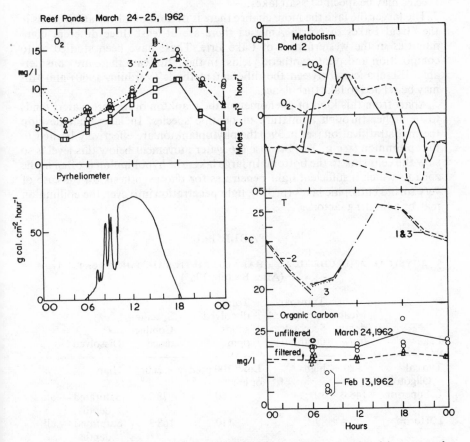

Fig. 19.9   Diurnal record of O_2 heat-energy, metabolic rate (CO_2 out and O_2 in), temp. and organic carbon in triplicate reef ponds, March 24–25, 1962. (after Odum)

heavy nannoplankton blooms. In such habitats careful measurement of the environmental factors can be maintained and these showed some resemblance to data obtained in the field (Fig. 19.9). Comparable systems can be established in the laboratory to enable microenvironments associated with algae.

FRESHWATER PHYTOPLANKTON

Methóds of collection are the same as for marine phytoplankton, and in both cases it is most desirable to examine the living material, even if subsequently there has to be preservation.

The components of freshwater phytoplankton differ from the marine in that there are a greater number of species of Cyanophyceae, abundant Desmids (completely lacking from salt water), Volvocales and Chlorococcales but fewer Pyrrhophyta. In inland salt lakes a few marine forms can be found, but although genera such as *Peridinium* and *Ceratium* are common, the species may be specific to salt lakes.

The larger the lake the more chance there is of horizontal variation. Thus in the Great Lakes of North America there are at least three different communities in the western part of Lake Erie. These have been studied as to composition and photosynthetic yields. In the latter case there are considerable discrepancies between the different methods (Verduin, 1960) and this may be related to lake turbulence.

Apart from this kind of phenomenon the plankton of inshore waters tends to be richer through admixture of benthic species. In lakes that develop thermal stratification (see p. 396) the phytoplankton are effectively limited to the epilimnion (see p. 396) because the water movement below this level is so slow the cells sink to the bottom. In large lakes the hypolimnion is often in the zone to which insufficient light penetrates for photosynthesis. In the case of very turbid eutrophic lakes (p. 397), light penetration into even the epilimnion may be a limiting factor.

TABLE 19.1

PHYSICAL AND CHEMICAL CHARACTERISTICS OF THE GREAT LAKES
(After Beeton, 1965)

Lake	Mean depth (m/)	Transparency (av. Seechi disc depth m)	Total dissolved solids ppm	Specific Conductance	Dissolved O$_2$
Typical Oligotrophic	>20	high	Low: 100 ppm or less	<200	High
L. Superior	148.4	10	60	78·7	Saturated — all depths
L. Huron	59·4	9·5	110	168·3	Saturated — all depths
L. Michigan	84·1	6	150	225·8	Near saturation
Typical Eutrophic	>20	Low	High >100	>200	Depletion in hypolim
L. Ontario	86·3	5·5	185	272·3	50–60% saturation in hypolim
L. Erie	17·7	4·5	180	241·8	10–50% saturation in hypolim

The effect of eutrophication of freshwater lakes on the phytoplankton is becoming of great importance because the ratio of desmids to other selected algal groups can be used as a measure of eutrophication. Despite this it is often very difficult to define the terms precisely as different investigators have used different approaches. There are also seasonal variations to be taken into consideration. A well established example of the eutrophication process is provided by the Great Lakes of the St. Lawrence basin (Table 19.1).

From these figures it will be seen that Lake Michigan is on the verge of passing from one type to the other. The changes that have been taking place over the years, particularly in Lakes Ontario and Erie, are illustrated in Fig. 19.10. In western Lake Erie the changes have been particularly pronounced since 1950 and these are reflected in the phytoplankton (Table 19.2) (Verduin, 1964).

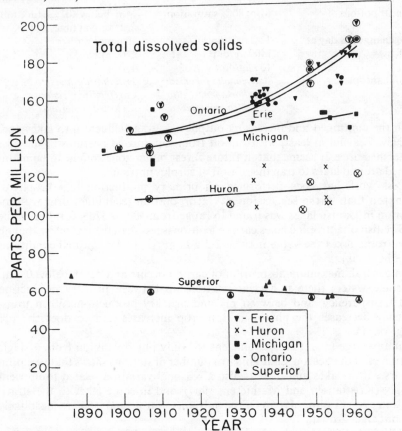

Fig. 19.10 Changes in concentration of total dissolved solids in the Great Lakes, 1890–1960. (O = 12 or more determ) (after Beeton)

Lake eutrophication in recent years is not restricted to the Laurentian lakes. In New Zealand the Rotorua lakes exhibit the same phenomenon and here it is associated also with excessive growth of the introduced aquatic phanerogam, *Lagarosiphon major*. This has reached nuisance proportions and is being presently controlled by diquat spraying. The elimination of the *Lagarosiphon* brings about some change in the phytoplankton which could be associated with the release of nutrients. This, however, is a feature that is being studied. The general process of eutrophication seems to be associated

TABLE 19.2

CHANGES IN THE CONDITIONS PREVAILING IN LAKE ERIE AND THE
RESULTING CHANGES IN ITS PHYTOPLANKTON BETWEEN
1950 AND 1960

	Before 1950	1960–61
pH	Max. 8·7	9·2
O_2 near bottom	Min. 80% saturation	Min. below 40% saturation near zero at times.
CO_2 change per day at	32 μ mol/dm^3	70 μ mol/dm^3
NO_3^-	1400 μg/dm^3	2200 μg/dm^3
PO_4^{---}	105 μg/dm^3	450 μg/dm^3
Dominant species	*Asterionella formosa*	*Fragillaria capucinus*
	Tabellaria fenestrata	*Coscinodiscus radiatus*
	Melosira ambigua	*Melosira binderana*

with the continued and increasing output of sewage effluent into lakes and, in New Zealand at least, with run-off from top-dressed pastures in the lake watershed area. It seems that in future forest buffer zones ought to be maintained around lakes to trap the run-off of surplus nutrients.

Less work has been performed on primary production in lake phytoplankton than in the sea. Using ^{14}C incorporation techniques primary production in Danish lakes was found to range from 300 to 370 g C m²/yr, while in Swedish oligotrophic lakes carbon fixation is less than 0·1 g C m²/yr though in eutropic lakes the value may be 2·2 g C m²/yr, under optimal conditions (Rodhe, 1958).

In general, maximum absorption of carbon occurs at a depth below 0·5 m. In some waters there is a clear relationship between primary production and transparency, and between pH and primary production. When transparency decreases the phytoplankton crop increases and so does the pH (Fig. 19.11).

In any large lake it is not sufficient to study phytoplankton from a single station or at one season. Sampling in a number of stations shows that variation occurs within lakes and there is also a seasonal variation related to nutrient supply (Kristiansen and Mathiesen, 1964), and in some lakes to the stratification cycle (Talling, 1966). Of the nutrients, nitrate and iron are most likely to limit algal growth.

Relatively little work has been done on the effect of temperature on carbon assimilation by phytoplankton. Some recent work by Aruga (1965) has shown a relationship between the optimum temperature for photosynthesis and the water temperature. In spring and winter the optimum temperature for the flora is around 20° C with actual temperatures of 8–15° C. In summer the optimum temperature and the actual water temperatures more or less coincide, reaching a maximum of 30° C in July and August.

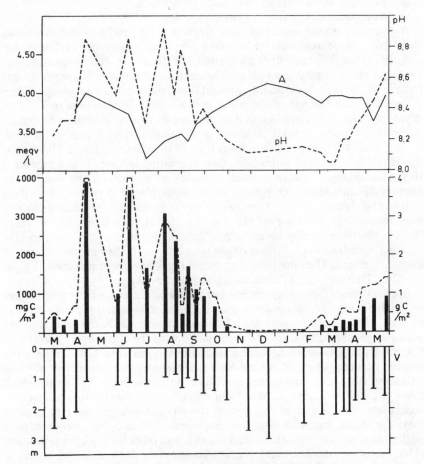

Fig. 19.11 Seasonal variation of primary production in Bavelse See (station 6), March 1961–May 1962, in relation to transparency, A (secchi disc), and acidity (pH). The black columns show gross production per day at the best depths, and the graph line shows gross production per unit area of lake surface per day. (after Kristiansen and Mathiesen)

The low temperature of high altitudes is obviously of paramount importance, but this factor has also to be considered in relation to the seasonal light intensity. The interaction of these two factors produces a plankton maximum immediately after the ice thaw and again in the autumn (Prescott, 1963).

Compared with the limnoplankton of lakes our knowledge of the potamoplankton of rivers is extremely meagre. Indeed it was not until nearly 1930 that the existence of a separate potamoplankton really became recognized.

Before then it was thought that river plankton originated from outside sources. Potamoplankton can be subdivided as follows:

(a) Eupotamic, thriving in the stream and its backwaters.

(b) Tychopotamic, thriving only in the backwaters.

(c) Autopotamic, thriving only in the stream.

A potamoplankton survey of three rivers in the Goros region showed that there were three maxima for diatoms and Chlorophyceae, and four for Cyanophyceae. The total flora comprized 155 species. It seems, however, that even though there may be a rich phytoplankton the component species do not form a distinctive association. In general the development of potamoplankton is dependent on the age of the water and whether the stream is in flood or drying up. The plankton of small streams appears to arise from the benthos.

Apart from the normal numbers of planktonic forms present in river waters, one or more species can from time to time develop into a bloom, but this is related to rate of water flow. One interesting feature of such blooms is their restriction to relatively small stretches of a river. In the case of the Danube blooms they even remain in the same place for two or three days (Claus and Reimer, 1961). At present there is no satisfactory answer to this phenomenon. In the case of the Danube, which has been studied fairly thoroughly, there does not seem to be a definite plankton cycle or any specific maxima. One important feature of potamoplankton is the high proportion of diatoms present. Thus the following percentages have been recorded: Volga, 45–53%; Dnieper, 48%; Rhine, 56%; Vistula, 55–74%.

In recent years increasing use has been made of the unicellular green algae, especially *Chlorella*, *Scenedesmus* and *Chlamydomonas*, to purify sewage effluent. Most of the pioneer work has been carried out in the United States where large lakes of effluent are innoculated with the algae, and these after a few days increase in such number that the oxygen they produce aids the aerobic decomposition of the remaining organic matter. When one of these oxidation lakes is operating, water is allowed to flow through at a rate which takes it about 6–7 days to pass from entry to exit, or else it is allowed to recirculate. At the end of this period the effluent emerges clear of organic matter and the bacterial count of pathogens has also been reduced to a negligible value. At present the largest oxidation lakes for sewage purification (1100 acres or 450 hectares) are at Auckland in New Zealand, where a detailed study of the phytoplankton has recently been carried out. A very substantial planktonic flora, especially in species of *Euglena* was found. Small lakes, such as those used in Israel, have a much more restricted flora and at times can be almost unialgal (e.g. *Prymnesium parvum*). The dominant species of algae are substantially the same in ponds in all countries. The large Auckland ponds, however, contained 36% of species not previously recorded in this habitat and there were two new species. A characteristic of the Auckland ponds was a paucity of Volvocales, especially *Chlamydomonas*, which are common elsewhere. The Volvocales are generally the first algae to develop in sewage treatment ponds but they are replaced later by *Chlorella* and *Scenedesmus*. There was a seasonal maximum in the warm weather which was probably related to the higher light intensities and temperatures, though there was also a heavier loading of the ponds. The flora was also more varied at this time.

Different types of treatment pond appear to carry different floras:

(a) Very lightly loaded ponds or polluted natural waters develop blooms of blue-green algae (*Microcystis*) in summer and a mixed Chlorococcalean flora at other times.

(b) Highly loaded ponds with low ammonia have a mixed Chlorococcalean flora with other types, e.g. *Closterium cornu*, *Cyclotella meneghiniana*, abundant at times.

(c) Ponds with a low biological oxygen demand (B.O.D.) and high ammonia are characterized by euglenoids.

(d) Ponds with a high B.O.D. have floras dominated by *Chlorella* and *Selenastrum miniatum*, with *Ankistrodesmus falcatus* var. *acicularis* or *Scenedesmus opoliensis* often becoming important.

In conclusion it is evident that much more work is needed in pure taxonomic phycology to find out which species are predominant in different areas and to catalogue, preferably statistically, their associations as well as the successions. The factors affecting the growth and distribution of phytoplankton are a complex of physical, e.g. light, temperature, turbidity, water movement; chemical, e.g. nutrients and excreted organic matter, and biological, of which grazing is perhaps the most important. The utilization of the nutrient varies with different species and is complicated. There is not present much data on the relationship between growth and concentration of nutrients, in particular nitrate and ammonia.

It has been known for some time that living planktonic cells, marine and freshwater, excrete organic matter, which may range from 0·5 to 34% of the total photosynthate. In many species the average amount is 3–6%. A few of the products are highly toxic (see p. 396), but we know very little about some of the other compounds. One important compound is glycollic acid (Fogg, 1966).

This liberation of a large proportion of photosynthetic product into the medium, whether sea or freshwater suggests that the water always contains material that can act as a source of energy and can exert an ecological effect. Excretion takes place in the light as well as in the dark, but recent work suggests that the phenomena are different (Fogg, 1966). Since there is a relationship with light intensity the amount excreted varies with water depth, increasing with depth and decreased light intensity.

The material excreted is not rich in carbohydrates, but may contain considerable quantities of amino acids and peptides, with lesser amounts of protein and chloroform — soluble substances. A few species excrete a single compound almost exclusively. There is also good evidence that some algae excrete substances that inhibit their own growth and that of other algae.

Finally, reference must be made to a peculiar type of freshwater plankton, the cryoplankton of the snow, e.g. red snow due to the presence of *Chlamydomonas nivalis*; this is by far the most common species, though in the Pacific north west of America *C. yellowstonensis* is also very important (Garric, 1965). The colour is due to carotenoids in the cells, especially in the resting spores.

Algae can also give other colours to snow; yellow snow associated with a flora of about twelve species, all containing much fat; green snow, principally caused by zoogonidia of green algae; brown snow or ice due to the presence of *Mesotaenium* and mineral matter; black snow caused by *Scotiella nivalis* and *Rhaphidonema brevirostre*; and a light, brownish-purple ice-bloom caused by a species of *Ancyclonema*.

Cryoplankton is normally only present where the snow or ice is stable for a considerable period.

REFERENCES

Allen, M. B. (1961). *Proc. Conf. Prim. Prod. Meas. Univ. Hawaii.*, U.S. Atomic Energy Comm., 7633, 58.

Anderson, G. C. and Banse, K. (1961). Ibid, 61.

Aruga, Y. (1965). *Bot. Mag., Tokyo*, **78**, 280.

Beeton, A. M. (1965). *Limnol and Oceangr.*, **10**, 240.

Bolin, R. L. and Abbott, D. P. (1963). *Cal. Co-op. Fish. Invest. Repts.*, **9**, 23.

Braarud, T., Graarden, K. R. and Grøntved, J. (1953). *Rapp. adm. Cons. Perm. Int. Explor. Mer.*, **133**, 1.

Brook, A. J. (1965). *Limnol. Oceanogr.*, **10**, 403.

Bunt, J. D. (1964). *Antarct. Res., Ser.* **I.**, 27.

Claus, G. and Reimer, C. I. W. (1961). *Revue Biol.*, **2**, 261.

Cassie, R. M. (1961). *Memoire Ist. Ital. Idrobiol.*, **13**, 157.

Fogg, G. E. (1966). *Oceanography and Marine Biology Ann. Rev.*, **4**, 195.

Garric, R. K. (1965). *Amer. J. Bot.*, **52** (1), 1.

Graham, H. W. and Bronikovsky, N. (1944). *Carnegie Institute Publication*, 565.

Grøntved, J. (1965). *Proceedings of the 6th Marine Biology Symposium, Bremerhavn.*

Kristiansen, K. and Mathiesen, H. (1964). *Oikos*, **15**, 1.

Leger, G. (1964). *Bull. Inst. Océanogr. Monaco*, **64**, No. 1326.

Odum, H. T., Siler, W. L., Beyers, R. J. and Armstrong, N. (1963) *Publs. Inst. mar. Sci. Tex.*, **9**, 373.

Pomeroy, L. R. (1959). *Limnol. Oceanogr.*, **4**, 386.

Prescott, G. W. (1963). *Trans. Am. Microsc. Soc.*, **82**, 83.

Rodhe, W. (1958). *Verh. int. Verein. theor. angew Limnol.*, **13**, 121.

Round, F. E. (1965). *Biology of the Algae* (London).

Strickland, J. D. H. (1965). *A. Rev. Microbiol.*, **19**, 127.

Subrahmanyan, R. and Sen Gupta, R. (1965). *Proc. Indian Acad. Sci.*, **61**, Ser. B, 12.

Talling, J. F. (1966). *Int. Revue ges. Hydrobiol. Hydrogr.*, **51**, 545.

Verduin, J. (1960). *Limnol. Oceanogr.*, **5** (4), 372.

Verduin, J. (1964). *Verh. Int. Verein. theor. angew. Limnol.*, **15**, 639.

Wood, E. J. Ferguson (1954). *Aust. J. mar. Freshwat. Res.*, **5**, 171.

20

ECOLOGICAL FACTORS

Any attempt to understand and interpret the zonation of a rocky coast or on a salt marsh must depend upon the part played by the different ecological factors. More than one factor will normally exert a controlling influence, but at one time of the year or at a certain phase in the life history of an organism one single factor may become paramount. A proper appreciation of some of the factors cannot be divorced from a study of the physiology of the algae and certain aspects of algal physiology must therefore be discussed in this connection.

The essential factors can be divided broadly into four groups: (1) physiographic (dynamic according to some writers), (2) physical, (3) chemical, (4) biological. Similarly, at any place it is possible to segregate the factors into those that are (1) causal, i.e. responsible, so far as one can see, for the basic zonation; (2) presence or absence, i.e. factors such as freshwater or the type of rock which determine whether a species shall be present or not, but do not affect the level at which the species grows; (3) modificatory, i.e. factors such as spray, cracks, shade which cause a variation in the level normally occupied.

PHYSIOGRAPHIC FACTORS

The two important factors here are topography and the tides. The former is

422   THE ALGAE

important because the nature of the coastline and the off-shore sea bed
determine the degree of shelter or exposure to wave action. Waves themselves
are determined by the stretch of water over which they are generated and by
the strength of the winds. In some places it is the prevailing winds that will be
most significant, e.g. trade-wind areas, while in other places it may be the
frequency and force of gale winds. Waves can be generated far out at sea and
result in a major swell when local conditions are quite calm, and this kind of
swell is not uncommon on Atlantic and Pacific coasts. The flatness or steep-
ness of the shore is also significant, the former giving relatively calmer conditions than the latter. The presence of immediate off-shore submarine
canyons, such as that of La Jolla in California, can also bring about local
modifications of sublittoral conditions. It is not easy to determine degrees of
exposure, but a system which appears fairly workable is that of Ballentine
(1961).

PHYSICAL DEGREE OF SHELTER OR EXPOSURE TO THE WAVES

This operates as a presence and absence factor, some species, *Durvillea*,
Postelsia, possessing holdfasts that enable them to withstand strong wave
action, or else it operates as a modificatory factor, as in elevating belts on
exposed shores. Even in the sublittoral there may be an effect, such as cover
degree (Fig. 20.1), where the maximum cover occurs at different levels under
the changing conditions; thus, with shelter the level is at 1 m, with moderate
wave action at 2 m, and on exposed coasts between 4–6 m. Under very exposed conditions it is evident that the sporelings of the algae must be able to

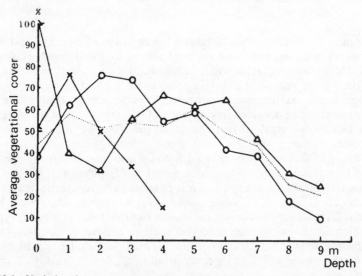

Fig. 20.1 Variation in average vegetation cover in relation to depth of water under
different conditions of exposure at Tsuyazake in Japan. (x————x = shelter, o————o
= moderate wave action, △————△ = exposed, ... = mean of all areas) (after Yoshida)

withstand severe wave action, and the attachment of spore or zygote to the rock surface must take place very rapidly. In the case of Laminarian species, e.g. *L. sinclairii, Postelsia*, the ovum must either be fertilized *in situ* or else there must be rapid adherence the moment a fertilized egg touches a rock surface. This factor will also operate on large freshwater lakes where the waves can be quite large. Such waves may indeed determine the type of algal community at the lake edge, e.g. where there is exposure to wave action Cyanophyceae tend to predominate, whereas in more sheltered places Chlorophyceae can play an important role (Kain, 1963).

WAVE ACTION AND TIDES

By wave action is meant the actual height of the waves and the degree of splash and spray that results from them. This is essentially a modifying factor, elevating the zones, especially the upper ones, often to a considerable degree.

Certain basic features of the tides are as follows:

(1) The maximum tidal range varies from place to place.

(2) The tidal rhythm can vary from place to place so that it is possible to observe the usual twice-daily tides (that are generally of unequal height) or the daily but less-common single high and low tides (diurnal) (page 425).

(3) The terminology used in respect of tides is as follows:

Max. tidal range	Min. tidal range		Mean range neap tides	Mean range spring tides
		Extreme high-water mark: spring tides		
		Mean high-water mark spring tides		
		Mean high-water mark neap tides		
		Extreme lowest high-water mark neap tides		
		Extreme highest low-water mark neap tides		
		Mean low-water mark neap tides		
		Mean low-water spring tides		
		Extreme low-water spring tides		

(4) The time of low water of spring tides varies from place to place. If it occurs around mid-day, especially in summer, this could have a profound effect upon any species that cannot tolerate high temperatures and consequent desiccation, or those in which photosynthesis is inhibited by high light intensities during exposure (see p. 434).

(a) Tidal Range

This is a modifying factor as it is primarily responsible for the actual width of the algal zones. Even with a very small tidal range, e.g. Caribbean, it is possible to observe a distinct zonation of the organisms. This factor also determines the extent to which rock pools may exist, the greater the tidal range the greater the variety among the rock pools. The different types of tidal variation are of great importance (Doty, 1957; Lewis, 1964) and may alter the position of an organism on the shore.

(b) Tidal Currents

In certain places there may be very pronounced currents close inshore where algae may grow, e.g. between Orkney Is. and the north of Scotland, at the mouth of the Bay of Fundy, etc. Such currents affect the growth of the plants but they can also operate as a presence or absence factor. Thus on the east coast of Scotland the strength of the current determines whether *Laminaria saccharina* or *L. hyperborea* is the dominant. In New Zealand increased current strength seems responsible for the presence of the so-called *var. richardiana* of *Ecklonia radiata*, which is therefore probably only an ecological growth form. The strength of the current flowing up salt marsh and mangrove swamp creeks determines the degree of erosion, and hence the extent to which creek bank algal communities can develop (Nienhuis, 1970).

(c) Silt Load

In some areas, e.g. near the mouth of the River Severn in England, Bay of Fundy in Nova Scotia, the load of silt carried by the sea exerts a distinct effect upon the fauna and flora. Certain organisms are unable to tolerate silt deposition, in the case of algae this may be due to inhibition of photosynthesis. Silt load can therefore operate as a presence or absence factor.

(d) Inter-tidal Exposure

This is the exposure suffered at any level between two consecutive tides and it varies with the type of tide, i.e. spring or neap. Exposure operates indirectly insofar as it affects water loss, temperature changes of the thallus (especially in tropics), and salinity of the cells. The degree of atmospheric humidity is also important during the periods of exposure, and it is this, together with the high temperature in the drier areas of the tropics that prevents algae from occupying the shore but enables them to form a dense covering at high moist latitudes.

The length of the inter-tidal exposure also affects the conditions in rock pools and salt pans, the diurnal changes being greater in the higher pools and pans where the length of exposure is longer. With one type of tide there will be certain levels on the shore where the inter-tidal exposure can be doubled, e.g. at 3·5 ft (Fig. 20.2). Similarly, the submergence period may be drastically altered over a short vertical range, e.g. just above 1·5 ft (Fig. 20.2) it is about 10 hr, whereas just below 1·5 ft it is about 30 hr. The effect of this phenomenon in relation to critical levels (p. 446) may well be profound.

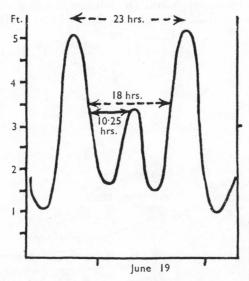

Fig. 20.2 Tide curves at San Francisco and the relevant hours of exposure at different levels. (after Doty)

(e) Continuous Exposure and Submergence

At levels lying between lowest high water of neap tides and extreme highest water of spring tides there will be periods of two or more days during the neap tides when the level will not be inundated at all. Similarly, between highest low water of neap tides and extreme lowest low water of spring tides there will be days during the neap tides when the levels will be continuously submerged. In the former case, desiccation, temperature and salinity changes will be far more profound than those that occur between two consecutive tides; in the latter case continual submergence may affect the metabolic activities and the compensation point, though this will depend upon the maximum tidal range (see p. 424). It is probable that the periods of continuous exposure are of far greater importance than inter-tidal periods, especially during the sporeling stages. Gail (1920) has suggested that it is the desiccation of young plants that prevents the appearance of algae outside their usual zones, and it is a remarkable fact that sporelings of fucoids are usually very strictly confined from an early stage to the level they occupy as adult plants.

A rather different exposure phenomenon can be observed around large lakes. Evaporation in the summer very often lowers the water level so that the winter and summer levels are different. This zone of exposure between the two seasonal levels may be occupied by a special algal community, or it may form a bare zone between the winter water-edge community and the lower, summer water-edge community.

Diurnal variations in temperature may affect the metabolic activities of the algae (see p. 432) and extreme variations in salinity can also bring about the death of some species (see p. 442). Water loss during periods of exposure is quite clearly an important factor. On salt marshes it is undoubtedly responsible for the occurrence of gelatinous Cyanophyceae at the higher levels

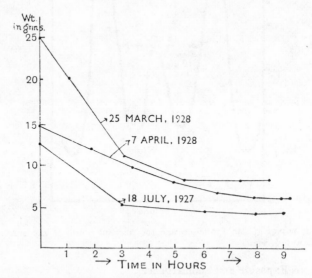

Fig. 20.3 Loss of water, as represented by loss in weight, in *Pelvetia canaliculata* during intertidal exposure. (after Isaac)

where the exposure periods are long. Among others, Pringsheim (1923), Zaneveld (1937), Isaac (1933), Kanwisher (1957), Brown and Johnson (1964) and Chapman (1962) have studied rate of water loss in intertidal algae. Much of the work has centered around the principal fucoids found on European shores. In *Pelvetia canaliculata* the main water loss occurs in the first six hours of exposure (Fig. 20.3) while in *Fucus spiralis* var. *platycarpus*, *F. vesiculosus*, *F. serratus* and *Ascophyllum* the maximum water loss can be spread over 18 h (Fig. 20.4): in *F. vesiculosus* as much as 90 per cent of the total initial water can be lost in 1·5 h. *Fucus spiralis* var. *platycarpus* loses its water at the slowest rate and it occurs at the highest level on the shore. Haas and Hill (1933) have also shown that the higher the alga grows the greater is the fat content (Table 20.1).

This fat is largely contained in the cell walls and examination has shown that the thickness of the cell wall is greater in plants growing higher on the shore.

These cell walls decrease in thickness when subjected to desiccating conditions, so it must be assumed that a large part of the water lost is contained there (Fig. 20.5). Those species which lose water most slowly also reabsorb it most slowly and, as a result, the growth rate of the highest species therefore tends to be the slowest. It would appear that an important factor controlling zonation, so far as these fucoids are concerned, is the biochemical nature and properties of the cell wall.

TABLE 20.1

FAT CONTENT OF SOME LITTORAL FUCOIDS

	Ether-extracted fat %	True fat %
Pelvetia canaliculata	4·88	3·6
Ascophyllum nodosum (mid-littoral)	2·87	
Fucus vesiculosus	2·60	1·9
Halidrys siliquosa (low littoral)	2·18	
Himanthalia lorea	1·21	
Laminaria digitata (sub-littoral)	0·46	0·46

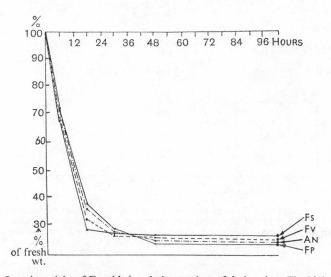

Fig. 20.4 Loss in weight of Fucoids in relation to time of desiccation. The higher an alga grows the slower it loses water and the greater the total loss. (Fs = *Fucus serratus*, Fv = *F. vesiculosus*, An = *Ascophyllum nodosum*, Fp = *Fucus spiralis* var. *platycarpus*) (after Zaneveld)

In the fucoid *Hormosira* the rate of water loss is correlated with bladder size (see below), the larger the bladder the slower the rate of water loss (Fig. 20.5). Plants with large bladders occur at the higher levels on the shore so that morphology provides protection in this species.

For some years now detailed studies have been made of water loss from selected New Zealand inter-tidal algal under varying conditions of exposure. In all cases the water loss is related to the maximum exposure time. A summary of results is set out in Table 20.2 (Chapman and Brown, 1966).

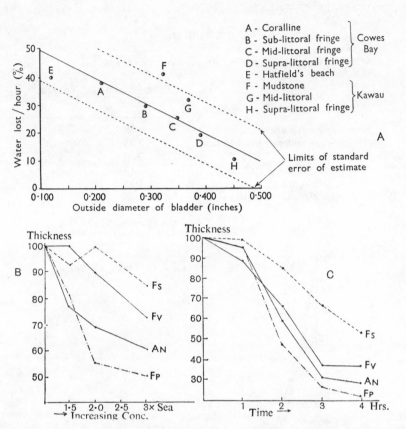

Fig. 20.5 A, rate of water loss per hour in *Hormosira*; B, decrease in diameter of cell walls when placed in sea water of increasing concentration; C, decrease in diameter of cell walls under normal conditions of exposure. (Fs = *Fucus serratus* Fv = *F. vesiculosus*, An = *Ascophyllum nodosum*, Fp = *Fucus spiralis* var. *platycarpus* (A, after Bergquist; B, C, after Zaneveld)

In the final column of Table 20.2 the probable water loss is that which is estimated to occur under summer conditions and it is believed this is the principal causal factor determining upper limits. With *Caloglossa* the high humidity in the mangrove swamps and in rock cracks reduces water loss to a level that the delicate thalli can tolerate. The high humidity of the mangrove swamps also reduces water loss in *Hormosira*. In the case of the two species of *Carpophyllum* successful transplants of adult plants to higher levels suggests that it is water loss from the sporelings that determines their upper limit.

A further feature that has received very little attention is the effect of desiccation upon the major metabolic processes, respiration and photosynthesis. A study upon *Hormosira* (Bergquist, 1957) has shown that the effects may be quite different between sporeling and adult (Fig. 20.6). This

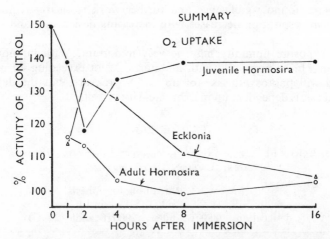

Fig. 20.6 Respiration rate of juvenile and adult *Hormosira banksii* and *Ecklonia radiata* after desiccation and re-immersion. Note that the adult *Hormosira* and *Ecklonia* return to the control value (non-dehydrated plants) but the juvenile *Hormosira* does not. (after Bergquist)

TABLE 20.2

Species	Max. Exp upper limit (Inter-tidal) (h)	Max. Water loss, upper limit (plant dead or barely viable) %	Probable water loss in nature at upper limit %
Caloglossa leprieurii	8·8	70	35–40
Gelidium caulacantheum	8·8	60	50
Hormosira banksii			
Estuarine	6	80	50–60
Mangrove	6	80	20–35
Rock	4–4·5	80	50–60
Pool	—	70–75	—
Scyothamnus australis	5–6	70	60–70
Carpophyllum maschalocarpum			
Sporeling	1·5	90	60
Adult		60	50
Carpophyllum plumosum			
Sporeling	1–5	65	40
Adult		40	30

may be of profound significance and further work is clearly required. In *Scytothamnus* respiration decreases with increasing desiccation and lowered humidity.

During exposure humidity may be very important. With humidities of 80 per cent r.h. or more and with a water loss not exceeding 40 per cent *Hormosira* will photosynthesise for up to 3 h exposure, but the degree of photosynthesis is dependent upon light intensity as well.

TABLE 20.3

RELATIONSHIP BETWEEN LIGHT INTENSITY, PHOTOSYNTHESIS AND DESICCATION IN *Hormosira*

	Rate of photosynthesis ml O_2/g.d.wt. in 1 h	
% desiccation	880 ft-candle	2000 ft-candle (20° C)
0	2·5	21·5
10	0·7	10·8
25	7·7	13·6

Similarly with the two species of *Carpophyllum* there is some photosynthesis at high humidities during exposure. *Scytothamnus*, on the other hand, shows assimilatory balance only at 100 per cent r.h. and high light intensities (Brown and Johnson, 1964). In *Caloglossa* there is no photosynthesis below 60 per cent r.h. or a desiccation value of 35 per cent.

Relatively little study has been made of the effect of subsequent reimmersion upon photosynthesis and respiration. In the case of adult *Hormosira* the immediate effect of submergence is to cause a rise in photosynthesis. This rise is dependent upon (1) the ecological form of the algae, (2) degree of desiccation, (3) intensity of illumination (Chapman and Brown, 1966). With *Caloglossa leprieurii* photosynthesis on resubmergence at 2000 ft-candle decreased with increasing desiccation up to 40 per cent water loss after which photosynthesis became disorganized (35–40 per cent water loss is probably the maximum sustained in nature, see Table 20.2). At lower light intensities a greater degree of water loss could be tolerated. *Gelidium caulacantheum*

TABLE 20.4

PHOTOSYNTHESIS OF RE-SUBMERGED *Scytothamnus* AT 2000 FT-CANDLE

% desiccation of submerged plant	O_2 output μl/g d.wt. in 1 hour	O_2 input μl/g d.wt. in 1 hour
	433	60
10	376	56
20	324	54
30	336	53
70	226	12

continues to photosynthesize after 8 h exposure and 50 per cent water loss. *Scytothamnus australis* shows a decrease in photosynthesis on re-immersion, but even up to 70 per cent desiccation the decrease is not more than half that of the control (Brown and Johnson, 1964).

The water relations of fucoids and other large algae may be compared with the *Enteromorpha* and Chrysophyceae belts of the Dover cliffs. Anand (1937) found that the *Enteromorpha* mat lost 25 per cent of its moisture in the first 3 h of exposure, whilst the Chrysophyceae belt lost 8·4 per cent (Fig. 20.7). The measurements were obtained by the simple method of weighing

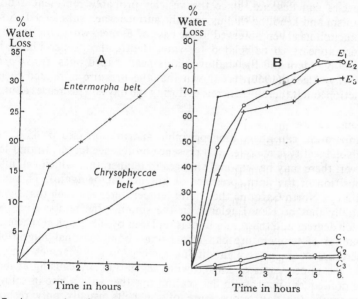

Fig. 20.7 A, water loss from samples of *Enteromorpha* and Chrysophyceae belts when exposed in their original position on the cliff face; B, water loss during drainage in nature from different levels in the *Enteromorpha* and Chrysophyceae belts during successive hours in winter. (E_1–E_3, C_1–C_3 = successive levels. Water loss in A and B expressed as % of that originally present.) (after Anand)

portions of the mat cut out so that they fitted into waterproof paper dishes which could be put back into position on the shore. The reduced loss from the Chrysophyceae belt as compared with the *Enteromorpha* is due to the gelatinous nature of the Chrysophyceae.

PHYSICAL FACTORS

(a) Substrate

The nature of the substrate, whether solid rock, boulders, pebbles, sand, mud or peat, is of fundamental importance in anchoring the plant and the general aspect of the flora is determined to a large extent by this factor. It is essentially a presence or absence factor. On rocky coasts the angle of slope and the presence of cracks may affect the local occurrence of some species, either operating as a presence or absence factor or as a modifying factor. The geological nature of the rock may also be important, e.g. soft rocks such as sandstone rarely carry large algae.

(b) Pressure

Normally this is not an important factor, but Damant (1937) has shown that in the case of vesicular fucoids there is a limit to the depth at which bladders will exist because the gas is forced out by the pressure of the water forces. This depth will therefore mark the approximate lower limit to which the species can descend, since the vesicles probably represent a flotation mechanism and keep the thallus near the surface when submerged. A similar phenomenon has been observed in the case of *Egregia* where the shape of the bladders appears to be related to water depth, the deeper plants having spherical bladders and the shallow plants pear-shaped ones. In *Ascophyllum* and *Fucus*, Aleem (1969) has shown that the pressure is related to oxygen production and it therefore increases during the day and decreases at night.

(c) Temperature

Temperature can affect biogeographic distribution (see p. 459), and for any given locality it operates as a presence or absence factor. In that locality, however, there may be seasonal changes in temperature which can affect the composition of the flora (presence or absence factor again). Thus at Cape Lookout in North Carolina the winter temperatures are such that the flora is essentially that of New England: in the summer the temperatures rise by eighteen degrees and the winter flora is replaced by sub-tropical species.

Changes of temperature can also bring about seasonal migrations, i.e. temperature operates as a modifying factor. On salt marshes temperature affects the upper levels of low-growing vegetation, by causing evaporation and a consequent increase in desiccation together with a rise in salinity. On rocky shores the day temperature of the belts usually only responds to changes of air temperatures during the summer, and then it is always less than that of the air. The temperature change is least in belts that retain more moisture, e.g. Cyanophyceae, Chrysophyceae. If, at high levels, the period of insolation is at all long in the summer then mats of Cyanophyceae, Chrysophyceae and *Enteromorpha* become cracked and fall off.

Most workers agree that temperature is a very important factor in rock pools and probably also in salt pans. While a pool is exposed the temperature of such a small body of water may rise considerably, especially in the case of pools at high levels, and then when the tide returns the cold sea water will lower the temperature very suddenly. An examination of pool floras has shown that Rhodophyceae tend to be more abundant in shaded pools whereas Chlorophyceae and Phaeophyceae are relatively more abundant in unshaded pools.

Rock pool temperatures probably rarely cause actual damage, except perhaps in high level pools, but respiration may be so speeded up that storage reserve breakdown is not made good by synthesis. Biebl (1937) found that warming up to 26° C over a period of 24 h has no effect on most Rhodophyceae and certain arctic Chlorophyceae (Biebl, 1969), and changes of 12° C could occur sharply without causing any damage. Tropical intertidal algae can tolerate up to 40° C (Biebl, 1962) and arctic algae up to 22° C (Biebl, 1970).

Work with *Fucus vesiculosus* and *Hormosira banksii* has shown that respiration rate decreases with temperature lowering but that with the former species it is still measurable at − 15° C (Kanwisher, 1957).*

Different species of *Fucus* exhibit varying temperature sensitivities broadly related to latitude (Biebl, 1970) and the belts they occupy, e.g. the order of decreasing sensitivity for Russian fucoids is: *Fucus filiformis* — *F. vesiculosus* (littoral) — *F. distichus* — *F. vesiculosus* (sub-littoral) — *F. serratus*. In at least three species (*F. vesiculosus, F. serratus* and *Ascophyllum*) heat resistance is higher in summer than in winter.

In laminarians temperature affects not only the macroscopic sporophyte, and parts of it, but also the microscopic gametophyte. The highly dissected appendages on the long 'boa' fronds of deep-water plants of *Egregia laevigata* (see p. 441) show a very marked response in apparent photosynthesis rate to temperature changes. Juvenile 'leaves' behave very differently as shown in Table 20.5 (Chapman, 1964).

TABLE 20.5

RELATIONSHIP BETWEEN PHOTOSYNTHESIS AND TEMPERATURE IN
LEAVES OF EGREGIA LAEVIGATA

Organ	10° C.	15° C.	20° C.	25° C.
Dissected 'leaves'	130	375	1175	630
Juvenile 'leaves'	670	725	700	965
Basal 'leaves'	530	540	265	145
Dissected 'leaves' (shallow plant)	600	415	500	707

The growth of *Laminaria digitata* in Norway is related to summer and winter temperatures as well as to light intensity (Sundene, 1964). Growth in *L. japonica* only occurs with sea temperatures between 0° C and 15° C. With *Alaria esculenta* the sporophyte cannot tolerate a sea temperature of 20° C and growth is poor at temperatures above 16° C, while below this temperature growth rate can be reduced by reductions in salinity. It must be emphasized that no one factor is solely responsible for changes that occur.

Sea-water temperature affects not only growth of dwarf gametophytes but also their fertility (see p. 227). At sea temperatures of 10° C the minimum light intensity for growth of gametophytes of *Laminaria hyperborea* is 20 lx whilst the saturating light intensity is 350 lx. At 17° C temperature the corresponding saturating light intensity is between 1000–2000 lx. In the Chordaceae growth of gametophytes of *Chorda tomentosa* is best at low temperatures and they do not become fertile at temperatures exceeding 10–12° C,

* Biebl (1969) has shown that *Enteromorpha (Blidingia) minima* and *Ulothrix zonata* tolerate periods at − 22° C without injury.

so that distribution of the adult is determined by these minimal winter temperature requirements.

It seems that algae of the Arctic (and Antarctic) can tolerate as much as 80 per cent of the internal water being frozen (with the consequent rise of internal concentration) and metabolic processes do not stop. After many months frozen into sea ice they must be capable of photosynthesis immediately upon being thawed out.

(d) Illumination

There are problems here involving seasonal variation in light intensity, diurnal variations in relation to times of high water, and the actual light intensity and spectrum at different depths. Practically nothing is known about the first two effects, and it is impossible therefore to assess their importance. It can, however, be pointed out that differences in illumination intensity at different latitudes appear to determine the maximum depth to which algae can descend. Thus in the North Sea the limit appears to be about 40–50 m, in the tropics it is about 100 m whilst in the clear waters of the Mediterranean it is about 130–180 m. Many of the deep-growing algae (and possibly others) are fully light-saturated at quite low values. Blades of *Macrocystis* are fully saturated at 1000 ft-candles ($\frac{1}{10}$ full sunlight of California) and parts of *Egregia* at 1400 ft-candles.

The third effect is undoubtedly important because measurements show that the incident light is cut down very considerably at even a depth of 1 m, while at 2 m only about 25 per cent of the surface light can penetrate. Waters of different turbidity may absorb light quite differently (Fig. 20.8A) and this in turn affects the relative photosynthetic rate and the compensation point level (Fig. 20.8B). During its passage through sea water the spectral composition of the light changes considerably, the red and yellow part of the spectrum being preferentially absorbed, such that the blue and green portions assume greater importance. Algae living in the sublittoral fringe and the upper portion of the sublittoral are subject to very different conditions as compared to algae living higher on the shore, so far as photosynthesis is concerned. There is also the different effect of submergence and emergence upon photosynthesis, the process being affected not only by the light change but also by the moisture and temperature changes. Algae can be found at 100 m depth in the arctic where the light intensity is extremely low and it has been suggested that such algae must be heterotrophic (Wilce, 1967).

It is also evident that temporary changes in light intensity as a result of cloudiness or chop on the water may affect photosynthesis. The outer edges of *Macrocystis* beds in California often occur at depths of 12–20 m in turbid water but at 27·5–30 m or more in clearer waters. Tschudy (1934) first showed that maximum photosynthesis occurred at the surface on choppy days but at about 5 m on calm days. Stocker and Holdheide (1937) used *Fucus spiralis* var. *platycarpus* and *F. serratus* where the difference between sunny and cloudy days was extremely pronounced (Fig. 20.9). It would seem that on exposure the fall off in assimilation rate on sunny days is correlated with water content because the exposed thalli quickly dry up and cease to assimi-

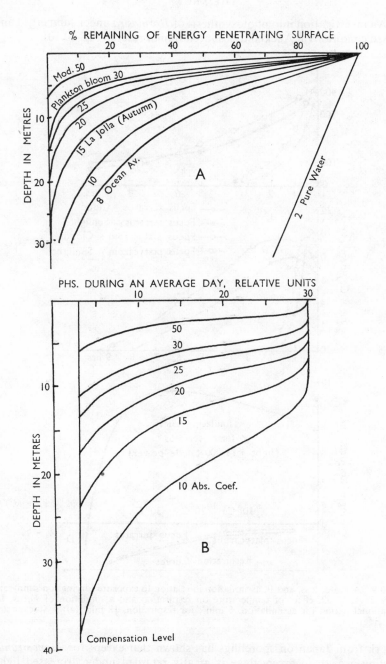

Fig. 20.8 A, % light absorption in sea water of different absorption co-efficients; B, relative photosynthesis in *Macrocystis* on an average day in waters of different absorption coefficients (A, above). (after North)

late. An investigation into photosynthesis of *Hormosira* under submerged and
exposed conditions at Auckland produced similar results (Fig. 20.10).

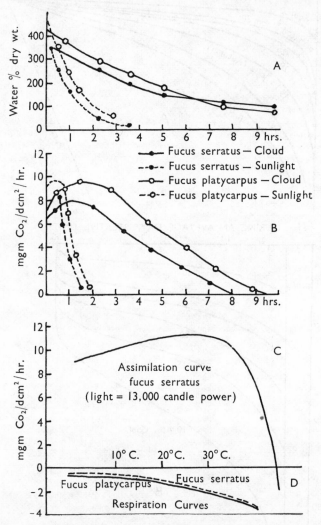

Fig. 20.9 A, water loss, and B, assimilation in relation to exposure (drying) on sunny and
cloudy days; C, D, effect of temperature on respiration and assimilation of *Fucus*. In-
vestigational period for assimilation, 5 min.; for respiration, 18 min. (after Stocker and
Holdheide)

Work from Japan on sporelings has shown that except for *Monostroma*,
sporeling growth in green algae is greatly retarded under decreased light,
while that of red and brown algae tends to be increased. Sporelings of green
algae also grew more rapidly under yellow, red or blue light than under green

light. Length of the light period may also be involved. Thus *Enteromorpha* sporelings grow better under long-day conditions, whereas those of *Monostroma* grow best at first under medium day length but that later short days are optimal.

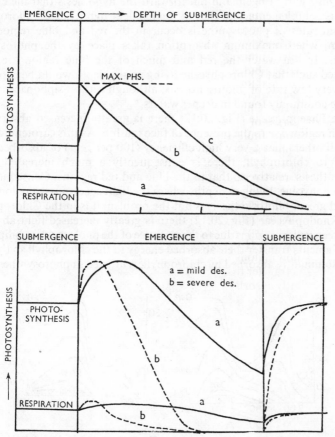

Fig. 20.10 Top: effect of light intensity upon photosynthesis of exposed and submerged *Hormosira banksii* in relation to depth. (a = low light intensity, b = optimum light intensity, c = supraoptimum light intensity); Bottom: effect of desiccation upon photosynthesis (full line) and respiration (broken line) of *Hormosira banksii* when exposed. (after Trevarthen)

At low light intensities there is little difference in photosynthesis between adult and sporeling estuarine *Hormosira banksii*. With increasing light intensity the juvenile plants respond very much more rapidly than do the adults. *Carpophyllum plumosum* sporelings, on the other hand, have a higher photosynthetic rate than the adults at comparable light intensities. Maximum efficiency occurs around 1500 ft. candles and this suggests that the species is essentially a shade alga. With *Laminaria hyperborea* the gametophytes and sporelings are distinct shade plants with light saturation at 80 μcal cm^{-2} sec^{-1}, while the adult sporophytes are not saturated until 400μcal cm^{-2} sec^{-1} (Kain, 1966).

There has been considerable discussion over many years in connection with the pigment composition of the algae in relation to photosynthesis and depth. Quite early Engelmann put forward the hypothesis that the colour of an alga is complementary to that of the incident light. In the Chlorophyceae maximum rates of photosynthesis occur in the red and blue region of the spectrum, where maximum absorption takes place by the photosynthetic pigments. In sea water the red and much of the blue region are rapidly eliminated such that Chlorophyceae living at any depth would be expected to have a very low rate of photosynthesis, although certain siphonaceous green algae are commonly found in deeper waters.*

In the Phaeophyceae (Fig. 20.12) there is greatly increased absorption in the green region due to the presence of fucoxanthin. As this carotenoid, unlike nearly all others, has a very high efficiency (100 per cent) of absorbed energy transfer to chlorophyll, there is consequently a much increased rate of photosynthesis, relative to that in the blue and red regions. Consequently for the Phaeophyceae, living at depths where the light spectrum is predominantly blue and green, photosynthesis is not the problem it is in the Chlorophyceae. In the Rhodophyceae (Fig. 20.12) there is greatly increased light absorption in the green yellow regions due to the presence of the photoreactive biliproteins. These pigments transfer their absorbed energy to the chlorophyll so that light, not predominantly absorbed by chlorophyll, is utilized in photosynthesis.

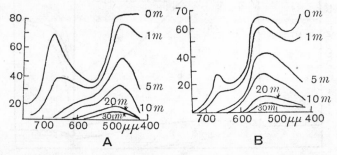

Fig. 20.11   Absorption curves of A, *Monostroma*, B, *Delesseria* at different depths. (after Seybold)

In all algal photosynthetic studies consideration must be given to the absorption curves and the action spectra. The latter is the more important because it indicates the extent to which the different portions of the absorbed spectrum are utilized in photosynthesis. Fig. 20.11 illustrates the absorption spectra of a green and red alga at different depths, and it will be seen that in *Monostroma* there is a rapid decrease in the absorption of the red wave lengths as depth increases. Fig. 20.12 illustrates absorption and action spectra for algae at different wave lengths.

* *Udotea* sp. attained max. photosynthesis at 15 m, and was found at 75 m. The compensation depth was 80 m (Drew and Larkham, 1967).

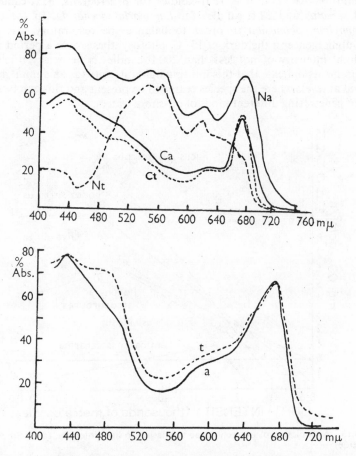

Fig. 20.12 Absorption (a) and action (t) spectra of marine algae. Top: *Coilodesme californica* (Ca, Ct) and *Smithora* (*Porphyra*) *naiadum* (Na, Nt); Below: *Ulva taeniata* (a, t). (after Haxo and Blinks)

Fig. 20.13 illustrates the rate of CO_2 assimilation by algae from all three groups under different light intensities. All exhibit an optimum light intensity for assimilation, which in the case of the littoral algae is in the same region as that of cormophytic land plants. In sub-littoral algae, such as *Laminaria saccharina*, the optimum is at a much lower light intensity. Because certain algae are more efficient metabolically at low light intensities or when the spectral composition of the light has been modified, it has been suggested that it is possible to recognize sun and shade algae. While there may be a definite ecological value in the distinction, nevertheless further experimental work is desirable before any final conclusions are reached. The compensation depth or light intensity at which photosynthesis just balances respiration may be very important. This differs for various species and is also dependent on

439

temperature. At 15° C it is 15 ft-candles for *Macrocystis*, 38 ft-candles for *Fucus serratus* and 32 ft-candles for *Laminaria saccharina*. In the case of *Carpophyllum plumosum* in order to balance the respiratory loss during 13 h/submergence in the dark at 13° C, photosynthesis over a period of 11 h at a light intensity of not less than 200 ft-candles is necessary. Field light measurements indicate that this limiting light intensity (as an annual mean) is reached at levels where the species ceases to be present and this can be related to light penetrating different kinds of off-shore waters.

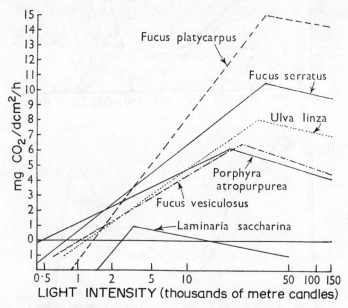

Fig. 20.13 Assimilation of different algae in relation to light intensity. (after Stocker and Holdheide)

In the case of large oarweeds, variations within the different organs may be considerable. Reference (p. 433) has already been made to the difference in

TABLE 20.6

PHOTOSYNTHESIS/RESPIRATION RATIOS FOR PARTS OF *Macrocystis* PLANTS (FOR CLENDENNING AND SARGENT, 1957)

	Photosynthesis/Respiration
Naked Stipe	4·1
Pneumatocyst (bulb)	7·1
Sporophyll	9·2
Apical blade to blade 20	11·1 (6·5–18·7 range)
Blade 21-blade 61	18·6 (10–29 range)
Blade 62-blade 120	22·0 (13–34 range)

apparent photosynthesis between juvenile, medium dissected and basal entire leaves of deep-water plants of *Egregia laevigata*. The same phenomenon, even more pronounced, is exhibited by different parts of *Macrocystis pyrifera* (Table 20.6).

For this species it is quite evident that the rate of photosynthesis increases with age of the blade. In the case of *Egregia* (p. 433) it is the middle-aged dissected appendages that exhibit the highest rate. Some recent work on *Macrocystis pyrifera* in California has shown that the net photosynthesis varies not only with depth, as evidenced by the rates exhibited by blades from different positions on the plant, but also with temperature.

TABLE 20.7

RELATIONSHIP BETWEEN PHOTOSYNTHESIS AND TEMPERATURE IN *Macrocystis* BLADES

Blade No.	Photosynthetic capacity ml O_2 per dcm³ per h					
	15° C			22·5° C		
	Plant 1	Pl. 2	Pl. 3	Pl. 4	Pl. 5	Pl. 6
					(inter-mediate depth)	(young plant at 50 ft.)
1 (apex)	1·33	1·52	1·45	·	·	1·74
9	·	·	·	0·89	·	·
20	2·62	2·83	1·83	0·64	·	1·74
40	2·94	2·66	3·28	0·68	2·4	2·38
50	·	·	·	·	·	2·85
60	3·44	3·04	4·02	1·35	3·08	·
80	4·09	3·60	4·46	1·99	3·42	·
99	4·77	2·74	4·92	2·44 (94)	3·83 (94)	·
109	5·14	·	·	·	·	·
120	5·28	3·4	5·12	·	·	·

Plants 1–4 all reached the water surface.

Ehrke (1931) suggested that there is a correlation between the temperature at which maximum assimilation takes place and the average temperature of the month of maximum development.

The problem, however, is not as simple as this. Thus Lampe (1935) found that in winter the assimilation rate of *Fucus serratus* plants rises when it is measured in sunlight under conditions of increasing temperature (Fig. 20.14); on the other hand, in the case of a red alga such as *Porphyra*, when the temperature is raised above 15° C the assimilation curve is lowered immediately in diffuse light and after seven days in sunlight. This suggests that *Fucus* is an eurythermal species, tolerating a wide range of temperature, while *Porphyra* is stenothermal, tolerating a narrow range. Comparable results for *Fucus serratus* were obtained by Hyde (1938), who found that between 15° C and 20° C the assimilation rate could be increased by raising the light intensity, and that there was a certain light value (2×500 lx) which yielded an optimum in the rate of assimilation (Fig. 20.15).

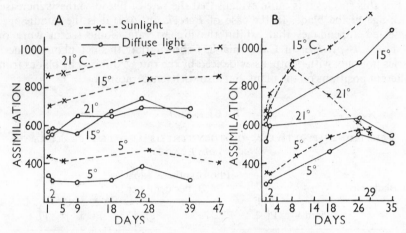

Fig. 20.14 Daily drift in assimilation of algae at different temperatures in sunlight and diffuse light. A, *Fucus serratus* winter plant; B, *Porphyra*. (after Lampe)

After a period of exposure there is the problem of return to normal assimilation rate, because experiments show that the assimilation rate is markedly reduced during exposure (see also p. 430). When reinundated, species such as *Ulva linza* and *Porphyra umbilicalis* take up water at once, and very soon are assimilating at their normal rate. The members of the Fucaceae behave differently and their behaviour can be correlated directly with the level they occupy on the shore and the periods of exposure they undergo.

There would seem little doubt that illumination and its effect on photosynthesis is an important causal factor, and probably is largely responsible for determining the lower limits of a number of algae. It is, however, a very complex factor, and much further work is necessary before its operation is fully unravelled.

CHEMICAL FACTORS

(a) Salinity

This may operate as a presence or absence factor in places where fresh water runs into the sea or where there are coastal lagoons as in The Gulf of Mexico (Conover, 1964). In such places Rhodophyceae and Phaeophyceae tend to be replaced by Chlorophyceae. In estuaries *Fucus spiralis* is replaced by *F. ceranoides*. Salinity changes due to loss of water also occur when the algae are exposed and this results in changes in osmotic pressure of the cell sap. Biebl (1938, 1970) found that algae could be placed in three groups according to their behaviour on exposure:

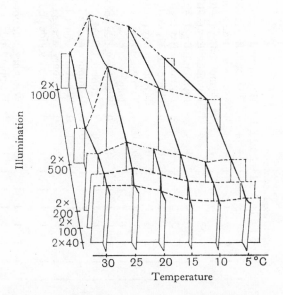

Fig. 20.15 Diagram of paper model to show the combined effects of light and temperature on the rate of apparent assimilation of *Fucus serratus*. (after Hyde)

   (i) Sublittoral algae which are never exposed to the air; they are resistant to a concentration of 1·4 times that of sea water.
  (ii) Algae of low-water mark and the lower littoral tide pools which never become dry; these are resistant up to a concentration of 2·2 times that of sea water.
 (iii) Algae of the littoral belt; these are often completely exposed and they can resist a concentration of 3·0 times that of sea water.

The influence of lowered salinity upon metabolic processes (see Doty, 1957) requires further study, especially with respect to the depth compensation point.

Salinity changes also occur in pools and salt pans, especially those at high levels, during neap tidal periods when there may be considerable evaporation. This factor can operate as a presence or absence factor in high-level pools. In pools, intertidal algae exhibit a greater tolerance to changes in salinity than do sublittoral algae (Table 20.8), and this may well explain why the latter are restricted to pools in the lower half of the littoral zone.

(b) Substrate

Generally, differences in chemical composition have little effect on the flora. An exception is the presence of chalk, the algal flora of chalk cliffs (Anand, 1937) differing distinctly from that of metamorphic and igneous rocks.

TABLE 20.8

OSMOTIC RESISTANCES OF RHODOPHYCEAE FROM DIFFERENT HABITATS (AFTER BIEBL)

ALGAE	Diluted sea water										Concentrated sea water											Habitat and depth
	00	·1	·2	·3	·4	·5	·6	·7	·8	·9	1·2	1·3	1·4	1·5	1·6	1·7	1·8	1·9	2·0	2·2	3·0	
Heterosiphonia plumosa	+	+	0	0	0	0	0	1	1	1	1	1	0	+	+	+	+	+	+	+	+	Plymouth Bay (8–10m.)
Heterosiphonia plumosa	+	+	+	0	0	1	1	1	1	1	1	1	1	0	+	+	+	+	+	+	+	Wembury, tide pool
Polyneura hilliae	+	+	+	+	0	1	1	1	1	1	1	1	0	0	+	·	·	·	·	·	·	Plymouth Bay (8–10m.)
Cryptopleura ramosa var. uncinatum	+	0	0	0	0	0	1	1	1	1	·	·	·	·	·	·	·	·	·	·	·	Plymouth Bay (8–10m.)
Brongniartella byssoides	+	0	0	0	1	1	1	1	1	1	1	1	·	·	0	0	0	?	?	?	+	Plymouth Bay (8–10m.)
Phycodrys rubens	+	0	0	0	0	0	1	1	1	1	1	1	1	1	0	+	+	+	+	+	+	Plymouth Bay (8–10m.)
Antithamnion tenuissimum	+	+	0	0	1	1	1	1	1	1	1	1	1	1	1	1	1	1	0	+	+	Plymouth Bay (8–10m.)
A. plumula	0	+	+	+	1	1	1	1	1	1	1	1	1	1	0	0	0	0	0	+	+	Tromsö, 30 cm.
A. cruciatum	+	0	0	1	1	1	1	1	1	1	1	1	1	1	1	1	1	1	1	1	+	Naples
Polysiphonia urceolata	+	0	0	1	1	1	1	1	1	1	1	1	1	1	1	1	1	1	?	?	0	Wembury, tide pool
Polysiphonia urceolata	+	+	1	1	1	1	1	1	1	1	1	1	1	0	0	0	0	0	1	1	?	Tromsö, L.T.M.
Membranoptera alata	0	0	0	0	1	1	1	1	1	1	1	1	1	0	1	1	1	0	0	0	0	Wembury, above L.W.M.
Ptilota plumosa	0	0	0	1	1	1	1	1	1	1	1	1	1	1	1	1	1	1	1	1	0	Wembury, above L.W.M.
Ceramium ciliatum	+	0	0	0	1	1	1	1	1	1	1	1	1	1	1	1	1	1	1	0	0	Wembury, above L.W.M.
Callithamnion tetragonum var. brachiatum	+	1	1	1	1	1	1	1	1	1	1	1	1	1	1	1	1	1	1	?	0	Wembury, tide pool
Spondylothamnion multifidum	+	+	0	1	1	1	1	1	1	1	1	1	1	1	1	1	?	0	0	+	+	Wembury, tide pool
Griffithsia flosculosa	+	+	+	0	0	0	0	0	1	1	1	1	1	1	1	1	1	1	0	0	+	Wembury, tide pool
Griffithsia furcellata	+	+	+	0	+	0	0	1	1	1	1	1	·	0	0	·	·	·	·	·	·	Naples
Griffithsia opuntioides	·	·	·	·	·	·	0	0	0	1	1	1	0	0	0	0	·	·	·	·	·	Naples, 60 cm. up.
Nitophyllum punctatum	+	+	+	0	0	0	0	1	1	1	1	1	0	0	0	0	0	0	0	+	+	Wembury, tide pool

(c) pH (Acidity)

The pH of sea water normally ranges from about 7·9 to 8·3 and the factor is of no great importance for littoral algae though it might be of significance in some of the higher rock pools where, during the day, the pH may rise to 10. Most algae that have been investigated tolerate a pH range of 6·8–9·6 and the pH of pools is rarely outside this range.

(d) Oxygen Content

The oxygen tension in sea water is normally low but sufficient for metabolic purposes. In pools, during the day the oxygen content may rise substantially.

(e) Nutrients

This factor is of importance in controlling the seasonal periodicity of marine and freshwater plankton (see p. 404). It is also of importance in lakes where oligotrophic and eutrophic water differ chemically, and this difference is reflected in the vegetation (see p. 397). High-level rock pools well supplied with bird droppings commonly support a very rich flora of microscopic algae.

BIOLOGICAL FACTORS

(a) Animals

Competition between plants and animals may in certain cases be sufficiently severe to eliminate some algae, it therefore operates as a presence or absence factor. If *Patella* (limpet) is present in abundance it may eat up the *Enteromorpha* felt and so stop *Fucus* sporelings from becoming established. The attacks of boring mollusca are probably responsible for the eventual detachment of *Durvillea* plants, and in Europe *Helicion pellucidum* is said to bring about the detachment of *Laminaria saccharina* plants. On salt marshes the mollusc *Hydrobia ulvae* can cause havoc among beds of *Ulva*. Sea urchins, if present in abundance, can completely clear rock pools of nearly all algae and, in conjunction with other factors, may help to set the lower limit, e.g. grazing of *Carpophyllum plumosum* by *Evechinus chloroticus* and *Lunella smaragda* in New Zealand probably helps to determine the lower limit in the clearer waters. Removal of *Echinus esculentus* increases the density of *Laminaria hyperborea* from 5·1/m² to 22·7/m² (Jones and Kain, 1967). Operating in the other direction is the occurrence of the red alga *Pachymenia himantophora* in relation to the prior presence of barnacles (Crossland, 1968).

(b) Plants

Many of the possible relationships here have not been worked out. There are, of course, the host-parasite and host-epiphyte relationships, because some of the epiphytes appear to be restricted in their host requirements. The establishment of *Fucus* sporelings also seems to depend very largely upon the pre-development of a green algal felt (*Enteromorpha*), whilst the establishment of *Porphyra tenera* sporelings on the bamboo poles or nets in Japanese harbours is dependent upon the development of a diatomaceous film.

In the case of epiphytes certain species at least may be restricted to specific parts of large host plants. Such epiphytes can be segregated into epithallic, epicaulic (stem), epiphyllic ('leaf'), epirhizoidic, ephihapteric (Beth and Linskens, 1964).

SUMMARY

Despite the great variety of factors that can and do operate on a maritime shore, there is nevertheless a uniformity in the belts occupied by the major organisms. Furthermore if the number of upper and lower limits occurring at any one level are calculated, it is found that there are certain levels at which the number of limits is greater than elsewhere. This suggests that there is some change in a major factor or factors at that level which affects a number of organisms. Such levels have been termed critical levels, and while a number have been established for different places there is general agreement that the following critical levels are widespread:

(a) between mean and extreme low-water marks of spring tide
(b) around mean low-water mark of neap tides
(c) around extreme high-water mark of neap tides.

This uniformity suggests that the physiographic or tidal factor is paramount. However, if the zones occupied by major algal species in different localities are compared, after allowance is made for splash and wave action, it is found that they do not always coincide (Fig. 20.2), so that even if the tidal factor is the major causal factor its effect is modified by other local factors.

One very important factor is the inter-relationship between respiration and photosynthesis. Normally the latter is in excess of the former, but depending on light intensity in the water, the two rates can coincide at different temperatures and at different depths. This is known as the compensation point and below that depth, if conditions remained uniform, an alga could not survive because breakdown would exceed build up. There will be, for each species, some level at which the periods below the compensation point exceed the periods above. This will effectively form a lower limit to the algal belt. It is likely that the lower levels of many of the species occupying the lower littoral and the infra-littoral are set by this phenomenon, but at present all too little work has been done upon it. An investigation of the littoral *Hormosira banksii*, has shown that under a low light intensity the compensation point is reached at about 1 m or less, at optimum light intensity (in relation to emergent conditions) the compensation point is reached at about 4 m, while at still higher light intensities it is not attained until nearly 6 m (Fig. 20.16). There is also evidence (Chapman and Brown, 1966) that the lower limits of *Scytothamnus australis* and two species of *Carpophyllum* are related to the compensation point.

Another means of approaching the problem was used by Klugh and Martin (1927), who studied the growth rates of various algae in relation to submergence by measuring plants and then tying them to floats which were suspended in the water at different depths. After some months the floats were pulled up and the plants remeasured. The curves (Fig. 20.17) show that maximum growth occurred between 1 and 2 m where normally under diffuse light maximum photosynthesis takes place (see p. 439).

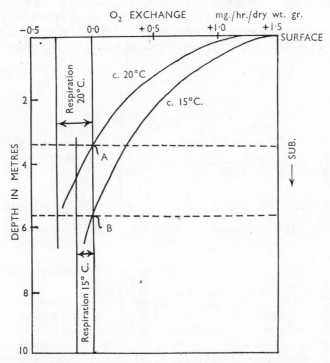

Fig. 20.16 Variation of compensation point with depth and sea temperature in *Hormosira banksii*. (after Trevarthen)

So far as the lower limit is concerned, it is possible than an alga may go deeper in the water the nearer its absorption and action spectra approach that of the shade type (i.e. algae are most efficient metabolically under low-light conditions).

Despite all that has been said in the previous pages, our knowledge of the factors controlling zonation is scanty. No real progress will be made until key species are studied exhaustively. Work of this nature has been commenced with *Macrocystis pyrifera* (Clendenning and Sargent, 1957; Clendenning and North, 1958), *Egregia laevigata* (Chapman, 1964), *Laminaria hyperborea* (Kain, 1962, 1963, 1966, 1967), *Hormosira banksii* and certain other New Zealand algae (Chapman and Brown, 1966). *Hormosira* occurs in Australasia where it is found within a range of 15° C. The belt that it occupies on the shore also rises with decreasing latitude, but the exact significance of this has yet to be worked out. Detailed studies have shown that it exists in a number of ecological forms and eventually each one of these forms will need to be studied so that the vertical limits reached under different conditions can be understood (cf. p. 253). There are several interesting free-living forms found in mangrove swamps and these occupy higher levels on the shore than do the attached counterparts on rocky shores. Shade, higher humidities during exposure and lower water loss may well be responsible for this phenomenon. A study of water loss that can occur during periods of exposure follow the trends that have been found for other fucoids. There is, however, in the form investigated, a relationship between degree of hydration and rate of respira-

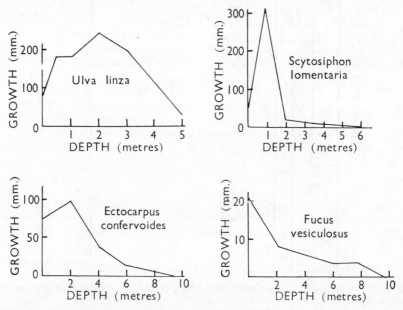

Fig. 20.17   Rate of growth of various algae at different depths in sea water, New Brunswick. (after Klugh and Martin)

tion, and it is evident that young plants differ from adult ones in this respect (Fig. 20.6). The difference is sufficient to suggest that certain metabolic paths in the young plants may be different to those in the adult. This is an aspect the implications of which will require to be explored more fully. The relationship of photosynthesis to depth and sea temperature has also been studied (Fig. 20.16) and the effect of periods of desiccation upon the photosynthetic rate (Fig. 20.10). Once the basic facts have been established, various refinements can be added. Thus the effects may vary depending upon whether high tide occurs at sunrise or whether it occurs in the middle of the day. The effects will also vary in different localities depending upon the nature of the tide and there will also be seasonal variations in quantity of light available at different depths. At the lower levels work has still to be carried out to determine the extent to which competition with other organisms may be important. At high latitudes where the alga is reaching its southern limit, competition from other algal species is probably more severe and as a factor it may be more important here than at its northern limit. It is not suggested that the final picture has yet been reached even in this case and there are further aspects that may or may not be of significance. Thus it has been found that *Hormosira*, like certain other

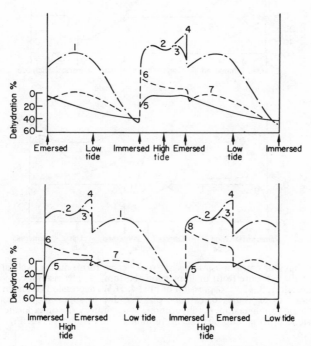

Fig. 20.18 Behaviour of *Hormosira banksii* (rocky coast form) during diurnal cycle with either low water midnt. (top) or high water (bottom). 1, phs rise on exposure; 2, phs depression at extreme H.W.; 3, phs rise prior to exposure; 4, extreme phs rise due to supra-optimal light conditions; 5, rehydration on re-immersion; 6, respiratory rise on re-immersion; 7, respiration during early phase of desiccation; 8, respiratory and phs rise on re-immersion. (phs = —·—·, water = ——, resp. = ---) (after Bergquist)

marine algae, possesses an Na^+efflux pump, removing excess sodium, and a K^+ influx pump which enables it to accumulate potassium (Bergquist, 1959). It is possible of course that either or both of these pumps may have a critical operating value, depending either on length of period of submergence or depth of submergence.

Using the data presently available it is possible to summarize the operation of the various factors schematically (Figs. 20.18, 20.19). Comparable diagrams have been prepared for other New Zealand algae: thus Fig. 20.20 shows the operation of such factors for high level algae with narrow vertical ranges (*Caloglossa leprieurii* and *Gelidium caulacantheum*), while Fig. 20.21 does the same for a low level alga (*Carpophyllum plumosum*) where the effect of

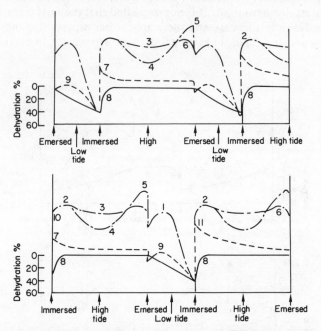

Fig. 20.19 Behaviour of *Hormosira banksii* (rocky coast form) during cycle at lower limit commencing with exposure (top) or immersion (bottom). 1, phs rise on exposure; 2, phs rise on re-immersion; 3, H.W. depression of phs; 4, H.W. depression of phs in silty water; 5, phs rise prior to exposure; 6, extreme phs rise due to supra-optimal conditions; 7, respiratory rise on re-immersion; 8, rehydration; 9, respiratory rise on exposure; 10, phs rise on re-immersion; 11, resp. and phs rise on re-immersion. (Symbols as for Fig. 20.18) (after Bergquist)

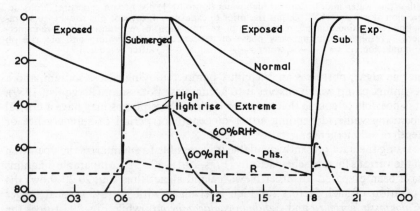

Fig. 20.20 Summarized behaviour of *Caloglossa leprieurii* during daily period. (Symbols as in Fig. 20.18)

450

mid-day exposure upon photosynthesis of the sporeling is a clear indication of the significance of this factor in setting the upper limit. At the lower limit the balance between maximum photosynthesis with a mid-day low tide and minimum photosynthesis with morning and evening low tides is probably important.

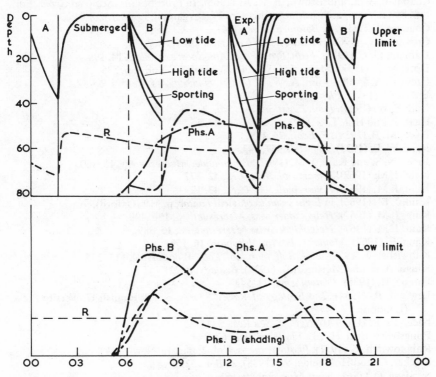

Fig. 20.21 Summarized behaviour of *Carpophyllum maschalocarpum* at upper and lower limits with low tide and mid-day (A) and early morning (B). (Symbols as for Fig. 20.18)

REFERENCES

Aleem, A. A. (1969). *Mar. Biol.*, **4** (1), 36.
Anand, P. (1937). *J. Ecol.*, **25**, 153, 334.
Ballentine, W. J. (1961). *Field Studies*, **1** (3).
Beth, K. and Linskens, N. F. (1964). *Naturw. Rdsch. Statt.*, **17**, 254.
Bergquist, P. (1957). *M.Sc. Thesis*, Auckland University.
Bergquist, P. (1959). *Physiologia, Pl.*
Biebl, R. (1937). *Bei. Bot. Ctbl.*, **57A**, 381.
Biebl, R. (1938). *Fb. Wiss. Bot.*, **86**, 350.
Biebl, R. (1962). *Protopl., Handb. d. Proto.-forsch.*, **12**, 344 p. (Springer, Vienna).
Biebl, R. (1969). *Mikroscopie*, **25**, 3.
Biebl, R. (1970). *Protoplasma*, **69**, 61.

Brown, J. M. A. and Johnston, A. (1964). *Botanica Marina*, **6** (3/4), 235–46.
Clendenning, K. A. and Sargent, M. C., (1957). *Ann. Rept. Kelp. Invest. prog.*, U. Cal. Inst. Mar. Res. 47–4.
Clendenning, K. A. and North, W. (1958). Ibid. 58–4.
Chapman, V. J. (1962). *Botanica Marina*, **3**, (3/4), 33–122.
Chapman, V. J. (1964). *Botanica. Mar.*, Vol. 1.
Chapman, V. J. and Brown, J. M. A. (1966). In *Proc. 5th Int. Seaweed Symp.* Edit. by G. Young and J. MacLachlan, p. 29 (Pergamon, London).
Conover, J. T. (1964). *Botanica. Mar.*, **7**, 4.
Crossland, C. (1968). *M.Sc. Thesis*, Auckland University.
Damant, G. C. (1937). *Fedn. Proc. Fedn. Am. Socs. exp. Biol.*, **14**, 198.
Dŏty, M. S. (1957). *Mem. geol. Soc. Am. Memoir* 67, Vol. **1**. p. 535.
Drew, E. A. and Larkum, A. W. D. (1967). *Underw. Ass. Rept. 1966/67.*
Ehrke, G. (1931). *Planta*, **13**, 221.
Gail, F. W. (1920). *Publ. Puget Sd. Bio. Sta.*, **2**.
Haas, P. and Hill, T. G. (1933). *Ann. Bot.*, **47**, 55.
Hyde, M. B. (1938). *F. Ecol.*, **26**, 118.
Isaac, W. E. (1933). *Ann. Bot.*, **47**, 343.
Jones, N. S. and Kain, J. M. (1967). *Helgoländer Meeresunters*, **15**, 460.
Kain, J. M. (1962). *J. mar. biol. Ass. U.K.*, **42**, 377.
Kain, J. M. (1963). *J. mar. biol. Ass. U.K.*, **43**, 129.
Kain, J. M. (1966). in *Light as an Ecological Factor*, p. 319 (Oxford).
Kain, J. M. (1967). *Helgoländer wiss. Meeresunters.*, **150**, 409.
Kain, J. M. (1967). *Helgoländer wiss Meeresunters.*, **15**, 489.
Kann, E. (1963). *Memoire. Ist. Ital. Idrobiol.*, **16**, 153.
Kanwisher, J. (1957). *Biol. Bull. mar. biol. Lab.*, Woods Hole, **113**, 275.
Klugh, A. B. and Martin, J. R. (1927). *Ecology*, **8**, 221.
Lampe, H. (1935). *Protoplasma*, **23**, 543.
Lewis, J. R. (1964). *The Ecology of Rocky Shores*, p. 150 English University Press, (London).
Nienhuis, P. H. (1970). *Neth. J. Sea Res.*, **5** (1), 20.
Pringsheim, E. G. (1923). *Fb. Wiss. Bol.*, **62**, 244.
Rabinowitch, E. (1951). *Photosynthesis*, Vol. **2**, part 1, (New York).
Stocker, O. and Holdheide, W. (1938). *Z. Bot.*, **32**, 1.
Sundene, O. (1961). *Nytt Mag. Bot.*, **9**, 155.
Sundene, O. (1964). *Nytt Mag. Bot.*, **11**, 33.
Tshudy, R. H. (1934). *Am. J. Bot.*, **21**, 546.
Wilce, R. T. (1967). *Bot. Marina.*, **10**, 185.
Zaneveld, J. S. (1937). *J. Ecol.*, **25**, 431.

21

GEOGRAPHICAL DISTRIBUTION, LIFE FORM

The problem of algal biogeography is difficult because of the unreliability of earlier records and the somewhat scanty literature, especially for tropical and sub-tropical areas. Those studies that have been published have established certain general features which are summarized below:

(1) There is a general resemblance between the algal floras of the West Indies and the Indo-Pacific. Vicarious pairs of species (i.e. two separate species closely related taxonomically and yet widely separated geographically) are known and even vicarious generic groups. Examples are the several vicarious pairs in the Dasycladaceous genus *Neomeris* (Fig. 21.1). Murray's suggestion that such discontinuities were caused by major changes of climate in former epochs, would only appear to explain some cases, e.g. certain species in the Laminariaceae (cf. below), but it is equally obvious that the present factors do not provide an adequate explanation. It seems best to postulate migration during an earlier epoch when there was a sea passage through the Panama isthmus, and this involves a migration not later than the Cretaceous.

(2) There are some species which are common only to the Western Atlantic and the western part of the Indian Ocean around Madagascar, e.g. *Chamaedoris peniculum* and three species of *Cladocephalus* (cf. Fig. 21.2).

453

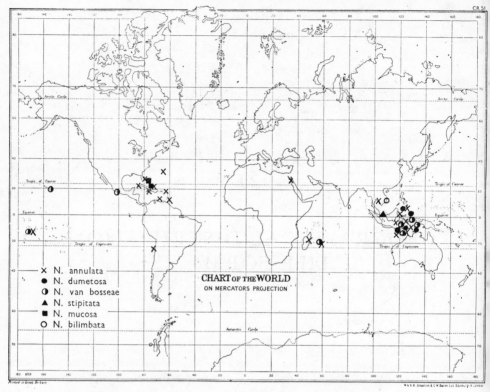

Fig. 21.1 Geographical distribution of species of *Neomeris*. (after Svedelius)

Although at present there is no very adequate explanation for this distribution, three possible hypotheses may be suggested:

(a) Migration via the Cape.

(b) Migration via the Pacific and Panama, the related species perhaps still
existing in the Pacific but not yet recorded.

(c) The related species or representatives in the interzone have died.

(3) There are some genera common to the Mediterranean and the Indo-
Pacific region, e.g. *Codium bursa* group, the vicarious pair *Halimeda tuna* in
the Mediterranean and *H. cuneata* in the Indo-Pacific, *Acetabularia mediterranea* and other species of *Acetabularia* in the Indo-Pacific (cf. Fig. 21.3).
In this case the only satisfactory explanation is the existence of a former
sea passage across the Suez isthmus. Species, e.g. *Caulerpa racemosa*,
Soliera dura, are in fact known from the eastern Mediterranean, which
must have migrated recently from the Indian Ocean via the Suez Canal. In
the flora of the northern part of the Arabian Sea, out of a total of 137
species and varieties, 22 per cent are endemic, 52 per cent are Indo-
Pacific and 60 per cent also occur in the Mediterranean and Atlantic
Ocean, the most striking example being *Cystoclonium purpureum* which

454

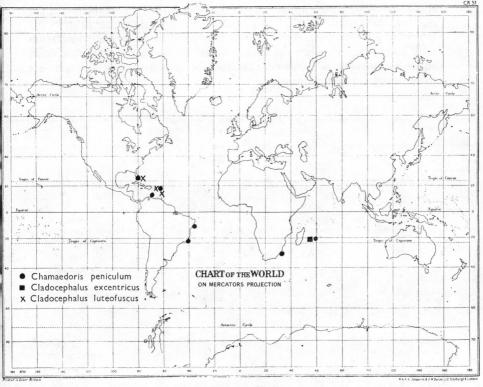

Fig. 21.2 Geographical distribution of *Chamaedoris* and *Cladocephalus*. (after Svedelius)

does not now exist between its widely separated stations along the southern shores of France and in the northern portion of the Arabian Sea.

(4) The Indo-Pacific region is more probable as the source of various tropical and subtropical genera which can be classified into

   (*a*) genera with no Atlantic representatives.

   (*b*) Genera with a few Atlantic species, e.g. *Halimeda, Caulerpa, Sargassum, Dictyota, Scinaia, Galaxaura.*

   The following genera are probably of Atlantic origin: *Dasycladus, Penicillus, Cladocephalus, Batophora.*

(5) Several families in the Laminariales, e.g. Laminariaceae, Alariaceae, are of Boreal Atlantic-Pacific discontinuity. These families must formerly have had a circum-arctic distribution but were pushed south by the onset of the Ice Age and then remained in their new habitat when the ice retreated. In this case, change of climate in a former epoch provides a satisfactory explanation of the present discontinuity. Other genera, however, e.g. *Lessonia, Macrocystis, Ecklonia,* are of Antipodes-Northern Pacific discontinuity; *Macrocystis* in particular is primarily circum-antarctic, after which it is absent from the tropics, to reappear again on the Pacific coast of North America and around the shores of South Africa. The two southern

species of *Macrocystis* appear to be identical with the two species in the northern hemisphere so that presumably they have disappeared from the intervening warm zone. Again, it must be concluded that their migration took place at a time when the temperatures of the ocean waters were more equable, unless it is assumed that the species have since become less tolerant towards temperature.

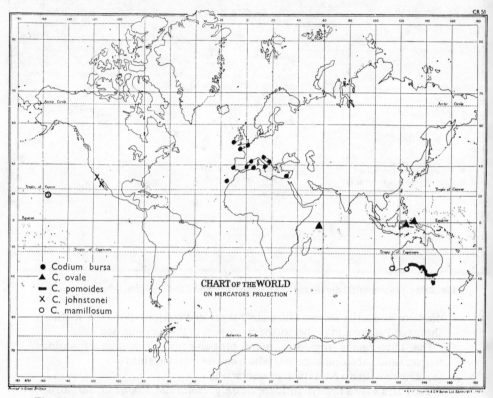

Fig. 21.3 Geographical distribution of *Codium* sect. *bursa*. (after Svedelius)

The existence in the Californian Miocene of extinct members (*Palaeo-cystophora*, see p. 322) allied to the extant southern genus *Cystophora* indicates a generic distribution almost identical with that of *Macrocystis*. Migration must have occurred from the south when temperatures were more equable and the species then disappeared with a rise in sea temperature. The absence of *Macrocystis* remains in this Miocene flora suggests that, if it came from the southern oceans, it must have migrated in post Miocene times.

(6) Stephenson (1946) has shown how there are different elements in South Africa based on temperature (see p. 457), the cold Antarctic current on the west being responsible for the presence of *Ecklonia, Laminaria pallida* and

Macrocystis. A similar cold upwelling on the coast of California enables cold-water kelps to grow much further south than they do at corresponding latitudes on the east coast. By contrast the impact of the Gulf Stream on Great Britain enables a number of Mediterranean and other warm-water genera (e.g. *Cystoseira*) to occur in south-west Ireland and on the coast of Cornwall.

In the southern hemisphere Bennett and Pope (1953) have studied the biogeographical provinces of temperate Australia and they consider that four components can be recognized, though more may prove to be desirable.

(*a*) tropical, the main species occurring north of 26° south; (*b*) warm-temperate; (*c*) cool-temperate; (*d*) universal.

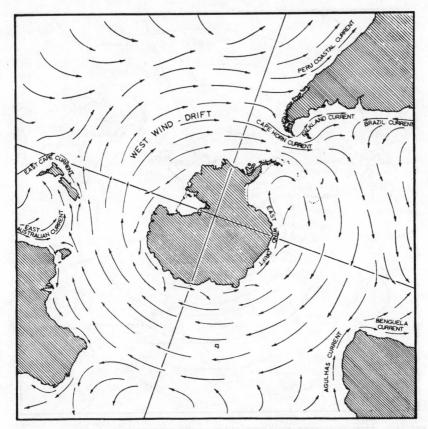

Fig. 21.4 Diagrammatic representation of the principal ocean currents of the southern oceans. (after Knox)

They suggested that the average winter sea temperature of 10° C, which Stephenson regarded as the limiting value between warm and cool temperate, may be too low as the Victorian flora is definitely cool temperate and the

average winter temperature is 11·8° C. The marine fauna and flora of
Australia as a whole fall into a number of provinces which agree very closely
with those established by Ferguson-Wood in his study of Australian dinoflagellates. In adjacent New Zealand there is a strong endemic element but
Australian and sub-Antarctic components are also represented, the latter
being found also in South Africa.

A recent review (Knox, 1964) of the biogeographical provinces of Australia
and New Zealand, based upon studies of animals, benthic algae and planktonic algae, has provided a much better understanding of the relationships in
this part of the world with modifications of earlier ideas. Knox (*loc. cit*)

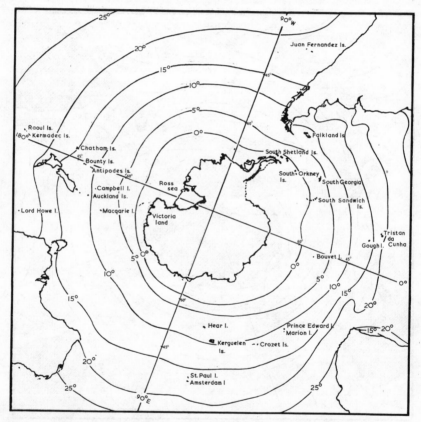

Fig. 21.5   Mean summer sea isotherms for the southern oceans. (after Knox)

introduced a transitional warm temperate region (Northern part of New
Zealand and S. Australia) but there seems no case for treating these areas as
other than warm temperate provided a subtropical zone is split off from the
tropical belt. On this basis the Peronian province of Australia and the Ker-

madecian province of New Zealand would be subtropical. It is also evident that the provinces are related broadly to currents and temperatures of the main water masses (Figs. 19.1, 21.4, 21.5). It is probable that further work on the Subantarctic islands is necessary before one can finally say that the Antipodean and Kerguelan provinces are justified.

(7) The peculiar distribution of certain algae in the southern hemisphere is significant in relation to Wegener's theory of Continental Drift. The presence of *Macrocystis*, *Splachnidium*, *Ecklonia* and *Ballia* in South Africa, widely distant from both Australia or South America, could conveniently be explained if at one time all three countries were part, with the Antarctic Continent, of a great southern continent which later broke up. If, instead, the permanence of the present land masses and oceans is accepted, then it must be assumed that fertile fragments of these algae must have drifted to South Africa, presumably from South America via the Antarctic current. Portions of algae can be found floating in the ocean, though it is very doubtful whether fertile fragments could survive such a long journey.

The most important factor concerned in biogeographical distribution is temperature, though locally presence or absence factors (see p. 421) operate as well. The only other major feature affecting distribution is that of salinity which, with lowered values, yields special floras such as the one found in the Baltic Sea.

Setchell (1920) divided the surface waters of the oceans into zones according to the course of the 10°, 15°, 20° and 25° C isotherms, and showed that the great majority of algal species are confined to only one zone, a considerable number occur in two, only a small number occur in three zones, while the number extending over four or five zones are very few indeed and their distribution is usually by no means certain. In New England many of the species are apparently separated by the 20° C isotherm, which approximates closely to the position of Cape Cod, so that the flora to the north of the Cape is essentially different to that of the south. Those species limited to one zone are called stenothermal while the wider-ranging forms are termed eurythermal. The former species are particularly characteristic of the warmer waters, but, even so, many apparent eurythermal species are found on examination to be essentially stenothermal. *Monostroma grevillei* and *Polysiphonia urceolata* are summer annuals in the cold waters of Greenland, but in the southern part of their range they develop in winter and early spring when the temperature will be the same as it is in the Greenland summer.

On the basis of these temperature studies Setchell recognised a number of climatic zones, e.g. Boreal, North Temperate etc., each of which contained species characteristic of the zone. An earlier study of Arctic algal floras by Borgesen and Jonsson indicated that the component species of the floras could be divided into five groups:

(1) The Arctic group; (2) A sub-Arctic group; (3) Boreal Arctic group; (4) A cold boreal group; (5) A warm boreal group.

Although Iceland is so far north, the flora is nevertheless predominantly boreal because 54 per cent belongs to the last three groups. If the different districts of Iceland are compared with neighbouring floras it is extremely interesting to see how the floras of the various parts of the Icelandic coast show resemblances of widely separated areas (Table 21.1).

TABLE 21.1

REGIONAL FLORA OF ICELAND COMPARED WITH FLORA OF OTHER COUNTRIES

	No. of species in Groups 1 and 2	No. of species in Groups 3 and 4
East Greenland	81	19
Spitzbergen	77	23
West Greenland	72	28
East Iceland	63	37
Finland	46	54
South-west Iceland	42	58
South Iceland	50	70
Faeroes	29	71
Nordhaven	27	73

A comparable analysis, in this classification, is Boreal and Austral, both of which represent warm temperate components. This classification has since been used as a basis for the analysis of the marine algal flora of Jamaica (Chapman, 1961) which showed that the principal element is the Caribbean, but, that while most of the remaining flora is tropical to warm temperate, there are also some cold temperate (Cold Boreal) species.

TABLE 21.2

ANALYSIS OF BOREAL AND ASTRAL FLORA

Region	No. of Species	Region	No. of Species
Caribbean	162	Cold Boreal Atlantic	23
Pan-tropical	52	Boreal Atlantic and Austral Pacific	10
Tropical Atlantic	34	North sub-tropical Atlantic	26
Indo-Pacific	11	South sub-tropical Atlantic	3
Cosmopolitan	57	Indeterminate	8

Studies in New Zealand (Chapman, 1956, 1961; Knox, 1964) have brought to light interesting distribution problems. It has been known for a long time that up to five distribution categories can be recognized:

(a) Species not extending much north of Dunedin in South Island, e.g. *Pachymenia lusoria, Apophloea lyallii.*

(b) Southern genera extending up to Kaikoura or Cook Strait, e.g. *Marginariella, Desmarestia*.

(c) Species only north of E. Cape, e.g. *Halopteris hordacea, Carpophyllum angustifolium*.

(d) Northern species extending south to the Cook Strait areas, e.g. *Caulerpa sedoides, Apophloea sinclairii*.

(e) Universal component, e.g. *Splachnidium, Ecklonia*.

In the Australasian region there is also a high degree of endemism, more so for Australia (71%) than for New Zealand (42%), though revision of the Rhodophyta of New Zealand will certainly increase the latter percentage (Table 21.3).

TABLE 21.3

COMPARISON OF ENDEMIC GENERA IN AUSTRALIA AND NEW ZEALAND

	No. & % Endemic Genera				No. & % Endemic Species			
	Australia		New Zealand		Australia		New Zealand	
Chlorophyceae	3	11%	2	7%	43	46%	67	38%
Phaeophyceae	12	19%	3	4·5%	134	70%	40	30%
Rhodophyceae	72	30%	8	6%	538	75%	174	43%
Total	87	26%	13	5·7%	715	71%	281	42%

In both regions it will be seen that there is a very high proportion of endemic species among the red algae.

Another interesting phenomenon that merits further study is the paucity of Phaeophyta in island groups in relation to Chlorophyta and Rhodophyta. Exactly why brown algae should be favoured by Continental or large island conditions is not immediately obvious. Table 21.4 illustrates this feature by making use of the Chlorophyta/Phaeophyta and Rhodophyta/Phaeophyta

TABLE 21.4

CHLOROPHYTA/PHAEOPHYTA AND RHODOPHYTA/PHAEOPHYTA RATIOS FOR CONTINENTAL AND LARGE ISLAND CONDITIONS

Continental	C/P	R/P	Island	C/P	R/P
West Coast Canada	0·6	2·3	Hawaii	2·7	4·4
Monterey Peninsula	0·5	2·9	Jamaica	2·7	3·6
South Vietnam	1·6	2·6	Costa Rica	1·7	5·0
Iceland*	0·8	1·1	Tonga	3·7	4·5
			Samoa	3·0	2·1
			Fiji	2·0	3·3
			Bahamas	2·7	4·6
			Bermuda	2·55	3·0

*These low values for a large island may be related to water temperature and further study is necessary.

ratios. It is true that in some cases the algal list requires revision and may not be complete but the trend is very obvious.

A study of life form has proved useful in the case of terrestrial phanerogams and there would seem to be justification for some sort of algal life form classification comparable to that of Raunkiaer's for the flowering plants. Such a system can be used to give a quantitative picture of the composition of the vegetation and thus raise the problem as to why some types are absent or reduced. Biological spectra, similar to those employed by Raunkiaer (1905), form a convenient way, if used with caution, of comparing floras from two different areas although they are subject to the limitation that they do not indicate the dominant types unless frequency figures are available.

Oltmann's scheme of 1905, which is one of the earliest, was based largely upon morphological criteria, as also was that of Funk's (1927) which applied particularly to the algae of the Gulf of Naples. He distinguished four major primary groups which are not readily separated, and each was subdivided on a morphological basis.

Gislèn in 1930 proposed yet another classification to include both plants and animals, the biological types referable to the plants being as follows:

Crustida (Crustaceous thallus):
  (1) Encrustida or encrusting forms, e.g. *Lithothamnion*
  (2) Torida or small cushion, e.g. *Rivularia*
Corallida (lime skeleton more or less developed):
  (1) Dendrida or tree-like forms, e.g. *Corallina*
  (2) Phyllida or leaf-like forms, e.g. *Udotea*
  (3) Umbraculida or umbrella-like forms, e.g. *Acetabularia*
Silvida (no lime skeleton):
  (a) Magnosilvida, or forms more than 10 cm high and with branches more than 1 mm thick.
    (1) Graminida, e.g. *Zostera* (a phanerogam)
    (2) Foliida, e.g. *Laminaria*
    (3) Sack-form, e.g. *Enteromorpha*
    (4) Palm form, e.g. *Lessonia*
    (5) Buoy form, e.g. *Nereocystis.*
    (6) Cord form, e.g. *Himanthalia*
    (7) Shrub-like form, e.g. *Chordaria*
    (8) *Sargassum* form
    (9) *Caulerpa* form
  (b) Parvosilvida (small delicate forms less than 100 mm high).

In 1926 Setchell propounded a scheme based primarily on the conditions found in tropical waters, with particular reference to coral reefs. For this reason the classification is restricted because it would require considerable extension if the flora of colder waters were to be included, but its basis was largely ecological:

Heliophobes:
(1) *Pholadophytes*. Forms nestling in hollows and avoiding excess light.
(2) *Skiarophytes*. Forms growing under rocks or in their shade.
Heliophiles:
(3) *Metarrheophytes* or attached flexible forms growing in moving water.
(4) *Lepyrodophytes* or encrusting forms.
(5) *Herpophytes*, composed of small creeping algae.
(6) *Tranophytes* or boring species.
(7) *Cumatophytes* or 'surf-loving' species.
(8) *Chordophytes*, where the thallus has the form of a cord.
(9) *Lithakophytes* or lime-encrusted species (Corallinaceae).
(10) *Epiphytes*.
(11) *Endophytes*.

The most workable scheme, so far, seems to be one proposed by Feldmann (1937) and modified later (1966) which is based on the same criteria, duration and perennation, that Raunkiaer used for the higher plants. The scheme is summarized as follows:

(1) Annuals

(a) Species found throughout the year. Spores or oospores germinate immediately, e.g.

EPHEMEROPHYCEAE: *Cladophora*
(b) Species found during one part of the year only.

(i) Algae present during the rest of the year as a microscopic thallus.

ECLIPSIOPHYCEAE
    (a) with microscopic gametophyte, *Sporochnus*.
    (b) with plethysmothallus, *Asperococcus*.
    (c) with microscopic sporophyte, *Ulothrix zonata*.

(ii) Algae passing the unfavourable season in a resting stage.

HYPNOPHYCEAE — Resting stage:
    (a) oospores, *Vaucheria*.
    (b) hormogones, *Rivularia*.
    (c) akinetes, *Ulothrix pseudoflacca*.
    (d) spores germinate and then become quiescent, *Dudresnaya*.

(2) Perennials

(a) Frond entire throughout the year.

(i) Frond erect. PHANEROPHYCEAE: *Codium tomentosum*.

(ii) Frond a crust. CHAMAEPHYCEAE: *Hildenbrandia*.

(b) Only a portion of the frond persisting the whole year.

(i) Part of the erect frond disappears. HEMIPHANEROPHYCEAE: *Cystoseira*.

(ii) Basal portion of thallus persists. HEMICRYPTOPHYCEAE.

    (a) basal portion a disc, *Cladostephus*.

    (b) basal portion composed of creeping filaments, *Acetabularia*.

This scheme is a great advance on the earlier ones but it does not take adequate account of the effect of environment and, furthermore, being primarily of use for the marine algae, it does not take into consideration the numerous freshwater and terrestrial species.

The real test will come if and when this scheme is employed to give biological spectra. If the spectra from different localities, e.g. temperate and tropical regions, show a distinct difference then it should prove possible to extend its use as a means of comparing the vegetation from different regions. Such differences may be expected to open up problems, the solutions to which should yield us valuable information concerning the general biology and ecology of the species concerned.

REFERENCES

Bennett, T. and Pope, E. C. (1953). *Aust. J. mar. Freshwat. Res.*, **4**, 105.
Chapman, V. J. (1956). *J. Linn. Soc. (Bot.)*, **55**, 333.
Chapman, V. J. (1961). *Nova Hedwigia*, **3**, 129.
Chapman, V. J. (1961). *Marine Algae of Jamaica*. **1** (Kingston).
Feldmann, J. (1937). *Revue Algol*, **10**, 1.
Feldmann, J. (1966). *Bull. Soc. Bot. Fr. Mem.*, 45.
Funk, G. (1927). *Pubbl. Staz. zool. Napoli*, **7**.
Gislèn, T. (1930). *Skr. K. Svensk. Vetensk.*, nos. **3**, **4**.
Knox, G. A. (1964). *Oceanogr. Mar. Biol. Ann. Rev.*, **I**, 341.
Oltmanns, F. (1905). *Morphologie und Biologie de Algen*, **12**, p. 276 (Jena, Stuttgart).
Setchell, W. A. (1920). *Am. Nat.*, **54**, 385.
Setchell, W. A. (1926). *Univ. Calif. Publs. Bot.*, **12**, 29.
Stephenson, T. A. (1964). *J. Linn. Soc. (Zool.)*, **45**.

22

ALGAL PHYSIOLOGY

It is impossible to separate algal physiology completely from some aspects of ecology, and many of the problems of algal physiology are those of plant physiology. It is essential to distinguish between physiological phenomena primarily found in algae (e.g. certain pigmentations, phototaxis, luminescence, hydrogen adaptation) and general plant physiology in which algae are experimental organisms. It is intended to give only a very brief outline, as algal physiology has been comprehensively discussed by Lewin (1962) and specific topics are annually discussed in the Annual Reviews of Plant Physiology and Microbiology, and Advances in the Microbiology of the Sea.

Apart from physiologico-ecological studies our knowledge of algal physiology stems largely from the study of bacteria-free algal cultures. Because unicellular organisms are much easier to handle than multicellular algae, the net result is that much of our knowledge is at present centred around these forms. The use of these forms as a means of extra food in arid and nutritionally poor areas may be important in the future and micro-algal cultures may also be significant in space vehicles (see Benoit, 1964). Some of the larger green algae, *Enteromorpha, Ulva, Caulerpa, Halimeda, Valonia, Acetabularia*, a few brown algae, e.g. *Pylaiella, Ectocarpus* and a few red algae, e.g. *Porphyra, Traillaiella, Callithamnion* have recently been grown and maintained successfully in culture.

PHOTOSYNTHESIS

Some mention has already been made of the considerable variation in pigment composition to be found in the different algal classes (p. 3). In the last few years considerable work has been carried out on these pigments in order to determine the part they play in the absorption of light for photosynthesis. In the larger marine algae this is important in the depth distribution of the green, brown and red algae. Fucoxanthin in the Phaeophyta, diatoms and Chrysophyta transfers absorbed light energy to chlorophyll with very high efficiency while peridinin in the Pyrrhophyta is also an efficient agent of energy transfer. Other carotenoids however have a low efficiency of energy transfer and β-carotene now appears to play an important role in protection against photosensitization. The phycocyanin and phycoerythrin biliproteins also show a very high efficiency of energy transfer in photosynthesis.

When light passes through water it is differentially absorbed and it seems that the red-orange and violet parts of the spectrum are more easily absorbed. In the sub-littoral depths, the green and green-orange portions are the principal penetrating light wavelengths. It is in this region that red and brown algae predominate while the Chlorophyceae, with maximum absorption in the blue and red-orange predominate near the surface. Thus pigmentation patterns do have definite ecological importance. The presence of biliproteins, fucoxanthin and peridinin is primarily genetically controlled, and these pigments are rarely, if ever, completely absent. It must be emphasised, however, that the pigment ratios are greatly affected by light intensity, light quality and nutrient availability.

Furthermore zonation is profoundly influenced by the general problem of exposure and submergence, e.g. many brown algae are intertidal. Engelmann's early views of chromatic adaptation are at least partially borne out. The phrase 'chromatic adaptation' should however be used with care and indeed 'chromatic control' might be a better term, restricted to such phenomena as the biliprotein variations in *Tolypothrix* (Fujita and Hattori, 1960).

There is no evidence that the mechanism of photosynthesis in the green algae differs materially from the mechanism in higher green plants. From the work that has been carried out it seems that the rate of photosynthesis in marine algae, such as fucoids, is lower than that of land plants. It is also evident that in the case of larger littoral algae the capacity to photosynthesise is greatly reduced during periods of exposure (Chapman, 1966).

Under normal conditions water acts as the electron donor for photosynthesis but certain algae are capable of photo-reduction using hydrogen after a period of adaptation under anaerobic conditions. The marine algae require a long period of adaptation.

The phenomenon is dependent upon the presence of a latent hydrogenase which is activated under dark anaerobic conditions.

There is also evidence that some algae, like certain bacteria, can utilize hydrogen sulphide as the hydrogen donor in place of water. *Oscillatoria*,

Synechococcus, *Pinnularia* (diatom), and *Scenedesmus* can all utilize H₂S, though *Synechococcus* and *Scenedesmus* use molecular hydrogen preferentially.

TABLE 22.1

ALGAE WHICH HAVE BEEN TESTED FOR ABILITY TO CARRY OUT PHOTO-
REDUCTION USING HYDROGEN

Class	Species adapted	Species not adapted
Chlorophyceae:	*Scenedesmus obliquus*	*Chlorella pyrenoidosa*
	Rhapidium sp.	
	Ankistrodesmus sp.	
	Chlamydomonas moewusii	
	Ulva lactuca	
Phaeophyceae:	*Ascophyllum nodosum*	
Bacillariophyceae		*Nitzschia*
Rhodophyceae:	*Porphyra umbilicalis*	
	Porphyridium cruentum	
Cyanophyceae:	*Synechococcus elongatus*	*Oscillatoria* sp.
	Synechocystis sp.	*Nostoc muscorum*
		Cylindrospermum sp.

Various factors, especially light (see p. 434) and temperature (p. 432) affect photosynthesis. With increasing light intensity there is light saturation and oxygen output then levels off. With shade algae, such as *Pterocladia capillacea* light saturation can be at quite low values, though even this point can be modified by temperature and CO_2 concentration.

When light is limiting, temperature exerts little effect, but once light is no longer limiting then increasing the temperature may raise the rate of photosynthesis at the level of light saturation. In the case of *Cystoseira barbata*, summer photosynthesis with maximum water temperature and light intensity of 0·16–0·20 cals cm⁻² min⁻¹ is around 80–135 mg O_2/100 g fresh alga in one day. From October to January, with reduced light intensity and low temperatures photosynthesis is much lower (Yatsenko, 1962). Work in New Zealand has shown comparable phenomena with local species (Chapman, 1966). Certain species of *Ulva*, *Chaetomorpha* and *Enteromorpha* that occur in the littoral can tolerate changes from 14° to 32° C and back to 14° C with little or no reduction in the rate of photosynthesis (Montfort, Ried and Ried, 1955, 1957).

RESPIRATION
Detailed studies of the respiratory process have shown that it follows the Emden-Meyerhof-Parnas glycolysis path and this is followed by the Krebs cycle, in *Chlorella*, *Scenedesmus*, *Ulva*, *Myelophycus* (Phaeophyceae) and *Gelidium* (Rhodophyceae). In the presence of light, respiration appears to follow a different path in *Nitella*, *Iridaea*, *Polyneura* and *Sargassum* (Brown and Tregenna, 1967) and this aspect wants further study. Respiration does

not appear to follow the normal path in *Cylindrospermum* (Cyanophyceae) nor does this alga appear to have the enzymes associated with the Krebs cycle. The respiration rate of some algae, especially fine and delicate species, is increased by lowered salinity and decreased by salinities in excess of sea water (Ogata and Takuda, 1968).

Anaerobic respiration (fermentation) has been reported for a number of algae, e.g. *Scenedesmus*, the colourless *Prototheca zopfii*.

CHEMOTROPHY

Chemotrophy involves the utilization of inorganic (-lithotrophy) or organic (-organotrophy) compounds as a source of energy. Certain algae appear to be obligate chemotrophs, e.g. *Gonium quadratum, Stephanosphaera pluvialis*, require acetate in the light for growth.

Many are facultative chemotrophs. In the dark they utilize sugars (e.g. glucose) organic acids (e.g. acetate) and occasionally alcohols and amino acids. Facultative chemotrophy (or dark heterotrophic growth) is established for members of the Chlorophyceae (e.g. *Volvocales*), Xanthophyceae (*Tribonema*), Bacillariophyta (e.g. *Navicula pelliculosa*), Chrysophyta (*Ochromonas, Prymnesium*), Euglenophyceae (*Euglena*), Cyanophyceae (*Tolypothrix*). Colourless algae such as *Prototheca* (colourless *Chlorella*), *Chilomonas* (colourless Cryptophyceae), *Polytoma* (colourless Chlamydomonadaceae) are of course obligate chemotrophs.

The green flagellate *Chlamydobotrys* is interesting because it can grow in red light when supplied with acetate, but not in green or blue light. This is an example of photochemo-organotrophy. *Ochromonas malhamensis* grows very slowly in the light, because of a very low rate of photosynthesis and an additional carbon source (e.g. glucose) is required for rapid growth.

At the other end of the scale is obligate phototrophy, which means the complete inability to grow in the dark. This phenomenon is manifested by certain *Chlamydomonas* spp; *Monodus* (Xanthophyceae), certain Cyanophyceae and numerous Bacillariophyceae and Dinophyceae.

Oxidative assimilation, in which a portion of the organic substrate (e.g. acetate) is synthesized into cell material, and part used as an energy source in respiration, has been demonstrated in the colourless *Prototheca*, and also in *Chlorella pyrenoidosa, C. vulgaris* and *Scenedesmus quadricauda*, and it is probably even more widespread.

Facultative chemotrophy may be important in the metabolism of deep growing marine algae and in situations where the available light may be negligible during long periods of time.

OTHER NUTRIENTS

Phosphorus is one of the essential elements for the algae but the amount necessary can vary very widely from less than 0.005 mg $P/10^{-3}$ m^2 for *Dinobryon* to 0·45 mg $P/10^{-3}$ m^2 for *Asterionella formosa*. In the presence of phosphate deficiency there can be an accumulation of fat. Certain algae, e.g. *Euglena, Volvox, Pandorina, Spirogyra* are capable of accumulating phosphates (Krumholtz, 1954).

Sulphur is probably assimilated as sulphate though under strong reducing conditions it may be present as hydrogen sulphide. There appears to be some connection between the presence of sulphur and the uptake of silica by diatoms. The uptake of sulphate by *Fucus vesiculosus* is apparently related to light intensity (Bidwell and Ghosh, 1962) and in this alga it is incorporated into fucoidin (see p. 184) and in other algae it appears generally in various organic compounds.

Silica is essential for diatoms and certain members of the Chrysophyta and Xanthophyta. It is normally present in natural waters in adequate supply but it can be limiting at certain times and this affects the development of some planktonic species (see p. 409). At least $0.5 \, mg/10^{-3} \, m^2$ are required for successful growth of *Asterionella formosa* and $25 \, mg/10^{-3} \, m^2$ for optimal growth of *Fragilaria crotonensis*.

Calcium and magnesium are normally both present in sufficient quantity. Certain algae, e.g. Calcareous Siphonales, Dasycladales, Dictyotales and Corallinaceae are capable of accumulating calcium and depositing it in the form of a calcareous skeleton (see p. 281).

Sodium is not normally a requirement for the growth of algae though it has been shown to be essential for certain Cyanophyceae, e.g. *Anabaena variabilis*, *Anacystis nidulans* and *Nostoc muscorum*. In the case of other algae excess sodium can be inhibitory and a sodium efflux pump has been demonstrated for *Hormosira banksii* (Bergquist, 1958), *Ulva lactuca*, *Valonia macrophysa*, *Halicystis ovalis* (Chapman, 1964) and *Acetabularia* (Saddler, 1970). Potassium is an essential element for all algae and in marine algae it is accumulated by an influx pump against a gradient, e.g. *Ulva lactuca* (Scott and Hayward, 1964). Under low concentrations of potassium both growth and photosynthesis are low and respiration is high.

Iron and manganese are essential elements because of the part they play in metabolic processes. Seasonal variations in manganese can be reflected in the composition of the phytoplankton. In the same manner that higher plants require certain trace elements so culture work has shown that these are essential for the successful growth of algae. Some of the trace elements concerned are molybdenum, copper, boron and cobalt. The red seaweed *Asparagopsis armata* grows very much better with both iodine and arsenic, and boron is important for Cyanophyta (Gerloff, 1968). Members of the Phaeophyceae and Rhodophyceae have the capacity to accumulate iodine in considerable quantities.

VITAMINS

Culture work has shown that three vitamins, cobalamin (B_{12}), thiamine and biotin are needed alone or in combination by many algae. Species requiring external growth factors, e.g. vitamins, are called auxotrophs. Many algal species, distributed among nearly all algal divisions other than the Phaeophyta are known to require an external source of vitamin B_{12} and at least 40 species are known to require thiamine or a part of the molecule. Biotin is required by only a few algae, all of which require one or both of the other two.

Growth rates of many auxotrophic algae are very sensitive to low concentrations and such algae (e.g. *Euglena gracilis, Ochromonas, Monochrysis*) have been used in the bio-assay of B_{12}. In general it may be said that algae with strong saprophytic and holozoic tendencies require vitamins.

In some of the higher algae, e.g. *Laminaria hyperborea, Ascophyllum*, the growth substance niacin varies seasonally, reaching up to 35 mg/g dry wt. in spring and falling to 15 mg/g dry wt. in autumn. Some species of algae contain as much vitamin C as do lemons, though the content is affected by season, temperature, location and pH. Vitamins D and E also appear to be present in at least some algae.

There is some evidence that the growth of marine algae may be dependent on organic substances present in sea water, though the nature of these substances is not fully understood. Surface sea water, for example, has substances necessary for the germination of sporelings of *Ulva* and *Enteromorpha*. Fucoids in the northern hemisphere are known to excrete phenolic compounds and so do *Hormisira* and *Ecklonia* in the South Pacific. Other organic substances are excreted by *Rhodymenia palmata, Spongomorpha arcta* and *Halosaccion ramentaceum* (Kailov, 1964). Not all marine algae excrete because there is no evidence of any compounds from *Pterocladia capillacea*.

An interesting phenomenon associated with some of the larger marine algae and the significance of which is not yet determined, is the production of carbon monoxide. In certain algae (*Nereocystis, Pelagophycus*) the carbon monoxide accumulates in the pneumatocyst. Algae in which carbon monoxide production has been demonstrated in fresh material include *Egregia menziesii*, and *Macrocystis* (Chapman and Tocher, 1966), it is also recorded for chopped material of *Laminaria, Calliarthron, Rhodomela larix* and *Iridophycus flaccidum* (Loewus and Delwiche, 1963).

NITROGEN METABOLISM

Algae utilize nitrate, ammonium salts and organic nitrogen as the main sources for nitrogen although ammonium ion appears to be preferentially absorbed over nitrate. Some algae can use nitrites and others are capable of fixing nitrogen from the atmosphere (see below). Certain marine algae have the capacity to accumulate nitrates against a gradient; thus *Valonia* accumulates nitrates up to 2000 times the concentration in sea water and *Halicystis* up to 500 times. Detailed studies of nitrogen absorption by various algae have led to the conclusion that irrespective of source the nitrogen is eventually incorporated into the metabolic pathways as ammonia. It appears that the amination of α-keto-glutaric acid, yielding glutamic acid, represents the major pathway of ammonia incorporation into amino acids with a lesser one via aspartic acid. The actual sites of protein synthesis vary in different species and it is not necessarily associated with the nucleus; thus enucleated *Acetabularia* continues to synthesise protein in both cytoplasm and plastids for at least three weeks (Weiz and Hammerling, 1959) (Fig. 22.1).

The establishment of bacteria-free cultures has enabled workers to show that some twenty or more species of Cyanophyceae are capable of 'fixing' atmospheric nitrogen (Pringsheim, 1968). This includes all genera in families that form heterocysts, e.g. *Nostoc, Anabaena, Cylindrospermum, Calothrix, Tolypothrix, Hapalosiphon, Anabaeniopsis* and *Mastigocladus*. Some species can fix nitrogen in the dark if supplied with sugars but others require light for the process. Fixation only takes place when there is little, if any, source of readily available combined nitrogen, e.g. nitrate, ammonium salts. The

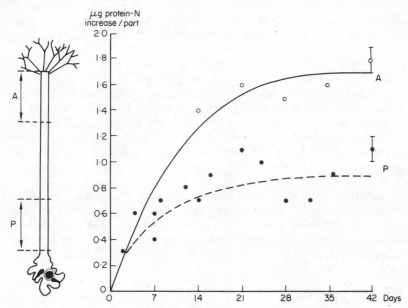

Fig. 22.1 The increase in protein nitrogen content of enucleate anterior segments of *Acetabularia* grown in the light. The diagram on the left shows the position from which the segments were taken. (from Hämmerling *et al.*, 1958)

process also apparently requires the presence of traces of molybdenum and is also more efficient in the presence of bacteria. In marine environments the phenomenon in *Calothrix scopulorum* and *Nostoc entophytum* is not affected by variations in salinity. The amount 'fixed' is highest in winter and lowest in summer (Stewart, 1964).

The fixation of free nitrogen by Cyanophyceae is of considerable importance in the culture of rice and continual crops are due to fixation by *Anabaena* and *Cylindrospermum gorahporense*. Watanabe (1957) has shown that *Tolypothrix tenuis* can fix nitrogen at a rate of 0·248 mg/m² in 12 h which could add as much as 2,900 lb of nitrogen per acre in a year. This alga is now grown in culture and is used to seed the rice fields; in India, for example, innoculation by Cyanophyceae is an effective substitute for fertilizing with ammonium sulphate (Subrahmanyan *et al.*, 1965).

GROWTH HORMONES

Auxin (indole-acetic acid) has been reported in *Valonia macrophysa*,

Laminaria agardhii and in a number of Pacific coast algae, including *Macrocystis, Desmarestia* and *Fucus evanescens*. I.A.A. promotes growth in *Codium decorticatum, Ulothrix subtilissima* and *Acetabularia mediterranea*. When associated with kinetin sporeling growth and initiation of normal filaments from green island cells in white thalli of *Ulva lactuca* (Chapman, 1964) was much better than kinetin by itself which only induced rhizoid formation. Gibberellic acid alone merely induces increased elongation of the thallus (Provasoli, 1958), though it promotes a 7-fold growth increase in *Ulothrix*. Not all algae appear to require growth hormones, e.g. *Prasiola stipitata*.

Growth substances have been detected in the eggs, sperm, and fruiting tips of European species of *Fucus*, and the presence of these may be partly responsible for the fact that the first rhizoid of a germinating zygote grows out on a side adjacent to neighbouring eggs. This is not the only factor as the appearance of the rhizoid is also affected by pH, and by unilateral illumination; furthermore blue rather than red light orientates the cleavage wall. When ova of *Fucus* or *Cystoseira* are passed through capillary tubing rhizoids are formed at a pole which suggests that pressure can play some part as well.

In addition to growth-promoting substances there may be growth-inhibiting compounds. Such substances are probably produced by *Nostoc punctiforme, Nitzschia palea* and *Chlorella vulgaris*. The inhibitor from *Chlorella* is known as 'Chlorellin' and appears to be a mixture of fatty acids. It is active against bacteria and this is one reason why *Chlorella* is effective in purifying sewage effluent.

SEX HORMONES

Since 1926 it has been known that the gametes or filaments of certain algae secrete sexual substances into the surrounding water. This has been demonstrated for *Dasycladus clavaeformis, Tetraspora lubrica, Sphaeroplea, Vaucheria, Oedogonium, Chlamydomonas eugametos, Volvox, Protosiphon botryoides, Stephanosphaera pluvialis, Ectocarpus* and *Botrydium granulatum*, (Smith, 1951; Hutner and Provasoli, 1964; Muller, 1968; Starr, 1968).

CHEMICAL COMPOSITION

As may be expected, the cell-wall material and the major storage products are composed mainly of polysaccharides. Apart from the Procaryota the principal wall polysaccharides are pectin (small amounts) uronic acid compounds, glucans, xylans, mannans and galactans, many being of commercial importance. In the larger Phaeophyceae alginic acid is present in the walls and in cells of certain Rhodophyceae can be found agar and carrageen which again are of commercial importance (Chapman, 1970).

Starch, (or starch-like compounds, with predominantly α 1 : 4 glucoside linkages) is the principal storage product of the Chlorophyta, Cryptophyceae, some Dinophyceae and Rhodophyceae (Floridean starch). The storage product of the Cyanophyceae is similar to the starches, but may be more closely related to glycogen. Storage reserves with principally β 1 : 3 glucoside linkages are found in the Phaeophyta (laminarin) Chrysophyta and Bacillario-phyta (chrysose or chrysolaminarin) and Euglenophyceae (paramylum). Other common reserve carbohydrates are trehalose, floridoside and manno-glycerates which are found in the Rhodophyceae, while polyhydric alcohols are common in the Phaeophyceae and Rhodophyceae. Dukitol is present in *Iridaea* and *Bostrychia* and mannitol is present in both groups. The presence of sorbitol (*Bostrychia*) and volemitol (*Pelvetia* and *Porphyra*) has also been established.

Because of the commercial uses of some of these storage compounds a great deal of work has been carried out on the detailed chemical composition of algae, especially the littoral rock weeds (Fucales) and sub-littoral oarweeds (Laminariales) of Europe. Regular analyses of plant populations have been made so that the results represent changes in the average composition rather than those in a single individual. During the period of rapid growth (spring), fresh weight, ash, protein content and alginic acid increase while laminarin and mannitol behave in a reverse fashion. The summer decrease in growth is correlated with the decrease of $PO_4^{\equiv}$ and NO_3^{-} in the sea at that time. During this period, however, the high photosynthetic rate permits the accumulation of mannitol and laminarin. There is also a tendency to more than one maxi-mum in mannitol and laminarin, due largely to the interaction of exposure, temperature and formation of reproductive organs. The low laminarin content in July–August of Scottish plants is almost certainly the result of a lowering of photosynthetic activity brought about by increased desiccation. The degree of the variations depends upon habitat, especially for *Laminaria digitata* and *L. saccharina*, being least in the open sea and greatest in locks or enclosed bodies of water. In *Laminaria hyperborea* the seasonal variations are almost entirely confined to the frond. With *L. saccharina*, gradients in mannitol and laminarin occur along the frond, the amounts appearing to be related to the age of the tissue. Mannitol and laminarin contents also vary with depth. In the case of littoral fucoids seasonal variations tend to increase with increasing frequency of immersion.

Outside Europe rather similar results have been reported for *Fucus vesiculosus*, *F. evanescens* and *Ascophyllum* from the Canadian Maritime provinces, except that laminarin tends to remain rather constant (Powell and Meeuse, 1964).

SAP, PERMEABILITY AND OSMOTIC PRESSURE

The Siphonales (*sensu lato*) and Charales have formed excellent material for studies in permeability, sap composition and osmotic pressure because of the size of their cells. There is considerable variation in the rate of penetration by the different ions; iodine, for example enters far more rapidly than chlorine, and potassium more rapidly than sodium but less rapidly than ammonia. The behaviour of salts and gases in the case of algae with a higher and more complex organization of tissues may well be different from the Siphonales (*sensu lato*) and Charales, but so far it has been but little studied. It is known that there are influx and efflux pumps operating for sodium and potassium (see p. 449). Dilution of sea water brings about the escape of chloride and potassium ions from *Chaetomorpha linum* but the sodium ion is little affected (Kesseler, 1964). The entry and exit of ions is related to the availability of free space, which in the case of *Ulva lactuca* and *Cladophora gracilis* may amount to 20 per cent of the wall volume. Very little work has been done on free space in algal cell walls and this field demands further investigation. The behaviour of the sulphate ion is also peculiar because it appears to be almost wholly excluded by both *Valonia* and *Halicystis* but not by *Ulva* and *Entero-morpha* (Polikarpov, 1960).

The cell sap of many freshwater algae has an osmotic pressure of about 5 atmospheres, whereas that of some marine algae, e.g. *Halicystis*, *Valonia*, *Nereocystis*, is only slightly higher than that of sea water. The osmotic pressure in Fucales would appear to be rather higher as is that of some of the Rhodophyceae. In general, however, the osmotic pressure in red algae is somewhat lower than that in green or brown algae. The tolerance of algae to changes in osmotic pressure (caused by water loss on exposure) may be important in the case of littoral seaweeds (cf. p. 442); some algae appear to respond almost instantaneously, e.g. *Monostroma*, *Dunaliella salina*. This suggests that cells of such species may be very permeable to the salts rather than impermeable to water. Changes in osmotic pressure affect respiration in *Laminaria*, *Hormosira banksii*, *Fucus serratus*, *Chaetomorpha linum* (Kesseler, 1962) and other algae (Chapman, 1966), but not in *Ulva*, *Porphyra* or *Fucus vesiculosus*. Photosynthesis is affected in *Fucus*, *Ulva* and *Hormosira*, being more than doubled in the first two genera when the sea water is diluted a third.

BIO-ELECTRIC PHENOMENA

Using large cells of the Charales, *Halicystis*, *Valonia* or *Hydrodictyon* it has been found that potential differences exist between the interior and the exterior of the cell, the latter usually being positive, e.g. *Chara*, *Nitella*, *Halicystis*, *Bryopsis*, *Hydrodictyon*, though in *Valonia*, *Ernodesmis* and *Chamaedoris* it is negative. It seems likely that the gradient of potassium ions is responsible for this potential in *Nitella* and *Hydrodictyon*. In other algae it seems probable that the potential is a result of an internal concentration gradient related to all the compounds within the protoplasm itself.

POLARITY AND MORPHOGENESIS

Polarity is a problem of considerable interest, especially in the case of coenocytic algae. Reference has already been made (p. 246) to determination of *Fucus* egg polarity, and spores of other algae, e.g. *Griffithsia*, are also polarized because rhizoid and shoot appear at opposite poles. Rhizoid formation can be induced by subjection to an electric current when rhizoids develop on the anode side of the thallus. Polarity also exists in *Enteromorpha* in relation to its powers of regeneration. Thus when apical sections are cut from the thallus papillae appear on the morphological upper surface and rhizoids and papillae on the morphological lower surface. It has been suggested that a gradient of substances is set up early in development and that it is not readily disturbed (Muller-Stoll, 1952).

In *Bryopsis* accumulation of auxin in the lower half of the thallus brings about rhizoid formation. If plants of *Bryopsis* are cut, inverted, and the cut end treated with I.A.A. rhizoid formation takes place at the lower tips where I.A.A. accumulates.

The formation of the cap in *Acetabularia* is also an example of polarity. Enucleate apical segments have a greater regenerating power than enucleate basal segments. In this alga there is now clear evidence of the production of morphogenetic substances by the nucleus in the rhizoid.

The genus *Acetabularia,* since 1931, has proved an admirable genus for experimental work on morphogenesis. Important summaries of the great volume of work can be found in Hammerling (1957), Keck (1961), Brachet (1968) and Puiseux-Dao (1970). Because of the large single nucleus in the rhizoid a study can be made of the influence of the nucleus on structure by its removal. It was found in 1931 that enucleate portions of *Acetabularia* could survive and regenerate and possessed a morphogenetic capacity related to the orginal nucleus.

When the nucleus is removed by cutting off the rhizoidal base, the stalk portion will grow a cap, or with *A. crenulata* more than one cap. If the stalk is cut into several portions normally only the apical one will elongate and it may form a cap at one or both ends. It is argued that the nucleus can produce a morphogenetic substance or substances which accumulate in the cytoplasm and exert their control even after the nucleus is removed. The greater regenerating power of the apical segments of the stalk suggests that there is a gradient of these substances.

Quite early on it was found possible to make grafts by inserting the cut stalk of one plant into the cut base of another. As a result grafts between species could be made and it also became possible to obtain bi-, tri- and quadrinucleate plants. Later it has been found possible to transfer single nuclei so that in the final plant there is no mixed cytoplasm. Using these grafts it has been shown that the form of the cap is determined by the nucleus present, and if more than two nuclei were present the cap morphology tended to be characteristic of the preponderant species nuclei, e.g. in a graft with two *crenulata* nuclei and one *mediterranea* nucleus the cap would tend to be of the *A. crenulata* type. The morphogenetic material is therefore species specific. It can be destroyed by ultra-violet radiation (Weiz and Hammerling, 1961) and inhibited by red light (Clauss, 1968). In the latter case there is no destruction because the effect can be reversed by exposure to blue light.

It now appears that there are two morphogenetic substances; one is concerned with the initiation of cap formation and the other with controlling its actual shape. It is only this latter that is species specific. More recent work has suggested that there are one or more proteins involved in cap formation as well as a set of enzymes, both being separate from the proteins and enzymes concerned in stalk development (Zetsche, 1966).

When cap formation was studied in relation to light intensity it was concluded that the precursors of the morphogenetic substances can be formed under nuclear control at quite low light intensities (500 lx). They are then activated by a higher light intensity which is not under nuclear control. At low values, e.g. 70 ft-candles, changes of intensity are more important than

Q 2

duration, while at higher values 225 ft-candles duration is more important (Terborgh and Thimann, 1964). The amount of the morphogenetic substance produced seems to be proportional to nucleus size, the larger the nucleus the greater the amount of substance.

The importance of the nucleus in producing these morphogenetic substances has directed attention to protein synthesis. In enucleate plants RNA synthesis ceases, though protein manufacture can continue for up to three weeks. In decapitated rhizoidal portions the synthesis of RNA bears a close relationship to rate of regeneration of stalk and cap. When enucleate basal parts of *A. mediterranea* are provided with a new nucleus, RNA synthesis takes place in subsequent new growth and not in the old stalk. Since RNA synthesis occurs in the plastids of enucleate segments it is clear that there are two sites for RNA synthesis and it is only the cytoplasmic one that is under nuclear control.

Growth in *Acetabularia* requires not only protein synthesis but also the morphogenetic substances. Because enzymes may be concerned with production of the morphogenetic substances attention has, in recent years, been turned to them. After enucleation invertase, acid phosphatase and phosphorylase all increase, and there is no lack of ribonuclease. Using interspecific grafts Keck (1961) found that acid phosphatase exists in two species specific forms, med and acic (*Acicularia schenkii*), and that both are produced under enucleate conditions. When med nuclei are transplanted to *Acicularia* stalks the acic phosphatase is replaced by med phosphatase within six days, though if no nucleus is introduced the acic phosphatase remains detectable for up to 25 days after enucleation. When an acic nucleus was transferred to a med stalk no acic phosphatase was subsequently produced. This may mean that production of med phosphatase is genetically dominant to that of acic. Further experiments showed that so long as med cytoplasm was present only med phosphatase was produced. It seems evident, therefore, that both nuclear and cytoplasmic factors may be involved in protein synthesis through species specific enzymes.

Up to the present there is no clue as to the chemical nature of the morphogenetic substances, but they can scarcely be I.A.A. or related compounds since these, although influencing growth in *Acetabularia*, are as a rule, primarily concerned in cell-wall extension.

CIRCADIAN RHYTHM

In recent years, there has been a lot of interest in the study of biological 'clocks' or circadian rhythm (Sweeney, 1963a, b, 1969). A number of such rhythms have been known for a long time among the algae, the older ones being primarily concerned with lunar rhythms and reproductive-body discharge, e.g. Dictyotales (see p. 197), *Oedogonium* and *Halicystis*. In *Hydrodictyon* an endogenous photosynthetic and respiratory rhythm operates during normal alternating light and dark periods and this continues when the cultures are transferred to continuous light. The rhythm is also induced in red or blue light but not in green light (Pirson and Schon, 1957). There is a comparable and even more pronounced rhythm in *Acetabularia* (Fig. 22.2) which continues for up to 40 days in enucleated plants, so that the phenomenon in this plant is not wholly under nuclear control (Schweiger, *et al.*, 1964). Similar rhythmic respiratory activity has been reported for *Fucus* though in this genus there has been no report of a photosynthetic rhythm.

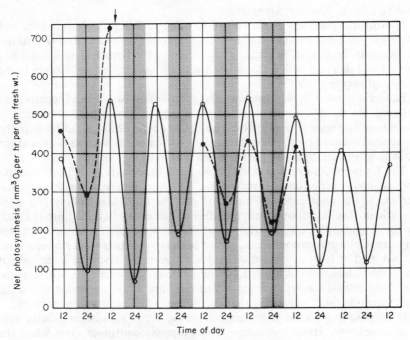

Fig. 22.2 Diurnal photosynthetic rhythm of *Acetabularia* in alternating day and night (shaded) and in continuous light. Dotted line shows the oxygen production by ten young plants whose caps have yet to form and the solid line is for one plant having a cap. i.e., relatively mature. The arrow indicates the time at which the nucleus was removed by amputating the rhizoids. (from Sweeney and Haxo, 1961)

Certain dinoflagellates (e.g. *Gonyaulax* spp.) are able to emit a flash of blue light resulting from the luciferase-catalysed oxidation of the substrate luciferin. When grown in dim light *Gonyaulax* demonstrates a diurnal rhythm in luminescence which is absent from those grown in bright light. The degree of luminescence is greatest at night. *Noctiluca* however does not show rhythm in luminescence. Other diurnal rhythms have been demonstrated for cell division in dinoflagellates, mitosis in *Spirogyra* and *Oedogonium*. Phototactic tidal rhythms are also known. *Euglena limosa* is positively phototactic but prior to high tide it burrows into the mud. The diatom *Hantzschia amphioxus* comes to the mud surface during low tide. These rhythms all appear to be temperature independent, but are affected by light. They are endogenous rhythms since they are maintained for some time after being brought into the laboratory and constant conditions.

In *Gonyaulax* the gradual decrease in the rhythm is due to energy exhaustion as *Gonyaulax* is an holophytic organism (Sweeney, 1963a, b). This particular rhythm does not depend on prior growth under alternating light and dark conditions, because cultures grown in continuous bright light for a considerable period exhibit this diurnal periodicity.

PHOTOTAXIS, PHOTOTROPISM

Phototropic phenomena are known to occur in *Bryopsis*, *Acetabularia* and *Derbesia* while rather weaker effects are found in *Cladophora* and *Griffithsia*. In *Bryopsis* the tips of branches grow towards the light and rhizoids away from it and this may well be associated with the accumulation of auxin in the basal region (see p. 104).

Phototaxis is the movement towards (positive) or away from (negative) an external source of light. The response is one of two types: topo-taxis is movement towards or away from the light source while photo-taxis is the movement induced by a sudden change in light intensity such as occurs at the boundary between light and dark in a beam of light.

The Cyanophyceae show both photo- and topo-taxis with both positive and negative responses, depending on the light intensity. In *Phormidium uncinatum* the photoreceptor pigments may be both carotenoids and phycoerythrin. Tactic responses are manifest in numerous flagellate groups such as the Euglenophyceae, Pyrrhophyta, Volvocales, *Ulva* gametes, Cryptophyceae, Chloromonadophyceae. Again both types of taxes and responses are known, and at least in *Platymonas subcordiformis* the response is dependent upon the external conditions. Low light intensity induces a positive taxis, high light intensity a negative taxis. The response is also dependent upon the light spectrum used, suggesting that two different mechanisms are involved.

In *Euglena gracilis* the phototactic response shows a daily periodicity with a peak reaction around mid-day. This response continues even when the *Euglena* is transferred to continuous darkness (Pohl, 1948).

The photoreceptor is not the eyespot. Flagellates which lack eyespots are still phototactic. The receptor is believed to be located near the flagellar base, and the eyespot is thought to play a role in shading. Diatoms of the order Pennales are also phototactic. The responses are not always clearly defined, being influenced by light intensity and light quality. The irregular motion of the gliding diatom further confuses the interpretation. The nature of the photoreceptor pigment is unknown, but action spectra studies indicate differences from class to class, and within the dinoflagellates, differences from genera to genera.

REFERENCES

GENERAL

Blinks, L. R. (1951). In *Manual of Phycology*. Edit. by G. M. Smith. p. 231. (Chronica Botanica, Waltham).

Fogg, G. E. (1966). *Oceanogr. Mar. Biol. Ann.-Rev.* **4**,

Hutner, S. H. and Provasoli, L. (1964). *A. Rev. Pl. Physiol.* **15**, 37.

Lewin, R. A. (1962). *Physiology and Biochemistry of Algae* (Academic Press, New York and London).

References Specific

Benoit, R. J. (1964). *Mass culture of microalgae for photosynthetic gas exchange in Algae and Man* (Pergamon Press, Oxford).

Bergquist, P. L. (1958). *Physiologia Pl.*, **11**, 760.

Bidwell, R. G. S. and Ghosh, N. R. (1962). *Can. J. Bot.*, **41**, 209.

Brachet, J. L. A. (1965). *Endeavour*, **24**, 155.

Brachet, J. L. A. (1968). *Current topics in Developmental Biology* (Acad. Press, N.Y.).

Brown, D. L. and Tregenna, E. B. (1967). *Can. J. Bot.*, **45**, 1135.

Chapman, D. J. and Tocher, R. D. (1966). *Canad. J. Bot.*, **44**, 1438.

Chapman, V. J. (1964). *Oceanogr. Mar. Biol. Ann.-Rev.*, **2**, 193.

Chapman, V. J. (1966). *Proc. 5th Int. Seaweed Symp.* pp. 29–54 (Pergamon Press, Oxford).

Chapman, V. J. (1970). *Seaweeds and Their Uses* (Methuen, London).

Clauss, H. (1968). *Protoplasma*, **65**, 49.

Fujita, Y. and Hattori, A. (1960). *Plant and Cell Phys.*, **1**, 293.

Gerloff, G. C. (1968). *Physiol. Plant.*, **21**, 369.

Hammerling, J. (1957). *8th Cong. Int. Bot. Paris. Compt. Rend.*, *Sect.* **10**, 87.

Kailov, K. M. (1964). *Ref. Zhur. Biol.*, No. 10D201.

Kanwisher, J. W. (1966). In *Some Contemporary studies in Marine Science*. Edit. by H. Barnes p. 407–420.

Keck, K. (1961). *Ann. N.Y. Acad. Sci.*, **94**, 741.

Kesseler, H. (1962). *Helgoländer wiss. Meeresunters*, **10**, 73.

Kesseler, H. (1964). *Helgoländer wiss. Meeresunters*, **8**, 1.

Krumholtz, L. A. (1954). *U.S. Atomic Energy Information Service*, ORO-132, 1.

Loewus, M. W. and Delwiche, C. C. (1963). *Pl. Physiol., Lancaster*, **38**, 371.

Montfort, C., Ried, A. and Ried, J. (1955). *Beit. Biol. Pflanz.*, **31**, 349.

Montfort, C., Ried, A. and Ried, J. (1957). *Biol. Zbl.*, **76**, 257.

Muller, D. G. (1968). *Planta*, **81**, 160.

Muller-Stoll, R. W. (1952). *Flora*, **139**, 148.

Ogata, E. and Takuda, H. (1968). *J. Shimonoseki Coll. Fish.*, **16** 2/3, 67.

Pirson, A. and Schon, W. S. (1957). *Flora*, **144**, 447.

Pohl, R. (1948). *Z. Naturf.*, **3B**, 367.

Polikarpov, G. G. (1960). *Nauk. Dokl. Vysshei Shkoly, Biol. Naux.*, 103.

Powell, J. H. and Meeuse, B. J. D. (1964). *Econ. Bot.*, **18**, 164.

Pringsheim, E. G. (1968). *Planta*, **79** (1), 1.

Provasoli, L. (1958). *Biol. Bull. mar. biol. Lab. Woods Hole*, **114**, 575.

Puiseux-Dao, S. (1970). *Acetabularia and Cell Biology* (Logos Press).

Richter, G. (1962). In *Physiology and Biochemistry of Algae*, Edit. by R. A. Lewin, p. 633 (Academic Press, London and New York).

Saddler, H. D. W. (1970). *J. expt. Bot.*, **21**, 605.

Scott, G. T. and Hayward, H. R. (1954). *J. gen. Physiol.*, **37**, 601.

Starr, R. C. (1968). *Proc. natn. Acad. Sci. U.S.A.*, **59**, 1082.

Stewart, W. D. P. (1964). *J. Gen. Microbiol.*, **36**, 415.

Subrahmanyan, R., Relwani, L. L. and Manna, G. B. (1965). *Proc. Indian Acad. Sci.*, **62**, (6), Ser. B. 252.

Sweeney, B. M. (1963a). In *Physiology and Biochemistry of Algae*, edit. by R. A. Lewin (Academic Press, London and New York).

Sweeney, B. M. (1963b). In *A. Rev. Pl. Physiol.*, **14**, 411.

Sweeney, B. M. (1969). *Rhythmic Phenomena in Plants* (Acad. Press, London).

Schweiger, E., Wallroff, H. G. and Schwerger, H. G. (1964). *Science, N.Y.*, **146**, 658.

Terborgh, J. and Thimann, K. V. (1964). *Planta*, **63**, 83.

Watanabe, A. (1951). *Archs. Biochem. Biophys.*, **34**, 56.

Weiz, G. and Hammerling, J. (1959). *Planta*, **53**, 145.

Weiz, G. and Hammerling, J. (1961). *Z. Naturf.*, **16b**, 829.

Yatsenko, G. K. (1964). *Referat. Z. Biologiya*, **11**, 29.

Zetsche, K. (1966). *Planta.* **68**, 360.

INDEX